Lithium-Related Batteries

Lithium-Related Batteries
Advances and Challenges

Edited by
Ngoc Thanh Thuy Tran
Wen-Dung Hsu
Jow-Lay Huang
Ming-Fa Lin

CRC Press
Taylor & Francis Group
Boca Raton London New York

CRC Press is an imprint of the
Taylor & Francis Group, an **informa** business

First edition published 2022
by CRC Press
6000 Broken Sound Parkway NW, Suite 300, Boca Raton, FL 33487-2742

and by CRC Press
4 Park Square, Milton Park, Abingdon, Oxon, OX14 4RN

ISBN: 978-1-032-20389-8 (hbk)
ISBN: 978-1-032-20489-5 (pbk)
ISBN: 978-1-003-26380-7 (ebk)

DOI: 10.1201/9781003263807

Typeset in Times
by codeMantra

Contents

Preface

Nowadays, the feverish scientific activity relating to the development of green energy materials is in urgent due to the non-stop increasing of air pollution. Significant efforts on exploiting new materials for energy storages were reported for potential applications to solve the environmental and energy issues. Among energy storages, the ion-based batteries have attracted a lot of theoretical and experimental studies. This book is focused on their diversified phenomena and special functionalities, as clearly illustrated in the following paragraphs.

This book is mainly focused on the further development of quasiparticle framework and the successful syntheses of cathode/electrolyte/anode materials mainly in lithium- and sodium-based batteries. Quasiparticles, electrons and polarons, exhibit the diverse phenomena through the Vienna Ab initio Simulation Package (VASP) calculations and molecule dynamics. The concise mechanisms, the multi-orbital hybridizations in various chemical bonds, are thoroughly identified from the essential physical/chemical/material properties. Furthermore, the experimental researches can greatly enhance the main functions of core materials and reduce the disadvantages. Only few theoretical predictions about the fundamental properties agree with the high-precision measurements. However, most of inconsistences come from the technical limits and intrinsic environments. A series of experimental examinations, which are absent up to now, need to be done in the near-future researches. Most importantly, a grand unified theory, corresponding to the stationary ion transports, is required for the basic and applied sciences.

The main-stream electrolyte/anode/cathode materials are outstanding candidates in verifying the quasiparticle viewpoints of the first-principles simulations. The significant multi-orbital hybridizations of different chemical bonds in solid-state crystals are responsible for the essential properties. The close relations between the various asymmetric environments, the time-dependent intermediate states, and the stationary ionic currents need to be fully established through the near-future basic researches. The complicated chemical reactions are successfully investigated by the method of molecule dynamics for the lithiation processes of silicon anode. Their direct combinations with the first-principles simulations will be very powerful in exploring the charging and discharging processes of ion-based batteries.

The experimental works focus on the enhanced functions and reduced dissipations of anode/electrolyte/cathode materials. A lot of summarized characteristics, which cover the sample growth, measurement, and design, will be very useful in comprehending the integrated systems and proposing the viewpoints of readers. Both lithium- and sodium-related batteries clearly show the diverse quasiparticle phenomena from certain significant differences. The second part could be verified by the first-principles simulations. It is also noticed that the X-ray diffraction spectra are available in revealing the spatial charge distributions. This will be very useful in verifying the theoretical predictions. Furthermore, how to develop the time-resolved X-ray scatterings in observing the various intermediate crystal structures becomes an emergent issue. The close relations among the electrode voltage, cycle number, and specific charge capacity are fully established through the high-precision measurements. On the other hand, the systematic investigations on the essential properties, as done for band structures, optical absorption spectra, and Coulomb excitations, are almost absent. Such studies will play the critical roles in greatly enhancing the battery performances.

Acknowledgments

This work was financially supported by the Hierarchical Green-Energy Materials (Hi-GEM) Research Center, from the Featured Areas Research Center Program within the framework of the Higher Education Sprout Project by the Ministry of Education (MOE) and the Ministry of Science and Technology (MOST 110-2634-F-006-017) in Taiwan.

Editors

Ngoc Thanh Thuy Tran earned a Ph.D. in Physics in 2017 at the National Cheng Kung University (NCKU), Taiwan. Afterward, she was a postdoctoral researcher and then an assistant researcher at Hierarchical Green-Energy Materials (Hi-GEM) Research Center, NCKU. Her scientific interest focuses on the fundamental (electronic, magnetic, and thermodynamic) properties of 2D materials and rechargeable battery materials by means of first-principles calculations.

Wen-Dung Hsu is an Associate Professor in the Department of Materials Science and Engineering, National Cheng Kung University (NCKU). He has expertise in utilizing computational materials science methods, including first-principles calculations, molecular dynamics simulations, Monte-Carlo methods, and finite-element methods to study materials issues. His research interests include the mechanical properties of materials from atomic to macroscale, lithium-ion batteries, solid-oxide fuel cells, ferroelectrics, a solid catalyst design for biodiesel, and processing design for single-crystal growth. He earned a Ph.D. in 2007 at the Department of Materials Science and Engineering, University of Florida. He served as a postdoc researcher in the Department of Mechanical Engineering at the University of Michigan, Ann Arbor. He joined National Cheng Kung University in 2008.

Jow-Lay Huang is a Chair Professor in the Department of Materials Science and Engineering, National Cheng Kung University (NCKU), Taiwan. He is the Director of the Hierarchical Green-Energy Materials (Hi-GEM) Research Center, NCKU, Taiwan. He earned a Ph.D. in materials science and engineering in 1983 at the University of Utah, Salt Lake City, Utah, USA. His research interests include the fabrication, development, and application of ceramic nanocomposites, piezo-phototronic thin films for photodetector devices, piezoelectric thin films for high-frequency devices, metal oxide/graphene and SiCx nanocomposites as anode materials for lithium-ion batteries, and 2D nanocrystal materials for photoelectrochemical applications.

Ming-Fa Lin is a Distinguished Professor in the Department of Physics, National Cheng Kung University (NCKU), Taiwan. He earned a Ph.D. in Physics in 1993 at the National Tsing-Hua University, Taiwan. His main scientific interests include the essential properties of carbon-related materials and low-dimensional systems. He is a member of the American Physical Society, the American Chemical Society, and the Physical Society of Republic of China (Taiwan).

Contributors

Sanjaya Brahma
Department of Materials Science and
    Engineering
National Cheng Kung University
Tainan, Taiwan

Chia-Chin Chang
Department of Green energy
National University of Tainan
Tainan, Taiwan

Vo Khuong Dien
Department of Physics
National Cheng Kung University
Tainan, Taiwan

Van An Dinh
Department of Precision Engineering
Graduate School of Engineering
Osaka University
Osaka, Japan

Nguyen Thi Han
Department of Physics
National Cheng Kung University
Tainan, Taiwan

Wen-Dung Hsu
Department of Materials Science and
    Engineering
Hierarchical Green-Energy Materials
    (Hi-GEM) Research Center
National Cheng Kung University
Tainan, Taiwan

Yuan-Shuo Hsu
Department of Chemical Engineering
National Cheng Kung University
Tainan, Taiwan

Jow-Lay Huang
Department of Materials Science and
    Engineering
Center for Micro/Nano Science and
    Technology
Hierarchical Green-Energy Materials
    (Hi-GEM) Research Center
National Cheng Kung University
Tainan, Taiwan

Thi My Duyen Huynh
Department of Physics
College of Sciences
National Cheng Kung University
Tainan, Taiwan

Febriana Intan
Department of Chemical Engineering
National Cheng Kung University
Tainan, Taiwan

Jeng-Shiung Jan
Department of Chemical Engineering
National Cheng Kung University
Tainan, Taiwan

Zhe-Yun Kee
Department of Materials Science and
    Engineering
National Cheng-Kung University
Tainan, Taiwan

Chin-Lung Kuo
Department of Materials Science and
    Engineering
National Taiwan University
Taipei, Taiwan

Kha M. Le
Viet Nam National University, Ho Chi
  Minh City (VNU HCM)
Applied Physical Chemistry Laboratory
Ho Chi Minh University of Science
  (HCMUS)
Department of Physical Chemistry
Ho Chi Minh University of Science
  (HCMUS)
Ho Chi Minh City, Viet Nam

Phung M.L. Le
Viet Nam National University, Ho Chi
  Minh City (VNU HCM)
Applied Physical Chemistry Laboratory
Ho Chi Minh University of Science
  (HCMUS)
Department of Physical Chemistry
Ho Chi Minh University of Science
  (HCMUS)
Ho Chi Minh City, Viet Nam

Ting-Yuan Lee
Department of Chemical Engineering
National Cheng Kung University
Tainan, Taiwan

Wei-Bang Li
Department of Physics
National Cheng Kung University
Tainan, Taiwan

Ming-Fa Lin
Department of Physics
Hierarchical Green Energy Materials,
  Hi-Research Center
National Cheng Kung University
Tainan, Taiwan

Shih-Yang Lin
Department of Physics
National Chung Cheng University
Chiayi, Taiwan

Huu Duc Luong
Division of Precision Science and
  Technology and Applied Physics
Graduate School of Engineering
Department of Theoretical
  Nanotechnology
The Institute of Scientific and Industrial
  Research
Osaka University
Osaka, Japan

Hoang V. Nguyen
Viet Nam National University, Ho Chi
  Minh City (VNU HCM)
Applied Physical Chemistry Laboratory
Ho Chi Minh University of Science
  (HCMUS)
Ho Chi Minh City, Viet Nam

Huong T.D. Nguyen
Viet Nam National University, Ho Chi
  Minh City (VNU HCM)
Department of Physical Chemistry
Ho Chi Minh University of Science
  (HCMUS)
Ho Chi Minh City, Viet Nam

Minh L. Nguyen
Viet Nam National University, Ho Chi
  Minh City (VNU HCM)
Applied Physical Chemistry Laboratory
Ho Chi Minh University of Science
  (HCMUS)
Ho Chi Minh City, Viet Nam

Thi Dieu Hien Nguyen
Department of Physics
National Cheng Kung University
Tainan, Taiwan

Li-Yi Pan
Department of Materials Science and
  Engineering
National Taiwan University
Taipei, Taiwan

Tai-Yu Pan
Department of Materials Science and
Engineering
National Cheng-Kung University
Tainan, Taiwan

Hai Duong Pham
Department of Physics
College of Sciences
National Cheng Kung University
Tainan, Taiwan

Thinh G. Phung
Viet Nam National University, Ho Chi
Minh City (VNU HCM)
Applied Physical Chemistry Laboratory
Ho Chi Minh University of Science
(HCMUS)
Ho Chi Minh City, Viet Nam

Hikari Sakaebe
Research Institute of Electrochemical
Energy
National Institute of Advanced
Industrial Science and Technology
(AIST)
Tokyo, Japan

Masahiro Shikano
Research Institute of Electrochemical
Energy
National Institute of Advanced
Industrial Science and Technology
(AIST)
Tokyo, Japan

Binh T. Tran
Department of Chemical Engineering
National Cheng Kung University
Tainan, Taiwan

Man V. Tran
Viet Nam National University, Ho Chi
Minh City (VNU HCM)
Applied Physical Chemistry Laboratory
Ho Chi Minh University of Science
(HCMUS)
Department of Physical Chemistry
Ho Chi Minh University of Science
(HCMUS)
Ho Chi Minh City, Viet Nam

Ngoc Thanh Thuy Tran
Hierarchical Green Energy Materials
(Hi-GEM) Research Center
National Cheng Kung University
Tainan, Taiwan

Nhan T. Tran
Viet Nam National University, Ho Chi
Minh City (VNU HCM)
Ho Chi Minh City, Viet Nam
Applied Physical Chemistry Laboratory
Ho Chi Minh University of Science
(HCMUS)
Ho Chi Minh City, Viet Nam

Thien Lan Tran
Department of Physics
Hue University of Education
Hue University
Hue, Vietnam

Tuyen T.T. Truong
Viet Nam National University, Ho Chi
Minh City (VNU HCM)
Applied Physical Chemistry Laboratory
Ho Chi Minh University of Science
(HCMUS)
Ho Chi Minh City, Viet Nam

Yu-Chao Tseng
Department of Chemical Engineering
National Cheng Kung University
Tainan, Taiwan

Shao-Chieh Weng
Department of Materials Science and
    Engineering
National Cheng Kung University
Tainan, Taiwan

Akira Yano
Research Institute of Electrochemical
    Energy
National Institute of Advanced Industrial
    Science and Technology (AIST)
Tokyo, Japan

1 Introduction

Ngoc Thanh Thuy Tran
National Cheng Kung University

Van An Dinh
Osaka University

Ming-Fa Lin
National Cheng Kung University

Hikari Sakaebe
National Institute of Advanced Industrial
Science and Technology (AIST)

Le My Loan Phung
University of Science, Viet Nam National
University – Ho Chi Minh city (VNU HCM)

Chin-Lung Kuo
National Taiwan University

*Jeng-Shiung Jan, Wen-Dung Hsu,
and Jow-Lay Huang*
National Cheng Kung University

CONTENTS

Green energy materials, which cover ion-based batteries [1–4], solar cells [5–8], and hydrogen storages [9–12], have become one of the mainstream condensed-matter systems in basic and applied science researches. They are capable of greatly diversifying the various properties and obviously promoting the potential applications. The first ones are the studying focus of this book, in which the current operations could be classified into five types according to the distinct ions and core materials (Tables 1.1–1.5). A lot of theoretical and experimental studies are done for achieving the high charging/discharging efficiencies. A series of systematic investigations on the critical mechanisms of the essential/physical/chemical/material properties are urgently required for the further development, mainly owing to the insufficient

DOI: 10.1201/9781003263807-1

TABLE 1.1
The Normal Lithium-Ion-Based Batteries under Science Researches with Merits and Drawbacks

a. **Advantages**

b. **Disadvantages**

- High energy density – potential for yet higher capacities.
  - Does not need prolonged priming when new. One regular charge is all that is needed.
  - Relatively low self-discharge – self-discharge is less than half that of nickel-based batteries.
  - Low maintenance – no periodic discharge is needed; there is no memory.
  - Specialty cells can provide very high current to applications such as power tools.

- Requires protection circuit to maintain voltage and current within safe limits.
  - Subject to aging, even if not in use – storage in a cool place at 40% charge reduces the aging effect.
  - Transportation restrictions – shipment of larger quantities may be subject to regulatory control. This restriction does not apply to personal carry-on batteries.
  - Expensive to manufacture – about 40% higher in cost than nickel–cadmium.
  - Not fully mature – metals and chemicals are changing on a continuing basis.

- **Cathodes**: V_2O_5 [119–121], $LiCoO_2$ [122,123], nano-$LiCoO_2$ [124], $LiMn_2O_4$ [125], $Li[Li_{0.20}Mn_{0.54}Ni_{0.13}Co_{0.13}]O_2$ [126], $LiNi_{1/3}Mn_{1/3}Co_{1/3}O_2$ [127,128], $LiMn_{1.5}Ni_{0.5}O_4$ [129], $0.5Li_2MnO_3 \cdot 0.5LiNi_{0.375}Mn_{0.375}Co_{0.25}O_2$ [130], $Li_{1.2}Ni_{0.2}Mn_{0.6}O_2$ [131], $FePO_4$ [132], $LiFe(Co/Ni)PO_4$ [133,134]
- **Anodes**: TiO_2 [135–138], graphite [139–144], patterned Si [145], Si film [146], carbon nanotubes [147,148], carbon nanofibers [149–151], Si nanowires [152,153], Si nanotubes [154], $Li_4Ti_5O_{12}$ [155–158], MoO_3 [159], SnO_2 [160], ZnO [161], Fe_3O_4/carbon foam [162], MnO [163], Co_3O_4 [164], GaS_x [165], MoS_2 [166]
- **Electrolytes**:
  - **Solid-state electrolytes**: Garnet ($Li_7La_3Zr_2O_{12}$) [167], perovskite ($Li_{3x}La_{2/3-x}TiO_3$) [168], Na superionic conductor (NASICON) [169], LIthium Super Ionic CONductor (LISICON) [170], LiMIV 2 (PO4)3 (MIV = Ti, Zr, Ge, and Hf) [171], $LiAlO_x$ [172], Li_3PO_4 [173], lithium silicate [174], Li (Ta/Nb)O_3 [175,176], Li_3N [177], $LiSiAlO_2$ [178], sulfide (Li_4GeS_4, $Li_{10}GeP_2S_{12}$, Li_2S-P_2S_5) [179], argyrodite (Li_6PS_5X (X = Cl, Br, I)) [180], anti-perovskite (Li_3OX (X = Cl, Br, I)) [181], LiSi/Ge/SnO [116,182–184]
  - **Liquid electrolytes**: 1-Ethyl-3-methylimidazolium bis-(trifluoromethylsulfonyl)-imide (EMITFSI) [185], 1-allyl-3-vinyl imidazolium bis(trifluoromethanesulfonyl)imide ([AVIm][TFSI]) [186], N, N-diethyl-N-methyl-N-(2-methoxyethyl)ammonium TFSA (DEME–TFSA) [187], N-methyl-N-butylpyrrolidinium bis(trifluoromethylsufonyl)imide ($PYR_{14}TFSI$) [188–194], N-methyl-N-propylpiperidinium bis(trifluoromethanesulfonyl) imide (PP_{13}–TFSI) [195–198], pyrrolidinium nitrate ($PYRNO_3$) [199], 1-allyl-3-methylimidazolium bis(trifluoromethanesulfonyl)imide and 1-methyl-3-propylimidazolium bis(trifluoromethanesulfone) imide (AMImTFSI and $Im_{13}TFSI$) [200], N, N-diethyl-N-methyl-N-(2-methoxyethyl)ammonium bis(trifluoromethylsulfonyl)imide (DMMATFSI) [201], 1-methyl-1-propylpiperidinium bis(trifluoromethanesulfonyl)imide ($Pip_{13}TFSI$) [202–204], methyl-methylcarboxymethyl pyrrolidinium bis(trifluoromethanesulfonyl) imide, N-butyl-N-methyl-pyrrolidinium bis(trifluoromethanesulfonyl) imide ($MMMPYRTFSIPYR_{14}TFSI$) [205], 1-methoxyethoxymethyl(tri-n-butyl)phosphonium bis(trifluoromethanesulfonyl)amide ($MEMBu_3PTFSI$) [206], 1-ethyl-3-methylimidazoliumbis(trifl uoromethanesulfonyl)imide ($Im_{13}TFSI$) [185,207]
  - **Gel polymer electrolytes**: Poly(vinylidene fluoride) (PVdF) [208], poly(ethylene oxide) (PEO) [209], poly(vinylidene fluoride) hexafluoropropylene (PVdF-co-HFP) [210,211], poly(acrylonitrile) (PAN) [212], poly(methyl methacrylate) (PMMA) [213,214], polystyrene-block-polymethyl methacrylate-block-polystyrene (PS-b-PMMA-b-PS) [215]

TABLE 1.2
Sodium-Ion-Based Batteries

c. Advantages	d. Disadvantages
• **Energy density**: Moderate or high energy density depending on the chemistry used for the sodium-ion battery. • **Safety**: High safety, especially when replacing flammable solvents with non-flammable solvents as co-solvents or as additives [216]. • **Cycle life**: Long cycle life (negligible self-discharge) • **Cost**: Due to the high natural abundance of sodium, commercial production of sodium-ion batteries might be extremely cheap. • Easy transportation	• Lower specific energy compared to LIBs (about one-half). • Less mature technology

• **Cathodes**:
  ◦ **Layered transition metal oxides**: $NaMO_2$ (M = Ti, Fe, Ni, Co, Cr, V) [217], $NaFe_{1/2}Co_{1/2}O_2$ [218], $NaNi_{1/3}Fe_{1/3}Co_{1/3}O_2$ [219], $NaNi_{1/3}Fe_{1/3}Mn_{1/3}O_2$ [220], $NaNi_{0.68}Mn_{0.22}Co_{0.10}O_2$ [221], $Na_{0.7}MO_{2+x}$(M = Co, Mn) [222,223], $Na_{2/3}Fe_{1/2}Mn_{1/2}O_2$ [224].
  ◦ **Metal phosphates and fluorophosphates**: $NaMPO_4$ (M = Co, Fe, Mn) [225], $Na_3M_2(PO_4)_3$ (M = Fe, V) [225], $Na_2FeP_2O_7$ [226], $Na_4Fe_3(PO_4)_2(P_2O_7)$ [227], Na_2FePO_4F [228], $Na_4Co_{2.4}Mn_{0.3}Ni_{0.3}(PO_4)_2P_2O_7$ [229], $Na_3V_2(PO_4)_2F_3$ [230].
  ◦ **Prussian blue and its analogues**: $Na_{1.92}Fe_2(CN)_6$ [231], $KFe_2(CN)_6$ [232], $MnFe(CN)_6$ [233], $Fe_2(CN)_6$ [234], $M_3(Co(CN)_6)_2$ (M = Mn, Fe, Co, Ni, Cu, Zn) [235]
• **Anodes**: Hard carbon [236], $Na_2Ti_3O_7$ [237–239], $Na_3Ti_2(PO_4)_3$ [240], $Na_4Ti_5O_{12}$ [241], $NaTiOPO_4$ [242], $NaTiO_2$ [243], alloys [244,245]
• **Liquid electrolytes**: Similar to Li-ion battery, in which lithium salt is replaced by sodium salt such as $NaPF_6$, $NaN(SO_2CF_3)_2$, and $NaClO_4$ [246]. Additives and solvents might include bis(2,2,2-trifluoroethyl) ether (BTFE), fluoroethylene carbonate (FEC), ethylene carbonate (EC), propylene carbonate (PC), dimethyl carbonate (DMC), diethyl carbonate (DEC), and ethyl methyl carbonate(EMC) [246,247].
• **Solid electrolytes**:
  ◦ Sulfide-based Na-ion conductors: Na_3PS_4 [248], Na_3SbS_4 [249], $Na_{10}SnP_2S_{12}$ [250], $Na_{2.88}Sb_{0.88}W_{0.12}S_4$ [251]
  ◦ NASICONs [250]: $NaZr_2P_3O_{12}$, $Na_3Zr_2PO_4(SiO_4)_2$, $Na_{3.35}La_{0.35}Zr_{1.65}Si_2PO_{12}$, $Na_{1.3}Al_{0.3}Ti_{1.7}P_3O_{12}$.
  ◦ Complex hydrides [250]: $NaAl_4$, Na_3AlH_6, $NaBH_5$, $NaNH_2$
• **Polymer electrolytes [252,253]**: P(VdF-HFP), poly(ethylene oxide) (PEO)

drawbacks of theoretical frameworks. It is well-known that certain ion-based batteries have become very popular commercial products. However, the high-precision measurements are frequently absent because of the technical limits and the complicated environments, such as the detailed examinations on lithium oxides of 3D intermediate crystal structures, electronic energy spectra, density of states, magnetic moments, frequency-dependent optical absorption/reflectance spectra, and photoluminescence spectra [1,13,14].

As for the theoretical progress of the emergent materials, the first-principles simulations and phenomenological models are frequently utilized to explore the essential material/physical/chemical properties and the applied sciences. Their direct combinations have been developed under the quasiparticle viewpoints, as fully done for the mainstream and emergent materials. For example, the systematic investigations

TABLE 1.3

Lithium–Sulfur Batteries

Advantages	Disadvantages
• **Energy density**: Combining lithium and sulfur delivers high energy density. • **Maintenance**: Need little maintenance compared to the other batteries such as lithium-ion batteries. • **Environment**: The most resilient batteries that are not easily damaged in harsh environments and last a few years longer than lithium-ion batteries. • **Self-discharge**: Do not discharge their own power itself therefore still keep charges. • **Safety**: Safety concerns because a negative electrode comprised of metallic lithium need protected by a passivating layer.	• **Transportation**: Due to the chemicals used, nearly every airline company restricts lithium–sulfur batteries to be carried in the airplane. • **Cost**: Expensive compared with other batteries, for example, with the same cost, you will buy nearly 2 to 3 lithium-ion batteries as much as you get a lithium–sulfur battery.

- **Cathodes**: Polyethylene glycol-coated [254], pitted mesoporous carbon [254], sulfur-coated [255,256], disordered carbon hollow carbon nanofibers [255,256], disordered carbon nanotubes made from carbohydrates [257], sulfur, copolymerized sulfur [258], porous TiO_2-encapsulated sulfur nanoparticles [259], sulfur [260], sulfur-graphene oxide nanocomposite with styrene-butadiene-carboxymethyl cellulose copolymer binder [261], sulfur/lithium–sulfide passivation layer [262], sulfur-copolymer (poly(S-co-DVB)) [263], carbon nanotube/sulfur [264–266], glass-coated sulfur with mildly reduced graphene oxide for structural support [267]
- **Anodes**: Lithium metal [254,255,261], silicon nanowire/carbon [257], lithiated hard carbon [263]
- **Electrolytes**: Li_2S/P_2S_5 [268], $Li_2S/P_2S_5/LiI$, $Li_2S/P_2S_5/Li_2O$, $Li_2S/P_2S_5/P_2O_5$, $Li_2S/Li_2O/P_2S_5$, $(0.7B_2S_3·0.3P_2S_5)/Li_2S$ [269], $(0.75:0.25\ Li_2S/P_2S_5)/LiBH_4$ [269], $Li_2S/P_2S_5,\ Li_7P_3S_{11-zz}$, solid polymer electrolyte: PEO–LiX (ClO_4, PF_6, $N(SO_2CF_3)$) [269]

cover 3D AA-, AB- and ABC-stacked graphites [13–15], layered graphene without/ with chemical modifications [16,17], graphene nanoribbons [18,19], and silicene-related materials [20,21]. The intrinsic quasiparticle interactions, which mainly arise from charge and spin distributions, are deduced to dominate the diverse phenomena. All the researches are focused to clarify the active orbital hybridizations of chemical bonds and the non-magnetic/ferromagnetic/anti-ferromagnetic spin configurations. Generally speaking, this most important goal can be achieved by thoroughly examining the geometric, electronic, magnetic, optical, and transport properties [1,22–25]. The quasiparticle viewpoints, being assisted by the Vienna Ab initio Simulation Package (VASP) and molecular dynamics, are able to provide the concise pictures in understanding the diverse phenomena of cathode/electrolyte/anode core components in ion-based batteries; that is, they can thoroughly clarify the critical mechanisms in pristine subsystems and even the composite systems with heterojunctions. Very interestingly, there exist a lot of theoretical predictions on the stationary ion transport. However, how to develop a unified theoretical framework suitable for the very complicated and active chemical environments in ion-related batteries remains to be urgently solved during the near-future studies.

TABLE 1.4
Aluminum-Related Batteries

Advantages

- Safe: The new battery is non-flammable and safer than LIBs, which can burst into flames.
- Environmentally friendly: It is safer for the environment than alkaline batteries.
- Flexible: The battery has great potential for use in flexible electronics because it can be bent or folded.
- Fast charging: It charges in just 1 minute
- **Long cycle life**: The aluminum battery can withstand more than 7,500 cycles without any loss of capacity, which is much better than a typical LIB, which lasts about 1,000 cycles.
- Inexpensive: Aluminum and graphite are relatively cheap materials
- **Cathodes**: Graphite [270], Mn_2O_4 [271], V_2O_5 [272–274], VO_2 [275], CFx [270], graphitic foam [276], Pyrolytic graphite [276], graphic carbon paper [277], $LiFePO_4$ [278], Mo_6S_8 [279], Ni_3S_2 [280], NiS [281], S [282], VCl3 [283], CuHCF [284], polythiophene [285], polypyrrole [285], FeS_2 [286,287], FeS [286,288], Al [289], Zn [290], Cd [290]
- **Anodes**: Al [270,271]
- **Electrolytes**: Aluminum chloride and 1-ethyl-3-methylimidazolium chloride ($AlCl_3$/[EMIm]Cl) [273,278], aluminum chloride and 1-butyl-3-methyl imidazole salt chloride and ($AlCl_3$/[BMIm]Cl) [272], aluminum chloride and 1,3-di-n-butylimidazolium bromide ($AlCl_3$/[BIM]Br) [270], $AlCl_3$–KCl–NaCl (acidic) [291] $AlCl_3$–NaCl (acidic) [292,293], $AlCl_3$–NaCl–Al_2S_3 (basic) [288], alkali tetrahaloaluminate salts (LiAlCl$_4$–NaAlCl$_4$–NaAlBr$_4$–KAlCl$_4$) [294], NaCl saturated NaAlCl$_4$ (acidic) [295], aluminum chloride and 1-methyl-3-ethylimidazolium chloride ($AlCl_3$–MEICl (acidic)) [274,289], $AlCl_3$–dipropylsulfone–toluene [296], $LiPF_6$–ethylene carbonate (EC)–ethyl methyl carbonate (EMC) ($LiPF_6$–EMC–VC) [297], aluminum chloride-1-butyl-3-methylimidazolium chloride benzyl sulfoxide ($AlCl_3$–BMICl–benzyl sulfoxide) [275], aluminum bromide and 1-ethyl-3-methylimidazolium chloride (AlBr$_3$–MEIBr (acidic)) [298]

Disadvantages

- Increased expansion
- Contraction of the reactive material

TABLE 1.5
Iron-Related Batteries

Advantages

- High energy density (because lesser numbers of ions are required to achieve the same electrochemical capacity)
- Abundance
- Very low tendency for dendrite formation on repeated cycling and easy recyclability
- **Cathodes**: V_2O_5 [299]
- **Anodes**: Fe [299]
- **Electrolytes**: Tetraethylene glycol dimethyl ether (TEGDME) [299]

Disadvantages

- The redox potential of Fe/Fe$_2$+ (-0.41 V vs SHE) is higher than that of Li/Li+ (-3.01 V vs. SHE), which may decrease its energy density.

Compared with other energy storage systems, lithium-ion batteries (LIBs) have attracted more and more attention recently due to their desirable features, such as lightweight, high energy density, fast charging rate, great cycle stability, long cycle life, and shape variability [26,27]. Besides LIBs, sodium-ion batteries (SIBs) have also attracted significant interest as large-scale energy storage candidates due to a naturally available sodium resource and low cost. Generally, ion-based battery is constructed by three main components, namely electrolyte, positive (cathode) and negative (anode) electrodes [26,27]. The electricity is stored mostly in the cathode and anode pair in the form of chemical energy, which is similar to an electrochemical cell. The cathode provides active component, namely the transition metal ions, while the anode stores the chemical energy during the charging session of battery. Ion-based batteries operate according to a "rocking chair" concept, in which ions are extracted from cathode and intercalated into anode when the battery is charged and then deintercalated from the anode and are reintroduced into the cathode under the discharge process. The physical/chemical/material properties of each component of ion-based battery are rather complex and directly related to its performance. In this book, the close relations among the electrode voltage, cycle number, specific charge capacity, and electrode/electrolyte interface stability are fully established through the high-precision experimental measurements. On the other hand, the essential properties, as done for band structures, optical absorption spectra and Coulomb excitations, diffusion mechanism, and anisotropic lithiation, are systematically investigated by means of first-principles calculations.

The organization of this book is as follows. Chapter 1 includes the general introduction, especially for the whole developments in the theoretical and experimental researches on ion-based batteries. Chapters 2–9 discuss the electrode materials, the main key components of batteries, in which various types of cathode and anode materials are investigated in Chapters 2–5 and Chapters 6–9, respectively. In particular, by means of first-principles calculations, Chapter 2 introduces a self-developed small polaron model to investigate the diffusion mechanism in transition metal-based cathode materials, while Chapter 3 focuses on the electronic, magnetic, and optical properties of $LiFeO_2$ cathode. On the experimental side, Al- and Zr-oxide-coated commercial cathodes $LiNi_{1/3}Mn_{1/3}Co_{1/3}O_2$ and $LiCoO_2$ are investigated in Chapter 4 to emphasize the coating effect on stability and Li-ion transfer kinetics at high-voltage region of LIBs, whereas Chapter 5 discusses the doping/substituting effect on phase transition and structural stability for sodium layered metal oxides cathode of SIBs. Chapters 6–8 show theoretical studies on promising anode materials, namely $FeCl_3$-intercalated graphite, silicon nanowire, and Li-intercalated HfX_2, providing a deep understanding of chemical bonding, charge transfer, anisotropic lithiation, and optical properties. Chapter 9 is experimental research on Mn-based oxide nanocomposite with reduced graphene oxide (rGO) anodes, covering four main types, including rGO, MnO_2 nanoneedle, MnO_2/rGO, and Mn_3O_4/rGO hexagonal-flat structure. Apart from electrode materials, electrolyte also plays an important role in the performance of batteries. To address the risk of fire issue posed by the use of flammable liquids, solid electrolytes have been developed and this book will introduce solid-state polymer electrolytes for LIBs and crystalline Li_2S and P_2S_5 for lithium–sulfur batteries (discussed later in Chapters 10–12). On the other hand, another challenge affecting the performance of the full cell battery is the interface issues between the electrode and electrolyte, which are addressed in Chapters 13 and 14 for LIBs and SIBs, respectively. Chapter 15 covers the concluding remarks of the aforementioned main chapters, while Chapter 16 discusses open issues and challenges. Finally, the battery-related practice problems are provided in Chapter 17.

In Chapter 2, we elucidate the formation mechanism of a quasiparticle named small polaron in the insulating cathode materials during the battery operation [28,29]. The basic characteristics of a typical small polaron such as strong binding with alkali vacancy/ion, or bound states caused by polaron formation in the density of states, are addressed in great detail. Moreover, we introduce an advanced diffusion model of small polaron–alkali vacancy, which can help theoretical scientists open a door to describe naturally and accurately the diffusion mechanism of alkali ions during the charge and discharge [3,30–37].

Previous theoretical studies have been conducted on the geometric and electronic properties of lithium-based materials. However, their results and analysis are insufficient to fully comprehend the fundamental properties of these compounds. For example, there is lacking of the critical optical excitation, the excitonic effect on optical properties, and the close connection of electronic, magnetic with optical excitation. That is to say, critical mechanisms (concise physical pictures) in comprehending the diversified phenomena are questions that have no answer up to now. The theoretical framework is based on the numerical calculations, and delicate analyses were developed and applied for the layered $LiFeO_2$ – a candidate for cathode compound. The fundamental features, the critical quasiparticle properties, and the significant orbital hybridizations in various chemical bonds are thoroughly examined in Chapter 3 from the first-principles simulations. Furthermore, the optical excitation, the effect of electron–hole coupling, and other optical properties will be illuminated. The charging and discharging of LIBs are expected to become complicated, owing to the variation of chemical bonds and thus orbital hybridizations. Our prediction provides certain meaningful information about the critical physical/chemical pictures in LIBs. Such state-of-the-art analysis is very useful for fully comprehending the diversified properties in anode/cathode/electrolyte and other emerging materials.

Layered rock-salt-structured positive electrode materials, such as $LiCoO_2$ and $LiNi_{1/3}Co_{1/3}Mn_{1/3}O_2$, exhibit a theoretical capacity as high as 270–280 mAh g^{-1} and an average discharging voltage of approximately 4 V (vs. Li$^+$/Li). However, their actual capacity in commercial LIBs is typically limited to approximately 60% of their theoretical capacity; further, their charging voltage does not exceed 4.4 V. This is because the electrodes rapidly degrade at higher charging voltages. In addition to the stability of the bulk of the active material, the stability of the interface between the electrode and the electrolyte is also of great importance. The application of a surface coating on layered rock-salt-active materials is an effective technique for stabilizing the electrode/electrolyte interface, especially for high-voltage charge/discharge. Furthermore, some oxide coatings have been reported to improve the current-rate and cycle performance, suggesting that the surface coatings not only stabilize the interface but also affect the Li-ion transfer kinetics at the interface. In Chapter 4, the surface structures of Al- and Zr-oxide-coated $LiNi_{1/3}Co_{1/3}Mn_{1/3}O_2$ and $LiCoO_2$ electrodes were described first; subsequently, the effect of surface coating on the charge/discharge characteristics in the high-voltage region was shown. Next, kinetic analysis using an epitaxial thin-film electrode was performed. The underlying mechanism of interface stabilization by surface coating was investigated along with the kinetics of Li-ion transfer at the interface.

In Chapter 5, we have systematically summarized the synthesis, applications, and electrochemical properties of layered cathode materials for SIBs. The common structure of sodium layer metal oxides can be divided into two main groups: O3-type and P2-type, whereas the sodium ions are occupied at octahedral and prismatic sites, respectively. However, both P2 and O3 layered phases commonly suffer from a series of phase transitions induced by the extraction of Na ions during electrochemical cycling. For in-depth studying of the structure evolution and phase transition of layered structure materials for SIBs, we also provide a comprehensive overview of ex situ (both static) and in situ (real-time) techniques. In addition, we have described the structure and electrochemistry of sodium insertion materials having a layered structure such as single metal layered materials (Na_xMO_2), two metals ($Na_xMM'O_2$), and three metals ($Na_xMM'M''O_2$). In particular, $NaNi_{1/3}Mn_{1/3}Co_{1/3}O_2$ material has been synthesized at different conditions with different phase components such as single-phase (O3, P2, and P3), bi-phase P2/O3, and multi-phase P2/O1/O3.

Graphite is the most well-known and important material in basic science research and potential applications [38]. The special geometric structures have attracted many theoretical and experimental [39,40] studies in past years. The 3D graphite systems might present the AA, AB, ABC, and turbostratic stackings [41–43]. All of them belong to semimetals under the interlayer hopping integrals of C-2pz orbitals. In general, such condensed-matter systems become the n- or p-type metals, depending on the kinds of intercalated atoms or molecules [44]. For example, the intercalation of alkali metal atoms [45] and $FeCl_3$ [46] into graphite, respectively, creates many free conduction electrons and valence holes, as observed in a pure metal. The great interlayer spacing provides a chemical environment in creating the chemical intercalations or deintercalations for the various atoms/molecules/ions [47], especially for the charging and discharging processes in ion-based batteries [48]. Interestingly, this stable system is usually utilized as the electrode materials of the lithium-/aluminum-/iron-based batteries. The rich essential properties have been studied for the alkali-metal-graphite intercalation compounds in previous works [49]. This study is mainly focused on stage-n $FeCl_3$-graphite intercalation compounds. Based on the first-principles simulations and calculations, the critical quasiparticle properties and the significant orbital hybridizations in various intralayer and interlayer chemical bonds are fully examined in Chapter 6. The intercalations and deintercalations of large molecules are expected to become more complicated, mainly owing to the enlarged Moiré superlattices [50]. This study is very useful in constructing the theoretical framework for $FeCl_3$-graphite intercalation compounds [42–51] and helpful for in-depth investigating the full features of graphite-related intercalation compound electrodes.

Silicon anode has gained more and more interest these days due to its tenfold higher capacity as 3,579 mAh g^{-1} than commonly used graphite anode (342 mAh g^{-1}) [52]. In a common LIB, an anode stores electrons from the external circuit by anode reduction and forming chemical bonds with lithium ion in electrolyte during the charging process. This means that electron from external circuit will be stored with lithium ion in anode. Similarly, silicon anode stores lithium ion by forming Li-Si alloy as a-Li$_{15}$Si$_4$, which is called the lithiation of the silicon anode. The far greater amount of Li incorporation with respect to silicon in a-Li$_{15}$Si$_4$ caused high volume expansion as about 400% in alloying process and pulverizes the bulk silicon. Therefore, the nanostructured silicon anode is used to solve this problem [53] by providing more space for the lithium–silicon alloy expansion. Unfortunately, the *anisotropic lithiation* of the crystalline silicon nanowire anode [54,55], which means that the lithiation rate differs significantly in each facet, cracks the silicon anode due to the stress concentration on the concaved region formed by slower moving facet. Up until now, several theoretical studies [56–58] have been performed to investigate the reason for the anisotropic lithiation, but the anisotropic lithiation cannot be fully reproduced. In Chapter 7, we will study the anisotropic lithiation process in classical molecular dynamics simulation with our newly developed ReaxFF model [59] from Ostadhossein et al. [60]. Some problems in the previous model are corrected, especially for Si [111] lithiation rate. We will reproduce the anisotropic lithiation process for Si [100], Si [110], Si [111], and Si [112] slab as in the experiment [54,55,61] and observe the lithiation behavior on an atomic scale. Furthermore, we will perform stress analysis to find out the reason for the anisotropic lithiation. Finally, we will conclude if the anisotropic lithiation in crystalline silicon is a thermodynamic- or kinetic-dominated process.

Recently, applications of electrochemical energy storage, particularly LIBs and SIBs, help to express the challenges of global climate change that have attracted concern in science community. The satisfactory results open up the opportunities to apply in a wide range of smart electrical grids and electronic devices [62–71]. 2D materials, especially graphene, are emerged as potential and promising candidates for batteries. To improve and enhance applications of 2D materials, the newborn materials based on graphene have been investigated that provide new suggestions for this field. Among 2D materials, transition metal dichalcogenides (TMDs) become well-known with the unique and essential properties [63,71–87] for applications in battery. Most importantly, group IV TMDs, especially HfX$_2$, are one of the newborn and promising 2D materials for the development of high performance in energy storage. However, the fundamental and connected information of electronic and optical properties might be still limited, which needs to be analyzed to provide the perspective of their properties. Clearly, it is useful to understand the connection between structural, electronic, and optical properties of these materials; thus, further investigations or applications can be recommended from these findings. We therefore focused on summary of the geometric and electronic properties that had been discussed in the previous work and investigated the optical properties of monolayer HfX$_2$ to verify the connection and consistence of their properties presented in Chapter 8. From these properties, the suggestion for further investigations could be revealed so that the applications in battery will be enlarged and enhanced.

Energy conversion and storage is one of the ongoing fundamental researches over the years, and the research in LIBs is in the forefront in this direction to satisfy the energy requirements in the future [88]. Graphite is the most commonly used anode in LIBs, but the low intrinsic capacity of 372 mAh g^{-1} hinders its industrial applications, and therefore, a lot of research is going to develop new materials or to modify the existing materials [89–92]. Metal oxides and their composites are considered as the possible candidates because of their high capacities (~1,000 mAh g^{-1}) and long cyclic stability. Here, we demonstrate the use of Mn-based oxides such as MnO_2, Mn_3O_4, and their composites with rGO as the potential anode materials in LIB [93]. Both the as-prepared composites are highly crystalline, whereas the microstructure varied significantly from nanorod-like morphology for MnO_2 to hexagonal plate-like morphology for Mn_3O_4 sheets. Among all the Mn-based composites, Mn_3O_4 grown over the rGO has delivered significant capacity (1,086.5 mAh g^{-1}) in the first cycle and the capacity is retained at 638.13 mAh g^{-1} after 250 cycles and has very good rate capability (549.35 mAh g^{-1} @ 2C current rate). A detailed investigation about the structure, microstructure, bond vibration, and electrochemical property was carried out in Chapter 9.

Currently, LIBs have become well-known power components of portable mobile devices such as laptops and smartphones. Meanwhile, LIBs will play a much more important role in our future life owing to their flourishing development in the field of household appliances and electric/hybrid vehicles. Electrolytes are most important in batteries, as they carry ions between a pair of electrodes to complete a circuit of charges. Poly(ionic liquid)s (PILs), which are polymerized from ionic liquid molecules, have emerged as a novel topic of polymer electrolytes because of their extraordinary combination of ionic properties and polymer natures. In this chapter, a thiol-ene click cross-linking reaction was employed to form a thiosiloxane/poly(ethylene glycol) diacrylate (PEGDA)/imidazolium-based cross-linker (C-VIm) co-network for application as a polymer scaffold in the electrolytes. And the thermal stabilities, the electrical properties as well as the charge/discharge test of the Li/LiFePO$_4$ cells based on these materials would be fully discussed in Chapter 10.

Very interestingly, the lithium-, phosphorous-, and sulfur-related elements [94], molecule gases [95], as well as compounds [96–99] can provide very active chemical environments [99,100], thus creating a plenty of emergent materials [98] and the highly potential applications [101–103]. Obviously, such condensed-matter systems are outstanding candidates in both basic and applied science researches. According to the previous works on the theoretical framework of quasiparticle viewpoints, the first-principles numerical simulations [16,104,105] and the phenomenological models are available in thoroughly exploring the critical mechanisms of the diverse physical/chemical/material phenomena. The close relations need to be established between the single- or multi-orbital hybridizations & even the spin configurations [16,104] and the essential properties (e.g., geometric [16,104], electronic [16,104], magnetic [16,104], and optical properties [106]). Specifically, the real current operations show that the multi-component lithium phosphorus sulfide compounds are frequently utilized as the solid-state electrolyte of Li$^+$-ion-based batteries because of the significant safety. Such complicated systems are expected to possess the unusual chemical bondings in the non-uniform and anisotropic environments [99,100].

On the experimental side, the multi-component 3D compounds have been successfully synthesized from the distinct experimental methods. For example, the mainstream materials cover the binary Li_2S [97] and P_2S_5 [98]; the ternary $Li_2P_2S_6$ [98], $Li_4P_2S_6$ [98], $Li_7P_3S_{11}$ [98], and $Li_3P_1S_4$ [98]; the quaternary LiPSCl [107]; and so on (details in Table 1.3), where they could be regarded as the direct combinations of the different elements and the distinct ratios of two, three, or four subsystems. Generally speaking, the frequently utilized growth manners are classified into three types: (1) wet chemistry [108], (2) solid-state route [108], and (3) mechanochemical synthesis [108]. Among the abovementioned materials, the ternary, quaternary, and even pentanary/hexanary Li-, P- and S-related compounds could serve as the high-performance solid-state electrolytes through the most commensurate cathode and anode materials [109,110]. Most importantly, the sensitive dependences among the specific capacity, voltage, and cycle number are thoroughly conducted under the charging and discharging processes [101,110]. These complicated condensed-matter systems should be very suitable for further developing the theoretical quasiparticle framework. However, the high-precision examinations on the essential properties of core materials are almost absent except the X-ray diffraction patterns [111], e.g., the experimental observations of the occupied band structures by angle-resolved photo-emission spectra [112]/the van Hove singularities through tunneling quantum currents [113]/the optical absorption structures [106]. This might be associated with the intrinsic limits [106] and/or the giant barriers of delicate analyses [114].

The first principles [114] have been used to predict the optimal geometric structures [115,116] and electronic energy spectra [115,116] for most electrolyte/cathode/abode materials of lithium-ion-based batteries. Their simulations are reliable even for Moiré superlattices with a lot of atoms in a primitive unit cell [99,117]. For example, the multi-component lithium phosphorus sulfide compounds [98] exhibit the diverse crystal symmetries and the unusual semiconducting behaviors through the VASP calculations [114]. However, the delicate analyses, which are conducted on the specific atom-dominance energy ranges, are almost absent. Furthermore, their direct combinations with the spatial charge density and the featured density of states are not conducted on the previously published works [97,99,100,117,118]. But according to the theoretical development of quasiparticle framework [16,104,105], the atom-, orbital-, and spin-projected van Hove singularities are available in accurately identifying the active orbital hybridizations/the magnetic spin configurations of chemical bonds. The rich and unique quasiparticle properties, which appear in various 3D Li-, P-, and S-related materials, are worthy of the systematic investigations using the VASP that will be addressed in Chapters 11 and 12.

In recent years, the focus of LIB research has shifted toward hybrid electric vehicles (HEVs) and full-functioning electric vehicles (EVs). Currently, the largest problem arises from the cyclability and capacity retention of the high operation voltage power battery. Most of the cathode-related LIB performance issues are related to the reaction occurring at the cathode/electrolyte interface (CEI). For example, surface phase transformation, transition metal dissolution, and electrolyte decomposition, as these are the most pronounced issues, cause unreliability that obstructs the transition from fossil fuel to LIB-based EVs. To improve the overall performance of LIB, the CEI initiation mechanism and surface reaction must be scrutinized. As the experimental observation of such atomic-scale reaction is unattainable, first-principles simulation method has become the most successful complementary approach to guide the experiment and to provide reasonable explanation for the observed result. This speeds up the development of novel materials and the modification of currently available materials, by providing either the theoretical performance of the target in interest or the possible mechanism of undesired side reaction. More details are shown in Chapter 13.

On the other hand, since the accelerated growth of the last decade, commercial requirements for SIBs have been focused on applications for energy storage systems. However, the lack of knowledge of the physical and chemical dynamics occurring at the electrode/electrolyte interface is the greatest obstacle to the advancement of existing battery technology and the creation of new ones. Similar to LIBs, the most important problem to improve lifetime, power capability, and even the safety of SIBs is understanding the chemical electrode surface. In Chapter 14, we begin with a discussion on the chemical composition and structure of the Na-ion solid electrolyte interphase (SEI). We also review the current understanding of the mechanism of SEI formation on different types of anodes and the existence of the SEI in ether-based electrolytes to emphasize how the enhanced stability of the SEI fundamentally affects the efficiency and reversibility of electrochemical reactions. Therefore, the buildup of the appropriate SEI layer is a necessary step in optimizing the combination of anode/electrolyte/cathode for Na-ion batteries, either focusing on finding the new materials or developing novel ones. Furthermore, the understanding of the analysis techniques used to characterize the SEI layer properties is also provided in this chapter. Finally, we discuss the SEI layer's perspective and challenges in the schedule for developing commercial Na-rechargeable batteries.

REFERENCES

1. Dien VK, Pham HD, Tran NTT, Han NT, Huynh TMD, Nguyen TDH, et al. Orbital-hybridization-created optical excitations in $Li_2 GeO_3$. *Scientific Reports*. 2021; 11(1):1–10.
2. Pan T-Y, Tran NTT, Chang Y-C, Hsu W-D. First-principles study on the initial reactions at $LiNi_{1/3}Co_{1/3}Mn_{1/3}O_2$ cathode/electrolyte interface in lithium-ion batteries. *Applied Surface Science*. 2020;507:144842.
3. Tran TL, Luong HD, Duong DM, Dinh NT, Dinh VA. Hybrid Functional Study on Small Polaron Formation and Ion Diffusion in the Cathode Material $Na_2Mn_3 (SO_4)$ 4. *ACS omega*. 2020;5(10):5429–35.
4. Tsai P-C, Nasara RN, Shen Y-C, Liang C-C, Chang Y-W, Hsu W-D, et al. Ab initio phase stability and electronic conductivity of the doped-$Li_4Ti_5O_{12}$ anode for Li-ion batteries. *Acta Materialia*. 2019;175:196–205.
5. Hu M, Chen M, Guo P, Zhou H, Deng J, Yao Y, et al. Sub-1.4 eV bandgap inorganic perovskite solar cells with long-term stability. *Nature Communications*. 2020;11(1):1–10.
6. Jeong M, Choi IW, Go EM, Cho Y, Kim M, Lee B, et al. Stable perovskite solar cells with efficiency exceeding 24.8% and 0.3-V voltage loss. *Science*. 2020;369(6511):1615–20.
7. Kim JY, Lee J-W, Jung HS, Shin H, Park N-G. High-efficiency perovskite solar cells. *Chemical Reviews*. 2020;120(15):7867–918.
8. Shabbir SA, Azher Z, Latif H. Comparison of efficiency analysis of perovskite solar cell by altering electron and hole transporting layers. *Optik*. 2020;208:164061.
9. Lang C, Jia Y, Yao X. Recent advances in liquid-phase chemical hydrogen storage. *Energy Storage Materials*. 2020;26:290–312.

10. Zhang X, Liu Y, Zhang X, Hu J, Gao M, Pan H. Empowering hydrogen storage performance of MgH_2 by nanoengineering and nanocatalysis. *Materials Today Nano*. 2020;9:100064.

11. Zhu Y, Ouyang L, Zhong H, Liu J, Wang H, Shao H, et al. Closing the loop for hydrogen storage: Facile regeneration of $NaBH_4$ from its hydrolytic product. *Angewandte Chemie International Edition*. 2020;59(22):8623–9.

12. Ouyang L, Liu F, Wang H, Liu J, Yang X-S, Sun L, et al. Magnesium-based hydrogen storage compounds: A review. *Journal of Alloys and Compounds*. 2020;832:154865.

13. Lin M-F, Hsu W-D, Huang J-L. *Lithium-Ion Batteries and Solar Cells: Physical, Chemical, and Materials Properties*: CRC Press.

14. Lin M-F, Hsu W-D. *Green Energy Materials Handbook*: CRC Press; 2019.

15. Li W-B, Lin S-Y, Tran NTT, Lin M-F, Lin K-I. Essential geometric and electronic properties in stage-n graphite alkali-metal-intercalation compounds. *RSC Advances*. 2020;10(40):23573–81.

16. Tran NTT, Lin S-Y, Lin C-Y, Lin M-F. *Geometric and Electronic Properties of Graphene-Related Systems: Chemical Bonding Schemes*: CRC Press; 2017.

17. Tran NTT, Lin S-Y, Lin Y-T, Lin M-F. Chemical bonding-induced rich electronic properties of oxygen adsorbed few-layer graphenes. *Physical Chemistry Chemical Physics*. 2016;18(5):4000–7.

18. Lin S-Y, Tran NTT, Chang S-L, Su W-P, Lin M-F. *Structure-and Adatom-Enriched Essential Properties of Graphene Nanoribbons*: CRC press; 2018.

19. Lin S-Y, Tran NTT, Fa-Lin M. Diversified phenomena in metal-and transition-metal-adsorbed graphene nanoribbons. *Nanomaterials*. 2021;11(3):630.

20. Lin S-Y, Liu H-Y, Nguyen DK, Tran NTT, Pham HD, Chang S-L, et al. *Silicene-Based Layered Materials*: IOP Publishing Limited; 2020.

21. Tran NTT, Gumbs G, Nguyen DK, Lin M-F. Fundamental properties of metal-adsorbed silicene: A DFT study. *ACS Omega*. 2020;5(23):13760–9.

22. Chettri B, Patra P, Vu TV, Nguyen CQ, Yaya A, Obodo KO, et al. Induced ferromagnetism in bilayer hexagonal boron nitride (h-BN) on vacancy defects at B and N sites. *Physica E: Low-dimensional Systems and Nanostructures*. 2021;126:114436.

23. Tran NTT, Dahal D, Gumbs G, Lin M-F. Adatom doping-enriched geometric and electronic properties of pristine graphene: A method to modify the band gap. *Structural Chemistry*. 2017;28(5):1311–8.

24. Tran NTT, Lin S-Y, Glukhova OE, Lin M-F. π-Bonding-dominated energy gaps in graphene oxide. *RSC Advances*. 2016;6(29):24458–63.

25. Van Hoang V, Tran NTT, Giang NH, Dong TQ. Two-dimensional FeC compound with square and triangle lattice structure–Molecular dynamics and DFT study. *Computational Materials Science*. 2020;181:109730.

26. Pistoia G. *Lithium-Ion Batteries*: Elsevier; 2013.

27. Yoshio M, Brodd RJ, Kozawa A. *Lithium-Ion Batteries*: Springer; 2009.

28. Emin D. Optical properties of large and small polarons and bipolarons. *Physical Review B*. 1993;48(18):13691.

29. Maxisch T, Zhou F, Ceder G. Ab initio study of the migration of small polarons in olivine $Li_x FePO_4$ and their association with lithium ions and vacancies. *Physical review B*. 2006;73(10):104301.

30. Bui KM, Dinh VA, Ohno T. Diffusion mechanism of polaron–Li vacancy complex in cathode material Li_2FeSiO_4. *Applied Physics Express*. 2012;5(12):125802.

31. Bui KM, Dinh VA, Okada S, Ohno T. Hybrid functional study of the NASICON-type $Na_3V_2(PO_4)3$: Crystal and electronic structures, and polaron–Na vacancy complex diffusion. *Physical Chemistry Chemical Physics*. 2015;17(45):30433–9.

32. Debbichi M, Debbichi L, Dinh VA, Lebègue S. First principles study of the crystal, electronic structure, and diffusion mechanism of polaron-Na vacancy of $Na_3MnPO_4CO_3$ for Na-ion battery applications. *Journal of Physics D: Applied Physics*. 2016;50(4):045502.

33. Dinh VA, Nara J, Ohno T. A new insight into the polaron–li complex diffusion in cathode material $LiFe1-yMnyPO_4$ for Li ion batteries. *Applied Physics Express*. 2012;5(4):045801.

34. Duong DM, Dinh VA, Ohno T. Quasi-three-dimensional diffusion of Li ions in $Li_3FePO_4CO_3$: First-principles calculations for cathode materials of Li-ion batteries. *Applied Physics Express*. 2013;6(11):115801.

35. Luong HD, Dinh VA, Momida H, Oguchi T. Insight into the diffusion mechanism of sodium ion–polaron complexes in orthorhombic P2 layered cathode oxide $NaxMnO_2$. *Physical Chemistry Chemical Physics*. 2020;22(32):18219–28.

36. Luong HD, Pham TD, Morikawa Y, Shibutani Y, Dinh VA. Diffusion mechanism of Na ion–polaron complex in potential cathode materials $NaVOPO_4$ and $VOPO_4$ for rechargeable sodium-ion batteries. *Physical Chemistry Chemical Physics*. 2018;20(36):23625–34.

37. Bui KM, Dinh VA, Okada S, Ohno T. Na-ion diffusion in a NASICON-type solid electrolyte: A density functional study. *Physical Chemistry Chemical Physics*. 2016;18(39): 27226–31.

38. Brownson DA, Banks CE. Graphene electrochemistry: An overview of potential applications. *Analyst*. 2010;135(11):2768–78.

39. Abergel D, Apalkov V, Berashevich J, Ziegler K, Chakraborty T. Properties of graphene: A theoretical perspective. *Advances in Physics*. 2010;59(4):261–482.

40. Abbott's IE. Graphene: Exploring carbon flatland. *Physics Today*. 2007;60(8):35.

41. Ho J-H, Lu C, Hwang C, Chang C, Lin M-F. Coulomb excitations in AA-and AB-stacked bilayer graphites. *Physical Review B*. 2006;74(8):085406.

42. Lin C-Y, Lee M-H, Lin M-F. Coulomb excitations in ABC-stacked trilayer graphene. *Physical Review B*. 2018;98(4):041408.

43. Lin C-Y, Wu J-Y, Chiu C-W, Lin M-F. *Coulomb Excitations and Decays in Graphene-Related Systems*: CRC Press; 2019.

44. Matsumoto R, Okabe Y, Akuzawa N. Thermoelectric properties and performance of n-type and p-type graphite intercalation compounds. *Journal of Electronic Materials*. 2015;44(1):399–406.

45. Natori A, Ohno T, Oshiyama A. Work function of alkali-atom adsorbed graphite. *Journal of the Physical Society of Japan*. 1985;54(8):3042–50.

46. Li Y, Yue Q. First-principles study of electronic and magnetic properties of $FeCl_3$-based graphite intercalation compounds. *Physica B: Condensed Matter*. 2013;425: 72–7.

47. Eklund PC. Optical studies of the electronic and lattice dynamical properties of graphite-intercalation compounds. *Bulletin of the American Physical Society*. 1981;26(3):265–.

48. Xu JT, Dou YH, Wei ZX, Ma JM, Deng YH, Li YT, et al. recent progress in graphite intercalation compounds for rechargeable metal (Li, Na, K, Al)-ion batteries. *Advanced Science*. 2017;4(10).

49. Toyoura K, Koyama Y, Kuwabara A, Oba F, Tanaka I. First-principles approach to chemical diffusion of lithium atoms in a graphite intercalation compound. *Physical Review B*. 2008;78(21).

50. Ryu YK, Frisenda R, Castellanos-Gomez A. Superlattices based on van der Waals 2D materials. *Chemical Communications*. 2019;55(77):11498–510.

51. Gao YR, Zhu CQ, Chen ZZ, Lu G. Understanding Ultrafast Rechargeable Aluminum-Ion Battery from First-Principles. *Journal of Physical Chemistry C*. 2017;121(13):7131–8.

52. Bourderau S, Brousse T, Schleich D. Amorphous silicon as a possible anode material for Li-ion batteries. *Journal of Power Sources*. 1999;81:233–6.

53. Chan CK, Peng H, Liu G, McIlwrath K, Zhang XF, Huggins RA, et al. High-performance lithium battery anodes using silicon nanowires. *Nature Nanotechnology.* 2008;3(1):31–5.

54. Liu XH, Zheng H, Zhong L, Huang S, Karki K, Zhang LQ, et al. Anisotropic swelling and fracture of silicon nanowires during lithiation. *Nano Letters.* 2011;11(8):3312–8.

55. Lee SW, McDowell MT, Choi JW, Cui Y. Anomalous shape changes of silicon nanopillars by electrochemical lithiation. *Nano Letters.* 2011;11(7):3034–9.

56. Kim S-P, Datta D, Shenoy VB. Atomistic mechanisms of phase boundary evolution during initial lithiation of crystalline silicon. *The Journal of Physical Chemistry C.* 2014;118(31):17247–53.

57. Choi D, Kang J, Park J, Han B. First-principles study on thermodynamic stability of the hybrid interfacial structure of $LiMn_2O_4$ cathode and carbonate electrolyte in Li-ion batteries. *Physical Chemistry Chemical Physics.* 2018;20(17):11592–7.

58. Chen S, Du A, Yan C. Molecular dynamic investigation of the structure and stress in crystalline and amorphous silicon during lithiation. *Computational Materials Science.* 2020;183:109811.

59. Van Duin AC, Dasgupta S, Lorant F, Goddard WA. ReaxFF: A reactive force field for hydrocarbons. *The Journal of Physical Chemistry A.* 2001;105(41):9396–409.

60. Ostadhossein A, Cubuk ED, Tritsaris GA, Kaxiras E, Zhang S, Van Duin AC. Stress effects on the initial lithiation of crystalline silicon nanowires: Reactive molecular dynamics simulations using ReaxFF. *Physical Chemistry Chemical Physics.* 2015;17(5):3832–40.

61. Liu XH, Wang JW, Huang S, Fan F, Huang X, Liu Y, et al. In situ atomic-scale imaging of electrochemical lithiation in silicon. *Nature Nanotechnology.* 2012;7(11):749–56.

62. Mei J, Zhang YW, Liao T, Sun ZQ, Dou SX. Strategies for improving the lithium-storage performance of 2D nanomaterials. *National Science Review.* 2018;5(3):389–416.

63. Salavati M, Rabczuluk T. Application of highly stretchable and conductive two-dimensional 1T VS2 and VSe2 as anode materials for Li-, Na- and Ca-ion storage. *Computational Materials Science.* 2019;160:360–7.

64. Bissett MA, Kinloch IA, Dryfe RAW. Characterization of MoS_2-graphene composites for high-performance coin cell supercapacitors. *ACS Applied Materials & Interfaces.* 2015;7(31):17388–98.

65. Huang X, Zeng ZY, Fan ZX, Liu JQ, Zhang H. Graphene-based electrodes. *Advanced Materials.* 2012;24(45):5979–6004.

66. Xiao LF, Cao YL, Xiao J, Wang W, Kovarik L, Nie ZM, et al. High capacity, reversible alloying reactions in SnSb/C nanocomposites for Na-ion battery applications. *Chemical Communications.* 2012;48(27):3321–3.

67. Chakravarty D, Late DJ. Microwave and hydrothermal syntheses of WSe2 micro/nanorods and their application in supercapacitors. *RSC Advances.* 2015;5(28):21700–9.

68. Slater MD, Kim D, Lee E, Johnson CS. Sodium-ion batteries. *Advanced Functional Materials.* 2013;23(8):947–58.

69. Zhao GX, Wen T, Zhang J, Li JX, Dong HL, Wang XK, et al. Two-dimensional Cr2O3 and interconnected graphene-Cr2O3 nanosheets: Synthesis and their application in lithium storage. *Journal of Materials Chemistry A.* 2014;2(4):944–8.

70. ten Elshof JE, Yuan HY, Rodriguez PG. Two-dimensional metal oxide and metal hydroxide nanosheets: Synthesis, controlled assembly and applications in energy conversion and storage. *Advanced Energy Materials.* 2016;6(23):1600355.

71. Gao YP, Wu X, Huang KJ, Xing LL, Zhang YY, Liu L. Two-dimensional transition metal diseleniums for energy storage application: A review of recent developments. *Crystengcomm.* 2017;19(3):404–18.

72. Zhou PS, Collins G, Hens Z, Ryan KM, Geaney H, Singh S. Colloidal WSe2 nanocrystals as anodes for lithium-ion batteries. *Nanoscale.* 2020;12(43):22307–16.

73. Eng AYS, Cheong JL, Lee SS. Controlled synthesis of transition metal disulfides (MoS$_2$ and WS2) on carbon fibers: Effects of phase and morphology toward lithium-sulfur battery performance. *Applied Materials Today*. 2019;16:529–37.

74. Barik G, Pal S. Defect induced performance enhancement of monolayer MoS$_2$ for Li- and Na-ion batteries. *Journal of Physical Chemistry C*. 2019;123(36):21852–65.

75. Zhao TF, Shu HB, Shen ZH, Hu HM, Wang J, Chen XS. Electrochemical lithiation mechanism of two-dimensional transition-metal dichalcogenide anode materials: Intercalation versus conversion reactions. *Journal of Physical Chemistry C*. 2019;123(4):2139–46.

76. Wang XQ, He JR, Zheng BJ, Zhang WL, Chen YF. Few-layered WSe2 in-situ grown on graphene nanosheets as efficient anode for lithium-ion batteries. *Electrochimica Acta*. 2018;283:1660–7.

77. Srinivaas M, Wu CY, Duh JG, Wu JM. Highly rich 1T metallic phase of few-layered WS2 nanoflowers for enhanced storage of lithium-ion batteries. *ACS Sustainable Chemistry & Engineering*. 2019;7(12):10363–70.

78. Dong YL, Xu Y, Li W, Fu Q, Wu MH, Manske E, et al. Insights into the crystallinity of layer-structured transition metal dichalcogenides on potassium ion battery performance: A Case Study of Molybdenum Disulfide. *Small*. 2019;15(15).

79. Song DX, Xie L, Zhang YF, Lu Y, An M, Ma WG, et al. Multilayer Ion Load and Diffusion on TMD/MXene Heterostructure Anodes for Alkali-Ion Batteries. *ACS Applied Energy Materials*. 2020;3(8):7699–709.

80. Thanh TD, Chuong ND, Hien HV, Kshetri T, Tuan LH, Kim NH, et al. Recent advances in two-dimensional transition metal dichalcogenides-graphene heterostructured materials for electrochemical applications. *Progress in Materials Science*. 2018;96:51–85.

81. Huang HH, Fan XF, Singh DJ, Zheng WT. Recent progress of TMD nanomaterials: Phase transitions and applications. *Nanoscale*. 2020;12(3):1247–68.

82. Wu L, Sun RM, Xiong FY, Pei CY, Han K, Peng C, et al. A rechargeable aluminum-ion battery based on a VS2 nanosheet cathode. *Physical Chemistry Chemical Physics*. 2018;20(35):22563–8.

83. Zhao CY, Wang DS, Lian RQ, Kan DX, Dou YY, Wang CZ, et al. Revealing the distinct electrochemical properties of TiSe2 monolayer and bulk counterpart in Li-ion batteries by first-principles calculations. *Applied Surface Science*. 2021;540.

84. Soares DM, Singh G. Superior electrochemical performance of layered WTe(2)as potassium-ion battery electrode. *Nanotechnology*. 2020;31(45).

85. Soares DM, Mukherjee S, Singh G. TMDs beyond MoS$_2$ for Electrochemical Energy Storage. *Chemistry-a European Journal*. 2020;26(29):6320–41.

86. Yang E, Ji H, Jung Y. Two-Dimensional Transition Metal dichalcogenide mono layers as promising sodium ion battery anodes. *Journal of Physical Chemistry C*. 2015;119(47):26374–80.

87. Lin LX, Lei W, Zhang SW, Liu YQ, Wallace GG, Chen J. Two-dimensional transition metal dichalcogenides in supercapacitors and secondary batteries. *Energy Storage Materials*. 2019;19:408–23.

88. Poizot P, Laruelle S, Grugeon S, Dupont L, Tarascon J. Nano-sized transition-metal oxides as negative-electrode materials for lithium-ion batteries. *Nature*. 2000;407(6803):496–9.

89. Li H, Liu Z, Yang S, Zhao Y, Feng Y, Bakenov Z, et al. Facile synthesis of ZnO nanoparticles on nitrogen-doped carbon nanotubes as high-performance anode material for lithium-ion batteries. *Materials*. 2017;10(10):1102.

90. Balaya P, Li H, Kienle L, Maier J. Fully reversible homogeneous and heterogeneous Li storage in RuO2 with high capacity. *Adv Funct Mater*. 2003;13(8):621–5.

91. Varghese B, Reddy M, Yanwu Z, Lit CS, Hoong TC, Subba Rao G, et al. Fabrication of NiO nanowall electrodes for high performance lithium ion battery. *Chem Mater*. 2008;20(10):3360–7.

92. Hou C-C, Brahma S, Weng S-C, Chang C-C, Huang J-L. Facile, low temperature synthesis of SnO₂/reduced graphene oxide nanocomposite as anode material for lithium-ion batteries. *Appl Surf Sci.* 2017;413:160–8.

93. Wang Z-H, Yuan L-X, Shao Q-G, Huang F, Huang Y-H. Mn₃O₄ nanocrystals anchored on multi-walled carbon nanotubes as high-performance anode materials for lithium-ion batteries. *Mater Letter.* 2012;80:110–3.

94. Anastas PT, Zimmerman JB. The periodic table of the elements of green and sustainable chemistry. *Green Chemistry.* 2019;21(24):6545–66.

95. Li X, Liang J, Lu Y, Hou Z, Cheng Q, Zhu Y, et al. Sulfur-rich phosphorus sulfide molecules for use in rechargeable lithium batteries. *Angewandte Chemie International Edition.* 2017;56(11):2937–41.

96. Yi Z, Su F, Huo L, Cui G, Zhang C, Han P, et al. New insights into Li₂S₂/Li₂S adsorption on the graphene bearing single vacancy: A DFT study. *Applied Surface Science.* 2020;503:144446.

97. Liang P, Zhang L, Wang D, Man X, Shu H, Wang L, et al. First-principles explorations of Li₂S@ V₂CTx hybrid structure as cathode material for lithium-sulfur battery. *Applied Surface Science.* 2019;489:677–83.

98. Kudu ÖU, Famprikis T, Fleutot B, Braida M-D, Le Mercier T, Islam MS, et al. A review of structural properties and synthesis methods of solid electrolyte materials in the Li₂S– P₂S₅ binary system. *Journal of Power Sources.* 2018;407:31–43.

99. Zhang L, Sun D, Feng J, Cairns EJ, Guo J. Revealing the electrochemical charging mechanism of nanosized Li₂S by in situ and operando X-ray absorption spectroscopy. *Nano letters.* 2017;17(8):5084–91.

100. Tahri Y, Chermette H, Hollinger G. Electronic structures of phosphorus oxide P4O10 and phosphorus sulfides P4S10 and P4S7. *Journal of electron spectroscopy and related phenomena.* 1991;56(1):51–69.

101. Ding B, Wang J, Fan Z, Chen S, Lin Q, Lu X, et al. Solid-state lithium–sulfur batteries: Advances, challenges and perspectives. *Materials Today.* 2020.

102. Li G, Chen Z, Lu J. Lithium-sulfur batteries for commercial applications. *Chem.* 2018;4(1):3–7.

103. Ould Ely T, Kamzabek D, Chakraborty D, Doherty MF. Lithium–sulfur batteries: State of the art and future directions. *ACS Applied Energy Materials.* 2018;1(5):1783–814.

104. Shih-Yang Lin H-YL, Nguyen DK, Tran NTT, Pham HD, Chang S-L, Lin C-Y, Lin M-F. *Silicene-Based Layered Materials: Essential Properties*: IOP Publishing 2020.

105. Lin SYT, NTT, Chang SL, Su WP, Lin MF. *Structure- and Adatom-Enriched Essential Properties of Graphene Nanoribbons*: CRC Press 2018.

106. Fox M. *Optical Properties of Solids*: Oxford University Press; 2001.

107. Wang S, Xu X, Zhang X, Xin C, Xu B, Li L, et al. High-performance Li 6 PS 5 Cl-based all-solid-state lithium-ion batteries. *Journal of Materials Chemistry A.* 2019;7(31):18612–8.

108. Rao C, Vivekchand S, Biswas K, Govindaraj A. Synthesis of inorganic nanomaterials. *Dalton Transactions.* 2007;(34): 3728–49.

109. Barghamadi M, Best AS, Bhatt AI, Hollenkamp AF, Musameh M, Rees RJ, et al. Lithium–sulfur batteries—the solution is in the electrolyte, but is the electrolyte a solution? *Energy & environmental science.* 2014;7(12):3902–20.

110. Kang W, Deng N, Ju J, Li Q, Wu D, Ma X, et al. A review of recent developments in rechargeable lithium–sulfur batteries. *Nanoscale.* 2016;8(37):16541–88.

111. Li J, Sun JL. Application of X-ray diffraction and electron crystallography for solving complex structure problems. *Accounts of Chemical Research.* 2017;50(11): 2737–45.

112. Damascelli A. Probing the electronic structure of complex systems by ARPES. *Physica Scripta.* 2004;2004(T109):61.

113. Bussolotti F, Chi D, Goh KJ, Huang YL, Wee AT. STM/STS and ARPES characterization—structure and electronic properties. *2D Semiconductor Materials and Devices*: Elsevier; 2020. p. 199–220.

114. Hafner J. Ab-initio simulations of materials using VASP: Density-functional theory and beyond. *Journal of Computational Chemistry*. 2008;29(13):2044–78.

115. Nguyen TDH, Pham HD, Lin SY, Lin MF. Featured properties of Li+-based battery anode: Li4Ti5O12. *Rsc Advances*. 2020;10(24):14071–9.

116. Khuong Dien V, Thi Han N, Nguyen TDH, Huynh TMD, Pham HD, Lin M-F. Geometric and Electronic Properties of Li2GeO3. *Frontiers in Materials*. 2020;7(288).

117. Ozturk T, Ertas E, Mert O. A Berzelius reagent, phosphorus decasulfide (P4S10), in organic syntheses. *Chemical Reviews*. 2010;110(6):3419–78.

118. Ertaş E, Öztürk T, Ösken I. Purification of phosphorus decasulfide (P4S10). Google Patents; 2018.

119. Liu D, Liu Y, Pan A, Nagle KP, Seidler GT, Jeong Y-H, et al. Enhanced lithium-ion intercalation properties of V_2O_5 xerogel electrodes with surface defects. *The Journal of Physical Chemistry C*. 2011;115(11):4959–65.

120. Karthik K, Phuruangrat A, Chowdhury ZZ, Pradeeswari K, Kumar RM. V_2O_5 nanoparticles as cathode for lithium-ion battery applications: Fabricated via microwave-assisted green synthesis using A. paniculata leaf extract. 2019.

121. Shevchuk V, Usatenko YN, Demchenko PY, Antonyak O, Serkiz RY. Nano-and microsize V_2O_5 structures. *Chemistry of Metals and Alloys*. 2011;4(1–2):67–71.

122. Donders M, Arnoldbik W, Knoops H, Kessels W, Notten P. Atomic layer deposition of $LiCoO_2$ thin-film electrodes for all-solid-state Li-ion micro-batteries. *Journal of the Electrochemical Society*. 2013;160(5):A3066.

123. He P, Yu H, Zhou H. Layered lithium transition metal oxide cathodes towards high energy lithium-ion batteries. *Journal of Materials Chemistry*. 2012;22(9):3680–95.

124. Scott ID, Jung YS, Cavanagh AS, Yan Y, Dillon AC, George SM, et al. Ultrathin coatings on nano-$LiCoO_2$ for Li-ion vehicular applications. *Nano Letters*. 2011;11(2):414–8.

125. Miikkulainen V, Ruud A, Østreng E, Nilsen O, Laitinen M, Sajavaara T, et al. Atomic layer deposition of spinel lithium manganese oxide by film-body-controlled lithium incorporation for thin-film lithium-ion batteries. *The Journal of Physical Chemistry C*. 2014;118(2):1258–68.

126. Jung YS, Cavanagh AS, Yan Y, George SM, Manthiram A. Effects of atomic layer deposition of Al_2O_3 on the Li [$Li_{0.20}Mn_{0.54}Ni_{0.13}Co_{0.13}$] O_2 cathode for lithium-ion batteries. *Journal of the Electrochemical Society*. 2011;158(12):A1298.

127. Riley LA, Van Atta S, Cavanagh AS, Yan Y, George SM, Liu P, et al. Electrochemical effects of ALD surface modification on combustion synthesized $LiNi_{1/3}Mn_{1/3}Co_{1/3}O_2$ as a layered-cathode material. *Journal of Power Sources*. 2011;196(6):3317–24.

128. Li X, Liu J, Banis MN, Lushington A, Li R, Cai M, et al. Atomic layer deposition of solid-state electrolyte coated cathode materials with superior high-voltage cycling behavior for lithium ion battery application. *Energy & Environmental Science*. 2014;7(2):768–78.

129. Deng H, Nie P, Luo H, Zhang Y, Wang J, Zhang X. Highly enhanced lithium storage capability of $LiNi_{0.5}Mn_{1.5}O_4$ by coating with Li_2TiO_3 for Li-ion batteries. *Journal of Materials Chemistry A*. 2014;2(43):18256–62.

130. Bettge M, Li Y, Sankaran B, Rago ND, Spila T, Haasch RT, et al. Improving high-capacity $Li_{1.2}Ni_{0.15}Mn_{0.55}Co_{0.1}O_2$-based lithium-ion cells by modifiying the positive electrode with alumina. *Journal of Power Sources*. 2013;233:346–57.

131. Zhang X, Belharouak I, Li L, Lei Y, Elam JW, Nie A, et al. Structural and electrochemical study of Al_2O_3 and TiO_2 coated $Li_{1.2}Ni_{0.13}Mn_{0.54}Co_{0.13}O_2$ cathode material using ALD. *Advanced Energy Materials*. 2013;3(10):1299–307.

132. Gandrud KB, Pettersen A, Nilsen O, Fjellvåg H. High-performing iron phosphate for enhanced lithium ion solid state batteries as grown by atomic layer deposition. *Journal of Materials Chemistry A.* 2013;1(32):9054–9.

133. Liu J, Banis MN, Xiao B, Sun Q, Lushington A, Li R, et al. Atomically precise growth of sodium titanates as anode materials for high-rate and ultralong cycle-life sodium-ion batteries. *Journal of Materials Chemistry A.* 2015;3(48):24281–8.

134. Fongy C, Gaillot A-C, Jouanneau S, Guyomard D, Lestriez B. Ionic vs electronic power limitations and analysis of the fraction of wired grains in LiFePO$_4$ composite electrodes. *Journal of the Electrochemical Society.* 2010;157(7):A885.

135. Cheah SK, Perre E, Rooth M, Fondell M, Hårsta A, Nyholm L, et al. Self-supported three-dimensional nanoelectrodes for microbattery applications. *Nano Letters.* 2009;9(9):3230–3.

136. Wang W, Tian M, Abdulagatov A, George SM, Lee Y-C, Yang R. Three-dimensional Ni/TiO$_2$ nanowire network for high areal capacity lithium ion microbattery applications. *Nano letters.* 2012;12(2):655–60.

137. Ban C, Xie M, Sun X, Travis JJ, Wang G, Sun H, et al. Atomic layer deposition of amorphous TiO$_2$ on graphene as an anode for Li-ion batteries. *Nanotechnology.* 2013;24(42):424002.

138. Kim S-W, Han TH, Kim J, Gwon H, Moon H-S, Kang S-W, et al. Fabrication and electrochemical characterization of TiO$_2$ three-dimensional nanonetwork based on peptide assembly. *ACS Nano.* 2009;3(5):1085–90.

139. Jung YS, Cavanagh AS, Riley LA, Kang SH, Dillon AC, Groner MD, et al. Ultrathin direct atomic layer deposition on composite electrodes for highly durable and safe Li-ion batteries. *Advanced Materials.* 2010;22(19):2172–6.

140. Wang X, Yushin G. Chemical vapor deposition and atomic layer deposition for advanced lithium ion batteries and supercapacitors. *Energy & Environmental Science.* 2015;8(7):1889–904.

141. Wang H-Y, Wang F-M. Electrochemical investigation of an artificial solid electrolyte interface for improving the cycle-ability of lithium ion batteries using an atomic layer deposition on a graphite electrode. *Journal of power sources.* 2013;233:1–5.

142. Goriparti S, Miele E, De Angelis F, Di Fabrizio E, Zaccaria RP, Capiglia C. Review on recent progress of nanostructured anode materials for Li-ion batteries. *Journal of power sources.* 2014;257:421–43.

143. Liu Y, Wang YM, Yakobson BI, Wood BC. Assessing carbon-based anodes for lithium-ion batteries: A universal description of charge-transfer binding. *Physical Review Letters.* 2014;113(2):028304.

144. Raccichini R, Varzi A, Passerini S, Scrosati B. The role of graphene for electrochemical energy storage. *Nature materials.* 2015;14(3):271–9.

145. He Y, Yu X, Wang Y, Li H, Huang X. Alumina-coated patterned amorphous silicon as the anode for a lithium-ion battery with high Coulombic efficiency. *Advanced Materials.* 2011;23(42):4938–41.

146. Xiao X, Lu P, Ahn D. Ultrathin multifunctional oxide coatings for lithium ion batteries. *Advanced Materials.* 2011;23(34):3911–5.

147. Chen S, Chen P, Wang Y. Carbon nanotubes grown in situ on graphene nanosheets as superior anodes for Li-ion batteries. *Nanoscale.* 2011;3(10):4323–9.

148. Liu C, Li F, Ma LP, Cheng HM. Advanced materials for energy storage. *Advanced materials.* 2010;22(8):E28–E62.

149. Yoon S-H, Park C-W, Yang H, Korai Y, Mochida I, Baker R, et al. Novel carbon nanofibers of high graphitization as anodic materials for lithium ion secondary batteries. *Carbon.* 2004;42(1):21–32.

150. Yan X, Teng D, Jia X, Yu Y, Yang X. Improving the cyclability and rate capability of carbon nanofiber anodes through in-site generation of SiOx-rich overlayers. *Electrochimica Acta*. 2013;108:196–202.

151. Yue H, Li F, Yang Z, Tang J, Li X, He D. Nitrogen-doped carbon nanofibers as anode material for high-capacity and binder-free lithium ion battery. *Materials Letters*. 2014;120:39–42.

152. Kohandehghan A, Kalisvaart P, Cui K, Kupsta M, Memarzadeh E, Mitlin D. Silicon nanowire lithium-ion battery anodes with ALD deposited TiN coatings demonstrate a major improvement in cycling performance. *Journal of Materials Chemistry A*. 2013;1(41):12850–61.

153. Lotfabad EM, Kalisvaart P, Cui K, Kohandehghan A, Kupsta M, Olsen B, et al. ALD TiO_2 coated silicon nanowires for lithium ion battery anodes with enhanced cycling stability and coulombic efficiency. *Physical Chemistry Chemical Physics*. 2013;15(32):13646–57.

154. Lotfabad EM, Kalisvaart P, Kohandehghan A, Cui K, Kupsta M, Farbod B, et al. Si nanotubes ALD coated with TiO_2, TiN or Al_2O_3 as high performance lithium ion battery anodes. *Journal of Materials Chemistry A*. 2014;2(8):2504–16.

155. Snyder MQ, Trebukhova SA, Ravdel B, Wheeler MC, DiCarlo J, Tripp CP, et al. Synthesis and characterization of atomic layer deposited titanium nitride thin films on lithium titanate spinel powder as a lithium-ion battery anode. *Journal of Power Sources*. 2007;165(1):379–85.

156. Leonidov I, Leonidova O, Perelyaeva L, Samigullina R, Kovyazina S, Patrakeev M. Structure, ionic conduction, and phase transformations in lithium titanate $Li_4 Ti_5 O_{12}$. *Physics of the Solid State*. 2003;45(11):2183–8.

157. Liu J, Li X, Cai M, Li R, Sun X. Ultrathin atomic layer deposited ZrO2 coating to enhance the electrochemical performance of Li4Ti5O12 as an anode material. *Electrochimica Acta*. 2013;93:195–201.

158. Ahn D, Xiao X. Extended lithium titanate cycling potential window with near zero capacity loss. *Electrochemistry Communications*. 2011;13(8):796–9.

159. Riley LA, Cavanagh AS, George SM, Jung YS, Yan Y, Lee SH, et al. Conformal Surface Coatings to Enable High Volume Expansion Li-Ion Anode Materials. *ChemPhysChem*. 2010;11(10):2124–30.

160. Wang D, Yang J, Liu J, Li X, Li R, Cai M, et al. Atomic layer deposited coatings to significantly stabilize anodes for Li ion batteries: Effects of coating thickness and the size of anode particles. *Journal of Materials Chemistry A*. 2014;2(7):2306–12.

161. Lee J-H, Hon M-H, Chung Y-W, Leu C. The effect of TiO_2 coating on the electrochemical performance of ZnO nanorod as the anode material for lithium-ion battery. *Applied Physics A*. 2011;102(3):545–50.

162. Kang E, Jung YS, Cavanagh AS, Kim GH, George SM, Dillon AC, et al. Fe_3O_4 nanoparticles confined in mesocellular carbon foam for high performance anode materials for lithium-ion batteries. *Advanced Functional Materials*. 2011;21(13):2430–8.

163. Lipson AL, Puntambekar K, Comstock DJ, Meng X, Geier ML, Elam JW, et al. Nanoscale investigation of solid electrolyte interphase inhibition on Li-ion battery MnO electrodes via atomic layer deposition of Al_2O_3. *Chemistry of Materials*. 2014;26(2):935–40.

164. Liu W-W, Lau W-M, Zhang Y. The electrochemical properties of $Co_3 O_4$ as a lithium-ion battery electrode: A first-principles study. *Physical Chemistry Chemical Physics*. 2018;20(38):25016–22.

165. Meng X, Libera JA, Fister TT, Zhou H, Hedlund JK, Fenter P, et al. Atomic layer deposition of gallium sulfide films using hexakis (dimethylamido) digallium and hydrogen sulfide. *Chemistry of Materials*. 2014;26(2):1029–39.

166. Li S, Liu P, Huang X, Tang Y, Wang H. Reviving bulky MoS_2 as an advanced anode for lithium-ion batteries. *Journal of Materials Chemistry A*. 2019;7(18):10988–97.
167. Aaltonen T, Alnes M, Nilsen O, Costelle L, Fjellvåg H. Lanthanum titanate and lithium lanthanum titanate thin films grown by atomic layer deposition. *Journal of Materials Chemistry*. 2010;20(14):2877–81.
168. Swamy DT, Babu KE, Veeraiah V. Evidence for high ionic conductivity in lithium-lanthanum titanate $Li_{0.29} La_{0.57} TiO_3$. *Bulletin of Materials Science*. 2013;19:1115–9.
169. Kimpa M, Mayzan M, Yabagi J, Nmaya M, Isah K, Agam M. Review on material synthesis and characterization of sodium (Na) super-ionic conductor (NASICON). *E&ES*. 2018;140(1):012156.
170. Deng Y, Eames C, Fleutot B, David R, Chotard J-N, Suard E, et al. Enhancing the lithium ion conductivity in lithium superionic conductor (LISICON) solid electrolytes through a mixed polyanion effect. *ACS Applied Materials & Interfaces*. 2017;9(8):7050–8.
171. Thangadurai V, Weppner W. Recent progress in solid oxide and lithium ion conducting electrolytes research. *Ionics*. 2006;12(1):81–92.
172. Comstock DJ, Elam JW. Mechanistic study of lithium aluminum oxide atomic layer deposition. *The Journal of Physical Chemistry C*. 2013;117(4):1677–83.
173. Hämäläinen J, Holopainen J, Munnik F, Hatanpää T, Heikkilä M, Ritala M, et al. Lithium phosphate thin films grown by atomic layer deposition. *Journal of the Electrochemical Society*. 2012;159(3):A259.
174. Tomczak Y, Knapas K, Sundberg M, Leskelä M, Ritala M. In situ reaction mechanism studies on lithium hexadimethyldisilazide and ozone atomic layer deposition process for lithium silicate. *The Journal of Physical Chemistry C*. 2013;117(27):14241–6.
175. Liu J, Banis MN, Li X, Lushington A, Cai M, Li R, et al. Atomic layer deposition of lithium tantalate solid-state electrolytes. *The Journal of Physical Chemistry C*. 2013;117(39):20260–7.
176. Østreng E, Sønsteby HH, Sajavaara T, Nilsen O, Fjellvåg H. Atomic layer deposition of ferroelectric $LiNbO_3$. *Journal of Materials Chemistry C*. 2013;1(27):4283–90.
177. Xu H, Li Y, Zhou A, Wu N, Xin S, Li Z, et al. Li_3N-Modified garnet electrolyte for all-solid-state lithium metal batteries operated at 40° C. *Nano Letters*. 2018;18(11):7414–8.
178. Perng Y-C, Cho J, Sun SY, Membreno D, Cirigliano N, Dunn B, et al. Synthesis of ion conducting Li x Al y Si z O thin films by atomic layer deposition. *Journal of Materials Chemistry A*. 2014;2(25):9566–73.
179. Kamaya N, Homma K, Yamakawa Y, Hirayama M, Kanno R, Yonemura M, et al. A lithium superionic conductor. *Nature Materials*. 2011;10(9):682–6.
180. Boulineau S, Courty M, Tarascon J-M, Viallet V. Mechanochemical synthesis of Li-argyrodite Li6PS5X (X= Cl, Br, I) as sulfur-based solid electrolytes for all solid state batteries application. *Solid State Ionics*. 2012;221:1–5.
181. Zhao Y, Daemen LL. Superionic conductivity in lithium-rich anti-perovskites. *Journal of the American Chemical Society*. 2012;134(36):15042–7.
182. Kuganathan N, Tsoukalas L, Chroneos A. Defects, dopants and Li-ion diffusion in Li_2SiO_3. *Solid State Ionics*. 2019;335:61–6.
183. Tang T, Chen P, Luo W, Luo D, Wang Y. Crystalline and electronic structures of lithium silicates: A density functional theory study. *Journal of Nuclear Materials*. 2012;420(1–3):31–8.
184. Musa N, Woo H, Teo L, Arof A. Optimization of Li_2SnO_3 synthesis for anode material application in li-ion batteries. *Materials Today: Proceedings*. 2017;4(4):5169–77.
185. Kerner M, Plylahan N, Scheers J, Johansson P. Ionic liquid based lithium battery electrolytes: Fundamental benefits of utilising both TFSI and FSI anions? *Physical Chemistry Chemical Physics*. 2015;17(29):19569–81.

186. Wang Z, Cai Y, Wang Z, Chen S, Lu X, Zhang S. Vinyl-functionalized imidazolium ionic liquids as new electrolyte additives for high-voltage Li-ion batteries. *Journal of Solid State Electrochemistry.* 2013;17(11):2839–48.

187. Seki S, Ohno Y, Miyashiro H, Kobayashi Y, Usami A, Mita Y, et al. Quaternary ammonium room-temperature ionic liquid/lithium salt binary electrolytes: Electrochemical study. *Journal of the Electrochemical Society.* 2008;155(6):A421.

188. Appetecchi GB, Montanino M, Zane D, Carewska M, Alessandrini F, Passerini S. Effect of the alkyl group on the synthesis and the electrochemical properties of N-alkyl-N-methyl-pyrrolidinium bis (trifluoromethanesulfonyl) imide ionic liquids. *Electrochimica Acta.* 2009;54(4):1325–32.

189. Yun YS, Kim JH, Lee S-Y, Shim E-G, Kim D-W. Cycling performance and thermal stability of lithium polymer cells assembled with ionic liquid-containing gel polymer electrolytes. *Journal of Power Sources.* 2011;196(16):6750–5.

190. Ivanov S, Cheng L, Wulfmeier H, Albrecht D, Fritze H, Bund A. Electrochemical behavior of anodically obtained titania nanotubes in organic carbonate and ionic liquid based Li ion containing electrolytes. *Electrochimica Acta.* 2013;104:228–35.

191. Kühnel R-S, Böckenfeld N, Passerini S, Winter M, Balducci A. Mixtures of ionic liquid and organic carbonate as electrolyte with improved safety and performance for rechargeable lithium batteries. *Electrochimica Acta.* 2011;56(11):4092–9.

192. Kim G-T, Appetecchi GB, Carewska M, Joost M, Balducci A, Winter M, et al. UV cross-linked, lithium-conducting ternary polymer electrolytes containing ionic liquids. *Journal of Power Sources.* 2010;195(18):6130–7.

193. Francis CF, Kyratzis IL, Best AS. Lithium-ion battery separators for ionic-liquid electrolytes: A review. *Advanced Materials.* 2020;32(18):1904205.

194. Bi S, Banda H, Chen M, Niu L, Chen M, Wu T, et al. Molecular understanding of charge storage and charging dynamics in supercapacitors with MOF electrodes and ionic liquid electrolytes. *Nature Materials.* 2020;19(5):552–8.

195. Theivaprakasam S, MacFarlane DR, Mitra S. Electrochemical studies of N-Methyl N-Propyl Pyrrolidinium bis (trifluoromethanesulfonyl) imide ionic liquid mixtures with conventional electrolytes in LiFePO₄/Li cells. *Electrochimica Acta.* 2015;180:737–45.

196. Montanino M, Moreno M, Carewska M, Maresca G, Simonetti E, Presti RL, et al. Mixed organic compound-ionic liquid electrolytes for lithium battery electrolyte systems. *Journal of Power Sources.* 2014;269:608–15.

197. Plylahan N, Kerner M, Lim D-H, Matic A, Johansson P. Ionic liquid and hybrid ionic liquid/organic electrolytes for high temperature lithium-ion battery application. *Electrochimica Acta.* 2016;216:24–34.

198. Li H, Pang J, Yin Y, Zhuang W, Wang H, Zhai C, et al. Application of a nonflammable electrolyte containing Pp13TFSI ionic liquid for lithium-ion batteries using the high capacity cathode material Li [$Li_{0.2}$ $Mn_{0.54}$ $Ni_{0.13}$ $Co_{0.13}$] O_2. *RSC Advances.* 2013;3(33):13907–14.

199. Böckenfeld N, Willeke M, Pires J, Anouti M, Balducci A. On the use of lithium iron phosphate in combination with protic ionic liquid-based electrolytes. *Journal of the Electrochemical Society.* 2013;160(4):A559.

200. Wang M, Shan Z, Tian J, Yang K, Liu X, Liu H, et al. Mixtures of unsaturated imidazolium based ionic liquid and organic carbonate as electrolyte for Li-ion batteries. *Electrochimica Acta.* 2013;95:301–7.

201. Hofmann A, Schulz M, Indris S, Heinzmann R, Hanemann T. Mixtures of ionic liquid and sulfolane as electrolytes for Li-ion batteries. *Electrochimica Acta.* 2014;147:704–11.

202. Ababtain K, Babu G, Lin X, Rodrigues M-TF, Gullapalli H, Ajayan PM, et al. Ionic liquid–organic carbonate electrolyte blends to stabilize silicon electrodes for extending lithium ion battery operability to 100C. *ACS applied Materials & Interfaces.* 2016;8(24):15242–9.

203. Xiang H, Yin B, Wang H, Lin H, Ge X, Xie S, et al. Improving electrochemical properties of room temperature ionic liquid (RTIL) based electrolyte for Li-ion batteries. *Electrochimica Acta*. 2010;55(18):5204–9.

204. Wang L, Byon HR. N-Methyl-N-propylpiperidinium bis (trifluoromethanesulfonyl) imide-based organic electrolyte for high performance lithium–sulfur batteries. *Journal of Power Sources*. 2013;236:207–14.

205. Cao X, He X, Wang J, Liu H, Röser S, Rad BR, et al. High voltage $LiNi_{0.5}Mn_{1.5}O_4/Li_4Ti_5O_{12}$ lithium ion cells at elevated temperatures: Carbonate-versus ionic liquid-based electrolytes. *ACS Applied Materials & Interfaces*. 2016;8(39):25971–8.

206. Usui H, Yamamoto Y, Yoshiyama K, Itoh T, Sakaguchi H. Application of electrolyte using novel ionic liquid to Si thick film anode of Li-ion battery. *Journal of Power Sources*. 2011;196(8):3911–5.

207. Hofmann A, Migeot M, Arens L, Hanemann T. Investigation of ternary mixtures containing 1-ethyl-3-methylimidazolium bis (trifluoromethanesulfonyl) azanide, ethylene carbonate and lithium bis (trifluoromethanesulfonyl) azanide. *International Journal of Molecular Sciences*. 2016;17(5):670.

208. Karuppasamy K, Rhee HW, Reddy PA, Gupta D, Mitu L, Polu AR, et al. Ionic liquid incorporated nanocomposite polymer electrolytes for rechargeable lithium ion battery: A way to achieve improved electrochemical and interfacial properties. *Journal of Industrial and Engineering Chemistry*. 2016;40:168–76.

209. Malik R, Zhou F, Ceder G. Kinetics of non-equilibrium lithium incorporation in $LiFePO_4$. *Nature materials*. 2011;10(8):587–90.

210. Xu K. Nonaqueous liquid electrolytes for lithium-based rechargeable batteries. *Chem Rev*. 2004;104(10):4303–418.

211. Singh VK, Singh RK. Development of ion conducting polymer gel electrolyte membranes based on polymer PVdF-HFP, BMIMTFSI ionic liquid and the Li-salt with improved electrical, thermal and structural properties. *Journal of Materials Chemistry C*. 2015;3(28):7305–18.

212. Tamilarasan P, Ramaprabhu S. Graphene based all-solid-state supercapacitors with ionic liquid incorporated polyacrylonitrile electrolyte. *Energy*. 2013;51:374–81.

213. Li Y, Wong KW, Dou Q, Ng KM. A single-ion conducting and shear-thinning polymer electrolyte based on ionic liquid-decorated PMMA nanoparticles for lithium-metal batteries. *Journal of Materials Chemistry A*. 2016;4(47):18543–50.

214. Rao M, Geng X, Liao Y, Hu S, Li W. Preparation and performance of gel polymer electrolyte based on electrospun polymer membrane and ionic liquid for lithium ion battery. *Journal of Membrane Science*. 2012;399:37–42.

215. Kitazawa Y, Iwata K, Imaizumi S, Ahn H, Kim SY, Ueno K, et al. Gelation of solvate ionic liquid by self-assembly of block copolymer and characterization as polymer electrolyte. *Macromolecules*. 2014;47(17):6009–16.

216. Che H, Chen S, Xie Y, Wang H, Amine K, Liao X-Z, et al. Electrolyte design strategies and research progress for room-temperature sodium-ion batteries. *Energy & Environmental Science*. 2017;10(5):1075–101.

217. Nanba Y, Iwao T, Boisse BMd, Zhao W, Hosono E, Asakura D, et al. Redox Potential Paradox in Na x MO2 for Sodium-Ion Battery Cathodes. *Chemistry of Materials*. 2016;28(4):1058–65.

218. Amaha K, Kobayashi W, Akama S, Mitsuishi K, Moritomo Y. Interrelation between inhomogeneity and cyclability in O_3-$NaFe_{1/2}Co_{1/2}O_2$. *Physica Status Solidi (RRL)– Rapid Research Letters*. 2017;11(1):1600284.

219. Tsubota T, Kitajou A, Okada S. O3-type Na (Fe_1/3Mn_1/3Co_1/3) O_2 as a cathode material with high rate and good charge-discharge cycle performance for sodium-ion batteries. *Evergreen*. 2019;6(4):275–9.

220. Sun L, Xie Y, Liao XZ, Wang H, Tan G, Chen Z, et al. Insight into Ca-substitution Effects on O_3-type $NaNi_{1/3}Fe_{1/3}Mn_{1/3}O_2$ cathode materials for sodium-ion batteries application. *Small*. 2018;14(21):1704523.

221. Li X, editor Development of High-performance Si Anodes and O_3-type Metal Oxide Cathodes for Next Generation Li-ion Batteries and Beyond. ECS Meeting Abstracts; 2020: IOP Publishing.

222. Pati J, Chandra M, Dhaka RS, editors. Electrochemical analysis of $Na_{0.7}Co_{1-x}Nb_xO_2$ (x = 0, 0.05) as cathode materials in sodium-ion batteries. AIP Conference Proceedings; 2020: AIP Publishing LLC.

223. Zhang Y, Pei Y, Liu W, Zhang S, Xie J, Xia J, et al. AlPO4-coated P2-type hexagonal $Na_{0.7}MnO_{2.05}$ as high stability cathode for sodium ion battery. *Chemical Engineering Journal*. 2020;382:122697.

224. Sui Y, Hao Y, Zhang X, Zhong S, Chen J, Li J, et al. Spray-drying synthesis of P_2-$Na_{2/3}Fe_{1/2}Mn_{1/2}O_2$ with improved electrochemical properties. *Advanced Powder Technology*. 2020;31(1):190–7.

225. Chiring A, Mazumder M, Pati SK, Johnson CS, Senguttuvan P. Unraveling the formation mechanism of $NaCoPO_4$ polymorphs. *Journal of Solid State Chemistry*. 2021;293:121766.

226. Ren L, Song L, Guo Y, Wu Y, Lian J, Zhou Y-N, et al. Magnesium-doped $Na_2FeP_2O_7$ cathode materials for sodium-ion battery with enhanced cycling stability and rate capability. *Applied Surface Science*. 2021;544:148893.

227. Cao Y, Xia X, Liu Y, Wang N, Zhang J, Zhao D, et al. Scalable synthesizing nanospherical Na_4Fe_3 (PO_4) 2 (P_2O_7) growing on MCNTs as a high-performance cathode material for sodium-ion batteries. *Journal of Power Sources*. 2020;461:228130.

228. Kirsanova MA, Akmaev AS, Aksyonov DA, Ryazantsev SV, Nikitina VA, Filimonov DS, et al. Monoclinic α-Na_2FePO_4F with Strong Antisite Disorder and Enhanced Na+ Diffusion. *Inorganic Chemistry*. 2020;59(22):16225–37.

229. Nose M, Shiotani S, Nakayama H, Nobuhara K, Nakanishi S, Iba H. $Na_4Co_{2.4}Mn_{0.3}Ni_{0.3}$ (PO_4) $2P_2O_7$: High potential and high capacity electrode material for sodium-ion batteries. *Electrochemistry Communications*. 2013;34:266–9.

230. Yi H, Lin L, Ling M, Lv Z, Li R, Fu Q, et al. Scalable and economic synthesis of high-performance Na_3V_2 (PO_4) $2F_3$ by a solvothermal–ball-milling method. *ACS Energy Letters*. 2019;4(7):1565–71.

231. Liu Q, Hu Z, Chen M, Zou C, Jin H, Wang S, et al. The cathode choice for commercialization of sodium-ion batteries: Layered transition metal oxides versus Prussian blue analogs. *Advanced Functional Materials*. 2020;30(14):1909530.

232. Lu Y, Wang L, Cheng J, Goodenough JB. Prussian blue: A new framework of electrode materials for sodium batteries. *Chemical Communications*. 2012;48(52):6544–6.

233. Li W-J, Chou S-L, Wang J-Z, Wang J-L, Gu Q-F, Liu H-K, et al. Multifunctional conducting polymer coated $Na_{1+x}MnFe$ $(CN)_6$ cathode for sodium-ion batteries with superior performance via a facile and one-step chemistry approach. *Nano Energy*. 2015;13:200–7.

234. Wu X, Deng W, Qian J, Cao Y, Ai X, Yang H. Single-crystal FeFe $(CN)_6$ nanoparticles: A high capacity and high rate cathode for Na-ion batteries. *Journal of Materials Chemistry A*. 2013;1(35):10130–4.

235. Kaye SS, Long JR. Hydrogen Storage in the Dehydrated Prussian Blue Analogues M3 [Co $(CN)_6$] 2 (M= Mn, Fe, Co, Ni, Cu, Zn). *Journal of the American Chemical Society*. 2005;127(18):6506–7.

236. Stevens D, Dahn J. High capacity anode materials for rechargeable sodium-ion batteries. *Journal of the Electrochemical Society*. 2000;147(4):1271.

237. Senguttuvan P, Rousse G, Seznec V, Tarascon J-M, Palacin MR. $Na_2Ti_3O_7$: Lowest voltage ever reported oxide insertion electrode for sodium ion batteries. *Chemistry of Materials*. 2011;23(18):4109–11.

238. Rudola A, Saravanan K, Mason CW, Balaya P. $Na_2Ti_3O_7$: An intercalation based anode for sodium-ion battery applications. *Journal of Materials Chemistry A.* 2013;1(7):2653–62.

239. Rudola A, Sharma N, Balaya P. Introducing a 0.2 V sodium-ion battery anode: The $Na_2Ti_3O_7$ to $Na_{3-x}Ti_3O_7$ pathway. *Electrochemistry Communications.* 2015;61:10–3.

240. Li Z, Ravnsbæk DB, Xiang K, Chiang Y-M. $Na_3Ti_2 (PO_4)$ 3 as a sodium-bearing anode for rechargeable aqueous sodium-ion batteries. *Electrochemistry Communications.* 2014;44:12–5.

241. Naeyaert PJ, Avdeev M, Sharma N, Yahia HB, Ling CD. Synthetic, structural, and electrochemical study of monoclinic $Na_4Ti_5O_{12}$ as a sodium-ion battery anode material. *Chemistry of Materials.* 2014;26(24):7067–72.

242. Galceran M, Rikarte J, Zarrabeitia M, Pujol MC, Aguiló M, Casas-Cabanas M. Investigation of NaTiOPO4 as Anode for Sodium-Ion Batteries: A Solid Electrolyte Interphase Free Material? *ACS Applied Energy Materials.* 2019;2(3):1923–31.

243. Wu D, Li X, Xu B, Twu N, Liu L, Ceder G. $NaTiO_2$: A layered anode material for sodium-ion batteries. *Energy & Environmental Science.* 2015;8(1):195–202.

244. Song K, Liu C, Mi L, Chou S, Chen W, Shen C. Recent progress on the alloy-based anode for sodium-ion batteries and potassium-ion batteries. *Small.* 2019:1903194.

245. Balogun M-S, Luo Y, Qiu W, Liu P, Tong Y. A review of carbon materials and their composites with alloy metals for sodium ion battery anodes. *Carbon.* 2016;98:162–78.

246. Komaba S, Murata W, Ishikawa T, Yabuuchi N, Ozeki T, Nakayama T, et al. Electrochemical Na insertion and solid electrolyte interphase for hard-carbon electrodes and application to Na-Ion batteries. *Advanced Functional Materials.* 2011;21(20):3859–67.

247. Zheng X, Gu Z, Liu X, Wang Z, Wen J, Wu X, et al. Bridging the immiscibility of an all-fluoride fire extinguishant with highly-fluorinated electrolytes toward safe sodium metal batteries. *Energy & Environmental Science.* 2020;13(6):1788–98.

248. Santhosha AL, Medenbach L, Palaniselvam T, Adelhelm P. Sodium-storage behavior of exfoliated MoS_2 as an electrode material for solid-state batteries with Na_3PS_4 as the solid electrolyte. *The Journal of Physical Chemistry C.* 2020;124(19):10298–305.

249. Zhang D, Cao X, Xu D, Wang N, Yu C, Hu W, et al. Synthesis of cubic Na_3SbS_4 solid electrolyte with enhanced ion transport for all-solid-state sodium-ion batteries. *Electrochimica Acta.* 2018;259:100–9.

250. Kim JJ, Yoon K, Park I, Kang K. Progress in the Development of Sodium-Ion Solid Electrolytes. *Small Methods.* 2017;1(10):1700219.

251. Kumaravel V, Bartlett J, Pillai SC. Solid electrolytes for high-temperature stable batteries and supercapacitors. *Advanced Energy Materials.* 2021;11(3):2002869.

252. Shi J, Xiong H, Yang Y, Shao H. Nano-sized oxide filled composite PEO/PMMA/P (VDF-HFP) gel polymer electrolyte for rechargeable lithium and sodium batteries. *Solid State Ionics.* 2018;326:136–44.

253. Fan L, Dang Z, Nan C-W, Li M. Thermal, electrical and mechanical properties of plasticized polymer electrolytes based on PEO/P (VDF-HFP) blends. *Electrochimica Acta.* 2002;48(2):205–9.

254. Ji X, Lee KT, Nazar LF. A highly ordered nanostructured carbon–sulphur cathode for lithium–sulphur batteries. *Nature Materials.* 2009;8(6):500–6.

255. Zheng G, Yang Y, Cha JJ, Hong SS, Cui Y. Hollow carbon nanofiber-encapsulated sulfur cathodes for high specific capacity rechargeable lithium batteries. *Nano Letters.* 2011;11(10):4462–7.

256. Keller SJ. Sulfur in hollow nanofibers overcomes challenges of lithium-ion battery design. *News* (Stanford, CA: Stanford University) Retrieved Feb. 2012;18.

257. Rosenberg S, Hintennach A. Laser-printed lithium-sulphur micro-electrodes for Li/S batteries. *Russian Journal of Electrochemistry.* 2014;50(4):327–35.

258. Chung WJ, Griebel JJ, Kim ET, Yoon H, Simmonds AG, Ji HJ, et al. The use of elemental sulfur as an alternative feedstock for polymeric materials. *Nature Chemistry.* 2013;5(6):518–24.

259. Seh ZW, Li W, Cha JJ, Zheng G, Yang Y, McDowell MT, et al. Sulphur–TiO_2 yolk–shell nanoarchitecture with internal void space for long-cycle lithium–sulphur batteries. *Nature Communications.* 2013;4(1):1–6.

260. Lin Z, Liu Z, Fu W, Dudney NJ, Liang C. Lithium polysulfidophosphates: A family of lithium-conducting sulfur-rich compounds for lithium–sulfur batteries. *Angewandte Chemie International Edition.* 2013;52(29):7460–3.

261. Dodson B. New lithium/sulfur battery doubles energy density of lithium-ion, gizmag. com. 2013.

262. Liu J, Lu D, Zheng J, Yan P, Wang B, Sun X, et al. Minimizing polysulfide shuttle effect in lithium-ion sulfur batteries by anode surface passivation. *ACS applied materials & interfaces.* 2018;10(26):21965–72.

263. Nguyen DT, Hoefling A, Yee M, Nguyen GTH, Theato P, Lee YJ, et al. Enabling high-rate and safe lithium ion–sulfur batteries by effective combination of sulfur-copolymer cathode and hard-carbon anode. *ChemSusChem.* 2019;12(2):480–6.

264. Zhang Y-Z, Zhang Z, Liu S, Li G-R, Gao X-P. Free-standing porous carbon nanofiber/carbon nanotube film as sulfur immobilizer with high areal capacity for lithium–sulfur battery. *ACS Applied Materials & Interfaces.* 2018;10(10):8749–57.

265. Peter KT, Vargo JD, Rupasinghe TP, De Jesus A, Tivanski AV, Sander EA, et al. Synthesis, optimization, and performance demonstration of electrospun carbon nanofiber–carbon nanotube composite sorbents for point-of-use water treatment. *ACS Applied Materials & Interfaces.* 2016;8(18):11431–40.

266. Yuan Z, Peng HJ, Huang JQ, Liu XY, Wang DW, Cheng XB, et al. Hierarchical free-standing carbon-nanotube paper electrodes with ultrahigh sulfur-loading for lithium–sulfur batteries. *Advanced Functional Materials.* 2014;24(39):6105–12.

267. Li C, Zhang H, Fu L, Liu H, Wu Y, Rahm E, et al. Cathode materials modified by surface coating for lithium ion batteries. *Electrochimica Acta.* 2006;51(19):3872–83.

268. Ohara K, Mitsui A, Mori M, Onodera Y, Shiotani S, Koyama Y, et al. Structural and electronic features of binary Li 2 SP 2 S 5 glasses. *Scientific Reports.* 2016;6(1):1–9.

269. Umeshbabu E, Zheng B, Yang Y. Recent progress in all-solid-state Lithium– Sulfur batteries using high Li-ion conductive solid electrolytes. *Electrochemical Energy Reviews.* 2019;2(2):199–230.

270. Rani JV, Kanakaiah V, Dadmal T, Rao MS, Bhavanarushi S. Fluorinated natural graphite cathode for rechargeable ionic liquid based aluminum–ion battery. *Journal of the Electrochemical Society.* 2013;160(10):A1781.

271. Menictas C, Skyllas-Kazacos M, Lim TM. Advances in batteries for medium and large-scale energy storage: Types and applications: Elsevier; 2014.

272. Wang H, Bai Y, Chen S, Luo X, Wu C, Wu F, et al. Binder-free V_2O_5 cathode for greener rechargeable aluminum battery. *ACS Applied materials & interfaces.* 2015;7(1):80–4.

273. Jayaprakash N, Das S, Archer L. The rechargeable aluminum-ion battery. *Chemical Communications.* 2011;47(47):12610–2.

274. Reed LD, Menke E. The roles of V_2O_5 and stainless steel in rechargeable Al–ion batteries. *Journal of the Electrochemical Society.* 2013;160(6):A915.

275. Wang W, Jiang B, Xiong W, Sun H, Lin Z, Hu L, et al. A new cathode material for super-valent battery based on aluminium ion intercalation and deintercalation. *Scientific Reports.* 2013;3(1):1–6.

276. Levitin G, Yarnitzky C, Licht S. Fluorinated graphites as energetic cathodes for non-aqueous Al batteries. *Electrochemical and Solid State Letters.* 2002;5(7):A160.

277. Sun H, Wang W, Yu Z, Yuan Y, Wang S, Jiao S. A new aluminium-ion battery with high voltage, high safety and low cost. *Chemical Communications.* 2015;51(59):11892–5.

278. Sun X-G, Fang Y, Jiang X, Yoshii K, Tsuda T, Dai S. Polymer gel electrolytes for application in aluminum deposition and rechargeable aluminum ion batteries. *Chemical Communications*. 2016;52(2):292–5.

279. Geng L, Lv G, Xing X, Guo J. Reversible electrochemical intercalation of aluminum in Mo6S8. *Chemistry of Materials*. 2015;27(14):4926–9.

280. Wang S, Yu Z, Tu J, Wang J, Tian D, Liu Y, et al. A novel aluminum-ion battery: Al/AlCl$_3$-[EMIm] Cl/Ni$_3$S$_2$@ graphene. *Advanced Energy Materials*. 2016;6(13):1600137.

281. Yu Z, Kang Z, Hu Z, Lu J, Zhou Z, Jiao S. Hexagonal NiS nanobelts as advanced cathode materials for rechargeable Al-ion batteries. *Chemical Communications*. 2016;52(68):10427–30.

282. Xia S, Zhang X-M, Huang K, Chen Y-L, Wu Y-T. Ionic liquid electrolytes for aluminium secondary battery: Influence of organic solvents. *Journal of Electroanalytical Chemistry*. 2015;757:167–75.

283. Suto K, Nakata A, Murayama H, Hirai T, Yamaki J-i, Ogumi Z. Electrochemical properties of Al/vanadium chloride batteries with AlCl$_3$-1-ethyl-3-methylimidazolium chloride electrolyte. *Journal of the Electrochemical Society*. 2016;163(5):A742.

284. Reed L, Ortiz S, Xiong M, Menke E. A rechargeable aluminum-ion battery utilizing a copper hexacyanoferrate cathode in an organic electrolyte. *Chemical Communications*. 2015;51(76):14397–400.

285. Hudak NS. Chloroaluminate-doped conducting polymers as positive electrodes in rechargeable aluminum batteries. *The Journal of Physical Chemistry C*. 2014;118(10):5203–15.

286. Das SK, Mahapatra S, Lahan H. Aluminium-ion batteries: Developments and challenges. *Journal of Materials Chemistry A*. 2017;5(14):6347–67.

287. Koura N. A preliminary investigation for an Al/AlCl$_3$-NaCl/FeS$_2$ secondary cell. *JElS*. 1980;127:1529–31.

288. Takami N, Koura N. Anodic sulfidation of FeS electrode in a NaCl saturated AlCl$_3$-NaCl melt. *Electrochimica Acta*. 1988;33(8):1137–42.

289. Dymek Jr C, Williams J, Groeger D, Auborn J. An aluminum acid-base concentration cell using room temperature chloroaluminate ionic liquids. *Journal of the Electrochemical Society*. 1984;131(12):2887.

290. Reynolds G, Dymek Jr C. Primary and secondary room temperature molten salt electrochemical cells. *Journal of Power Sources*. 1985;15(2–3):109–18.

291. Holleck GL. The reduction of chlorine on carbon in AlCl$_3$-KCl-NaCl melts. *Journal of the Electrochemical Society*. 1972;119(9):1158.

292. Li Q, Bjerrum NJ. Aluminum as anode for energy storage and conversion: A review. *Journal of Power Sources*. 2002;110(1):1–10.

293. Jiang T, Brym MC, Dubé G, Lasia A, Brisard G. Electrodeposition of aluminium from ionic liquids: Part I—electrodeposition and surface morphology of aluminium from aluminium chloride (AlCl$_3$)–1-ethyl-3-methylimidazolium chloride ([EMIm] Cl) ionic liquids. *Surface and Coatings Technology*. 2006;201(1–2):1–9.

294. Hjuler H, Von Winbush S, Berg RW, Bjerrum N. A novel inorganic low melting electrolyte for secondary aluminum-nickel sulfide batteries. *Journal of the Electrochemical Society*. 1989;136(4):901.

295. Berrettoni M, Tossici R, Zamponi S, Marassi R, Mamantov G. A cyclic voltammetric study of the electrochemical behavior of NiS2 in molten NaCl saturated NaAlCl$_4$ melts. *Journal of the Electrochemical Society*. 1993;140(4):969.

296. Chiku M, Takeda H, Matsumura S, Higuchi E, Inoue H. Amorphous vanadium oxide/carbon composite positive electrode for rechargeable aluminum battery. *ACS Applied Materials & Interfaces*. 2015;7(44):24385–9.

297. Zhang X, Tang Y, Zhang F, Lee CS. A novel aluminum–graphite dual-ion battery. *Advanced Energy Materials*. 2016;6(11):1502588.

298. Tsuda T, Kokubo I, Kawabata M, Yamagata M, Ishikawa M, Kusumoto S, et al. Electrochemical energy storage device with a lewis acidic $AlBr_3$– 1-ethyl-3-methyl-imidazolioum bromide room-temperature ionic liquid. *Journal of the Electrochemical Society.* 2014;161(6):A908.
299. Saroja A, Samantaray SS, Sundara R. A room temperature multivalent rechargeable iron ion battery with an ether based electrolyte: A new type of post-lithium ion battery. *Chemical Communications.* 2019;55(70):10416–9.

2 Small Polaron–Li-Ion Complex Diffusion in the Cathodes of Rechargeable Li-Ion Batteries

Huu Duc Luong
Osaka University

Thien Lan Tran
Hue University of Education, Hue University

Van An Dinh
Osaka University

CONTENTS

2.1 FORMATION MECHANISM OF THE SMALL POLARON

In principle, the working of typical rechargeable batteries consists of two reversible processes: the charging and discharging processes, as shown in Figure 2.1. During the charging process, the Li ions deintercalate the cathode, then pass through the electrolyte, and finally destine for the anode, while the released electrons travel along the external circuit from the cathode to the anode. In contrast, during the discharging process, Li atoms release its electron to become Li ions. The free electrons move from the anode, go through the external circuit, and come back to the cathode, while Li ions diffuse back and intercalate the cathode material. The chemical reactions inside the cathode can be described as follows:

DOI: 10.1201/9781003263807-2

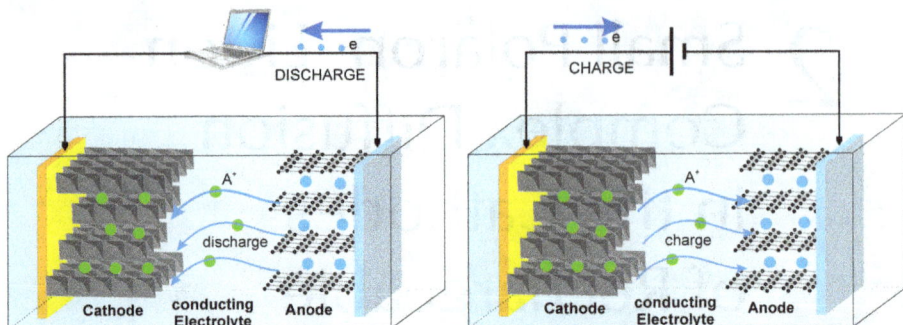

FIGURE 2.1   The operation of typical rechargeable batteries.

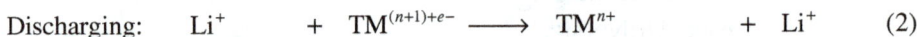

Charging: Li^+ + TM^{n+} $\longrightarrow$ $TM^{(n+1)+e-}$ + Li^+ (1)

Discharging: Li^+ + $TM^{(n+1)+e-}$ $\longrightarrow$ TM^{n+} + Li^+ (2)

where TM denotes the transition metal (TM = Ti, V, Mn, Fe, Ni, Co), and n stands for the oxidation number of transition metal ions.

During the battery operation, the removal/insertion of electrons may cause the formation of a small polaron at transition metal sites. Now, we describe how the polaron forms in cathode materials. As can be seen from Figure 2.2a, when the charging process proceeds, electrons leave the cathode and travel through the external circuit to the anode, resulting in holes introducing the cathode. Simultaneously, the Li ions leave their sites to diffuse inside the cathode and then go through the electrolyte to the anode. These introduced holes favorably distribute at the TM sites, causing the change of oxidation of TM^{n+} ions to $TM^{(n+1)+}$. A hole with a positive charge attracts strongly to the negative ions X ($X = O^{2-}$, F^-) surrounding TM ion by its Coulomb attraction so that TM–X bonds are forced to shrink, consequently, causing a local distortion around the TM site, and hence, the holes would be trapped inside the $TM^{(n+1)+}O_6$ octahedrons. Therefore, it can be concluded that when a Li is removed from the material, a hole is introduced and self-trapped by a local distortion at a transition metal site and a small polaron forms at a TM site. This small polaron comes from a hole with a positive charge; therefore, it is often called *hole polaron* or *positive polaron*.

Reversibly, when the discharging process proceeds, electrons and Li ions are introduced into the cathode material, and transition metal ions $TM^{(n+1)+}$ receive electrons. The insertion of an additional electron repels the surrounding anions further from the TM ion due to strong repulsion Coulomb interaction. After a self-deformation of the crystal framework, the TM–X bonds would be elongated so that the $TM^{n+}O_6$ polyhedron of transition metal ion receiving an additional electron becomes different from $TM^{(n+1)+}O_6$ octahedrons in the surrounding. In other words, a local distortion appears at the $TM^{n+}O_6$ octahedron where the additional electron is self-trapped. Consequently, similar to the charging case, a small polaron would form at the transition metal site as a Li ion is introduced. In this case, because the small

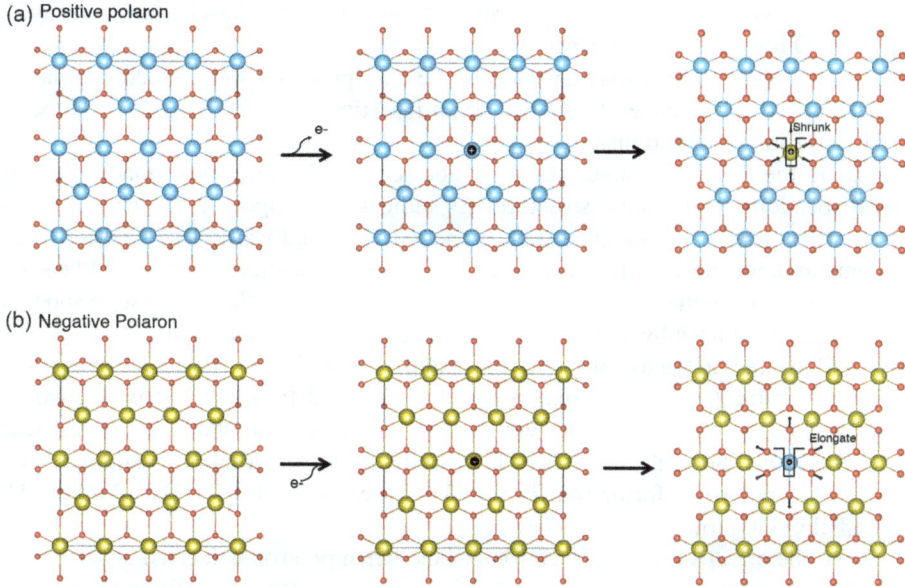

FIGURE 2.2   Positive (a) and negative (b) polaron formation. Light blue, yellow, and red balls represent TM^{n+} and TM$^{(n+1)+}$, and X ions. $\oplus$ and $\ominus$ symbols indicate the additional hole and electron, respectively.

polaron originates from the electron with the negative charge, this small polaron can be called *negative polaron* or *electron polaron*.

Depending on the sign of the charge of the quasiparticle, the small polarons can be classified into two opposite groups, including negative polaron and positive polaron. They may appear and behave differently at various contents of alkali ions in cathode materials. In the next section, we will describe the typical characteristics of small polaron in cathode materials.

2.2   TYPICAL CHARACTERISTICS OF SMALL POLARON

First of all, as mentioned in the previous section, a small polaron would form when a charge carrier is self-trapped at a locally distorted site. As a result, the significant change in the bond lengths would be a stringent signature of the polaron formation. For the negative polaron, it is observed a substantial elongation of TM–X bonds at the transition metal site where a polaron forms. In stark contrast to the negative polaron case, a shrinkage of TM–X bonds can be seen when a positive polaron forms. In order to address the bond length change, the density functional theory (DFT) is widely used to relax the crystal structure and evaluate the bond length difference before and after polaron formation. In general, the change of a unit charge induces an average bond length change of about 0.1–0.2 Å depending on the transition metal types and the crystal frameworks. For example, when positive (negative) polaron is formed in olivine phosphate LiFePO$_4$, the average Fe–O bond length decreases (increases) by 0.09 Å (0.11 Å) [2]. Similarly, for LiMnPO$_4$, the Mn–O bond change is

0.13 Å (0.11 Å) for the positive (negative) case [3]. For layer oxides $LiMnO_2$, the average Mn–O bond length difference is about 0.1 Å [4].

Secondly, the local distortion plays a role as a potential well, which confines a hole or an electron inside. A removal or an insertion of Li atom results in a localized distortion around transition metal sites, and then, newborn bound states would appear in the density of states (DOS) of the semiconductors, corresponding to the flat dispersions in the band structure. As indicated in Figure 2.3a, when a positive polaron forms, an occupied state in the valence band would lose its electron to become an unoccupied state. As a result, a sharp and localized peak would be seen in the band gap. In the case of a negative polaron formation, the bound state appears below the Fermi level by an extra electron. The new bound state is also sharp, localized and located on the top of the valence band, right before the Fermi level, as illustrated in Figure 2.3b. When polaron appears, the local framework structure would be deformed so that Jahn–Teller effect might cause several stabilized states near the bound state. Several examples of Jahn–Teller effect resulting in several newborn peaks when polaron is formed can be found in $LiFePO_4$, $LiMnPO_4$ [3], Li_2FeSiO_4 [5], or $Li_3FePO_4CO_3$ [6].

Next, the effective mass of electron or hole when polaron is formed is very large [7]. As a quasiparticle, polaron by itself can hop between the adjacent transition metal sites. The free polaron migration requires a substantial activation energy.

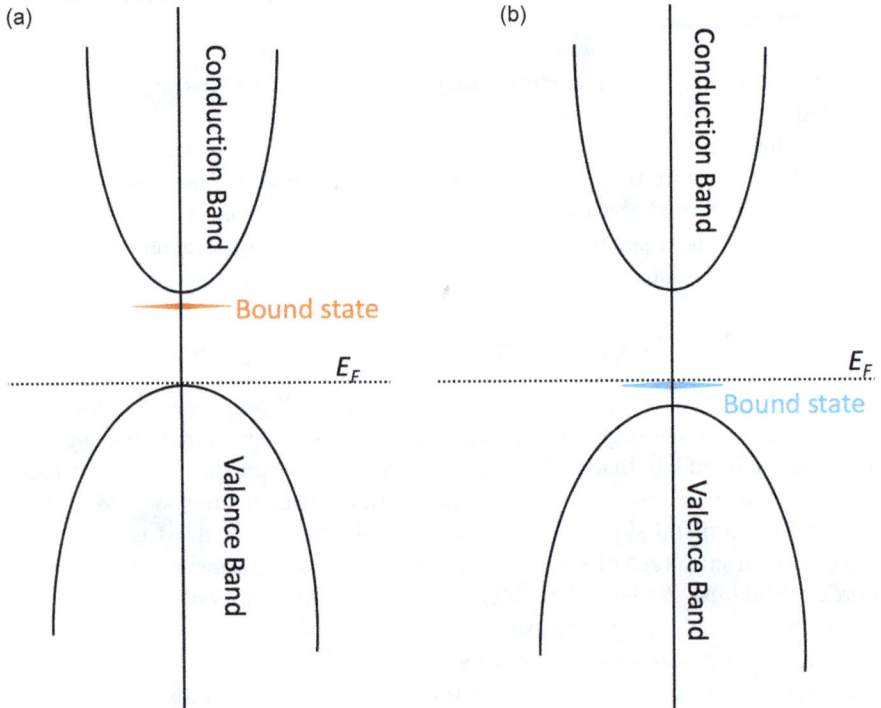

FIGURE 2.3 Bound states in the density of states appeared when a positive (a) or negative (b) polaron is formed.

TABLE 2.1

Activation Energy (meV) for Free Polaron Migration in Some Cathode Materials

Cathode Materials	Negative Polaron	Positive Polaron	References
$LiCoO_2$	100	20	[8]
$LiNiO_2$	210, 280	260, 280	[8]
$LiMnO_2$	390, 480	300, 340	[8]
$Li(NiCoMn)_{1/3}O_2$	310	400	[8]
$Li(NiCoAl)_{1/3}O_2$	0	300	[8]
Li_2MnO_3	-	330	[8]
$Li_xV_2O_5$	-	121	[9]
$LiFePO_4$	175	215	[2]
$LiMnPO_4$	196	303	[3]

When the positive (negative) polaron hops from TM^A site to the adjacent TM^B site, two reversible redox reactions coincide, which is reduction (oxidization) at TM^A site while oxidization (reduction) at TM^B site. Using nudged elastic band (NEB) calculation, the activation energy for free polaron migration can be estimated. Table 2.1 represents some examples of activation energies for free polaron migration. In general, it is noted that the free polaron migration needs activation energy from 100 up to 400 meV depending on the different transition metal sites and crystal framework [8,9]. For example, for free positive (negative) polaron migration, the activation energy is 250 meV (190 meV) in Li_xFePO_4, while it is slightly higher in Li_xMnPO_4 [2,3].

The polaron formation seems to be temperature-dependent. The experimental temperature-dependent behavior of small polaron can be characterized based on Arrhenius relations:

$$\sigma = AT^{-3/2}\exp\left(-\frac{E_\sigma}{k_BT}\right) = T^{-3/2}\exp\left(-\frac{E_S + W_H}{k_BT}\right)$$

where A is the pre-exponential factor, E_σ is the activation energy of electrical conductivity, k_B is the Boltzmann constant, E_S is the activation energy of thermopower, and W_H is the polaron hopping energy for non-adiabatic small polaron. A linear relation between ionic conductivity and $\log(1/T)$ indicates the polaron-type insulating behavior. The hoping energy of small polaron can also be estimated from the linear fitting relation on experimental data. Figure 2.4a indicates an example of the temperature-dependent behavior of small polaron [10]. The Seebeck coefficient shifts its sign from positive values at Li-rich phase to negative values at Li-poor phase, illustrating the change from positive polaron formation at high Li's concentration to negative polaron formation at low Li's concentration during the charge and discharge.

Finally, it is also noted that the positive/negative small polaron can bind strongly to Li vacancy/ion in proximity. For instance, the binding energy between positive (negative) small polaron and Li vacancy (ion) is about 500 meV (370 meV) [2]. When replacing Fe atoms by Mn, Co, and Ni, the binding energy between positive polaron and Li vacancy becomes higher of 650, 570, and 590 meV [11]. At the different

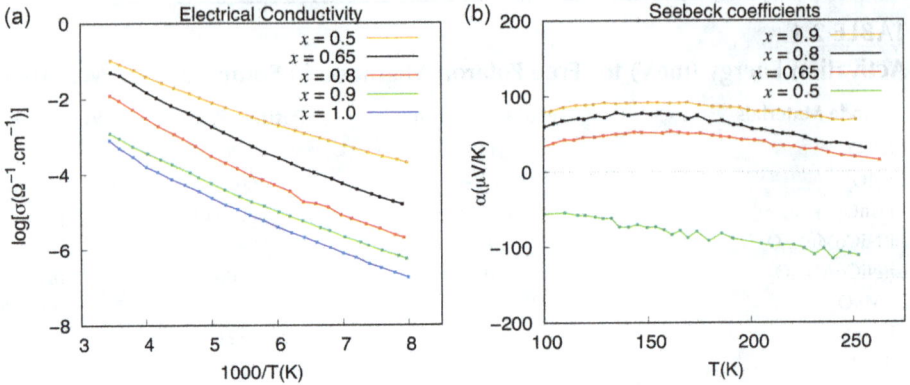

FIGURE 2.4 Electrical conductivity and Seebeck coefficients of $Li_xNi_{0.8}Co_{0.2}O_2$ phase. (Data were extracted from reference [10].)

TABLE 2.2
Energy Difference (meV) of the Systems Corresponding to the Four Different Polaron Sites in $Li_3FePO_4CO_3$ [6]

	1st NN	2nd NN	3rd NN	4th NN
V_{Li1}	0	61	168	210
V_{Li2}	98	0	146	168
V_{Li3}	215	206	198	0

distances from polaron to Li ion/vacancy in proximity, the binding energy of small polaron–Li ion/vacancy may not be the same. For example, in $LiFePO_4$, the Li vacancies are preferable binding with positive polarons at the first nearest neighbor Fe sites [12], while in Li_2FeSiO_4, the polaron most favorably forms at the third nearest neighbor of Li vacancy [5]. However, the binding between polaron and Li vacancy in $Li_3FePO_4CO_3$ is complexed because there are three distinguished Li sites that behave differently in binding with polaron. As indicated in Table 2.2 [6], it is noted that Li vacancy at the first site is preferable binding with the polaron formed at the first nearest neighbors of distance of 3.03 Å, while the second type is found at the second nearest neighbor site with a distance of 3.53 Å, and the last one is binding stronger to the polaron at the fourth nearest neighbors. Overall, with stronger binding energy, when a Li ion/vacancy diffuses, the polaron would simultaneously accompany the Li ion/vacancy.

2.3 SMALL POLARON–Li-ION/VACANCY COMPLEXES DIFFUSION MODEL

In order to evaluate the kinetic properties and ionic conductivity of cathode materials, it is necessary to investigate the diffusion mechanism of the charge carriers. The NEB method is often employed to predict the diffusion pathway and overall activation energy. Although the evidence of small polaron is clearly addressed, the investigations of the polaron migration are often separated from Li-ion diffusion. However, because of a strong binding energy between Li vacancy/ion and positive/negative polaron, when Li vacancy/ion migrates, polaron would simultaneously diffuse and accompany Li vacancy/ion. Thus, diffusion mechanisms in which the polaron migration and Li-ion diffusion are separated would not present the overall picture of the diffusion mechanism in the cathode materials.

Since the small polaron and Li ion/vacancy are strongly binding, they create two types of complexes: Li vacancy–positive polaron complex and Li ion–negative polaron complex. These types of complexes are dominant at different concentrations of Li ion. An advanced diffusion model of small polaron–Li-ion/vacancy complex was proposed by Dinh et al. [12] in 2012, which clearly distinguishes the feasible diffusion path and analyzes the effect of small polaron migration to diffusion of Li ions.

Because the three-dimensional frameworks of cathode materials are complicated, several diffusion pathways can be found. In this model, the diffusion of Li ion/vacancy is considered as a process so-called the polaron–Li-ion/vacancy complex diffusion in which Li ion/vacancy diffuses with its accompanying polaron along the elementary diffusion paths. The diffusion pathways in whole material can be found by combining the most favorable elementary diffusion paths. There are three types of elementary diffusion processes (EDPs) that are named based on the relation between the migration direction of small polaron and Li vacancy/ion trajectory and can be defined as follows:

i. **Single process**: This EDP occurs when the polaron stays at the same transition metal site during Li-ion/vacancy diffusion.
ii. **Parallel process**: This EDP happens if the polaron hopping direction is parallel to the Li-ion/vacancy diffusion trajectory.
iii. **Crossing process**: This EDP happens if the polaron hops in the direction that crosses to the Li-ion/vacancy diffusion trajectory.

In general, because the hoping movements of polaron are geometrically different, the activation energy for each EDP might not be tantamount. The binding Li-ion/vacancy–polaron complex diffusion might require a higher activation energy in the parallel and crossing processes. Therefore, such EDPs of the binding vacancy (ion)–polaron complex must be depicted as a crucial part of the transition metal-based cathode materials for Li-ion batteries.

2.4 APPLICATIONS OF THE DIFFUSION MODEL TO REAL CATHODE MATERIALS

In this final part, some examples in which Dinh's model was successfully applied to investigate the diffusion mechanism of Li ion inside some typical cathode materials would be depicted. First of all, the commercial olivine phosphate $LiFe_{1-y}Mn_yPO_4$ has been in the spotlight as a cheap, high-energy-density cathode material. In 2012, Dinh et al. [12] used the hybrid functional method to prove the formation of localized bound polaron by bond length change of 0.15 Å. EDPs were well-described

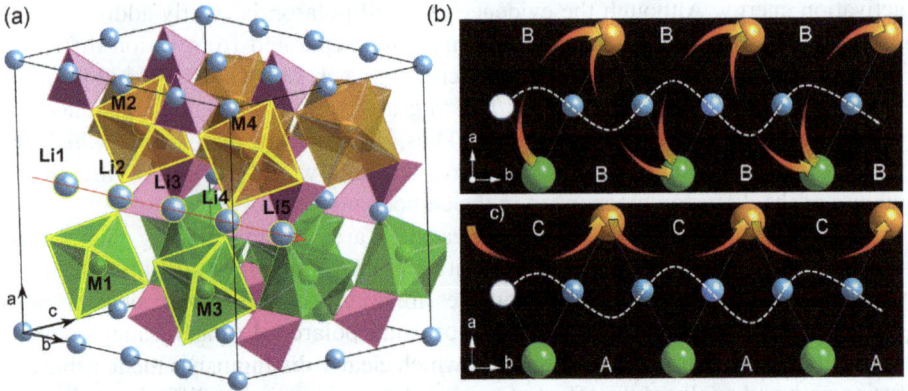

FIGURE 2.5 (a) Structure of LiFe$_{1-y}$Mn$_y$PO$_4$. Cyan balls are Li ions. (b) Crossing process. (c) Parallel process [12]. (Copyright (2012) The Japan Society of Applied Physics.)

TABLE 2.3
Activation Energy E_a (meV) of the Polaron–Li Vacancy Complexes in LiFe$_{1-y}$Mn$_y$PO$_4$ [12] and the Corresponding Experimental Values [13,14]

y	Single	Crossing	Parallel	Exp.
0	380	600	643	590, 630
1	411	684	623	
1/2	380		635	630

in Figure 2.5. The polaron in the crossing process has a hoping distance of 5.5 Å shorter than that in the parallel one. The activation energies for polaron–Li vacancy complex are summarized in Table 2.3. The diffusion of Li ion is preferable along [010] direction. For different y values, it is noted that the influence of polaron migration on Li vacancy diffusion is significant because the crossing and parallel processes (containing polaron migration) always need activation energy of more than 1.5 times the activation energy required for the single process (no polaron migration). It implies that the effect of polaron migration is substantial. The process with the lower activation energy would be preferable. For $y = 0$, the crossing process would be favorable with a lower activation energy of 600 meV. Therefore, the polaron diffusion along the zigzag pathway would be prominent with an activation energy of 600 meV, which is in good agreement with experimental data [13]. When $y = 1$, the parallel polaron diffusion pathway with E_a of 623 meV is more preferable than the crossing process. For $y = ½$, the crossing process in which polaron hops from Fe site to Mn site or vice versa is blocked; thus, only parallel diffusion with activation energy of 635 meV is allowed. The calculated activation energy is well consistent with the experiment result [14].

The silicate material has also been considered as a promising alternative for phosphate compounds. The orthorhombic Li_2FeSiO_4 (space group $Pmn2_1$) is a potential cathode material, which exhibits an initial capacity of 165 mAh g^{-1} and keeps 140 mAh g^{-1} after several cycles [15]. GGA+U method can well describe the formation of small polaron by indicating the sharp bound states appeared in the band gap together with the significant shrinkage in the bond length from 2.05 to 1.92 Å [5]. The diffusion mechanism was also described by the combination of four possible EDPs, as indicated in Figure 2.6. The accompanying polaron migration in this material has a small effect on the Li-ion diffusion along [100] direction, while the crossing process along [001] direction has a barrier slightly higher than others. Along the [100] direction, the combination of parallel and single processes requires activation energy of 840 meV. Li vacancy may also diffuse in zigzag trajectory in [001] direction by repeating only crossing process with energy cost of 880 meV. Unlike olivine phosphate with preferable diffusion along [010] direction only, the Li diffusion in the lithium iron silicate illustrates multidimensional character, which benefits the ionic conductivity. It is also found that the diffusion of Li ion–polaron complex is not affected by the spin configuration of Fe ions (ferromagnetic or antiferromagnetic) in the magnetic material.

Another example of applying Dinh's model is to investigate the diffusion of Li ion in $Li_3FePO_4CO_3$ (space group $P1$ triclinic). This material shows high specific capacity of 876 Wh kg^{-1} and a high discharge voltage of 4.6 V [16]. Similarly, the polaron formation can be characterized by the bond shrinkage and newborn bound state appeared in

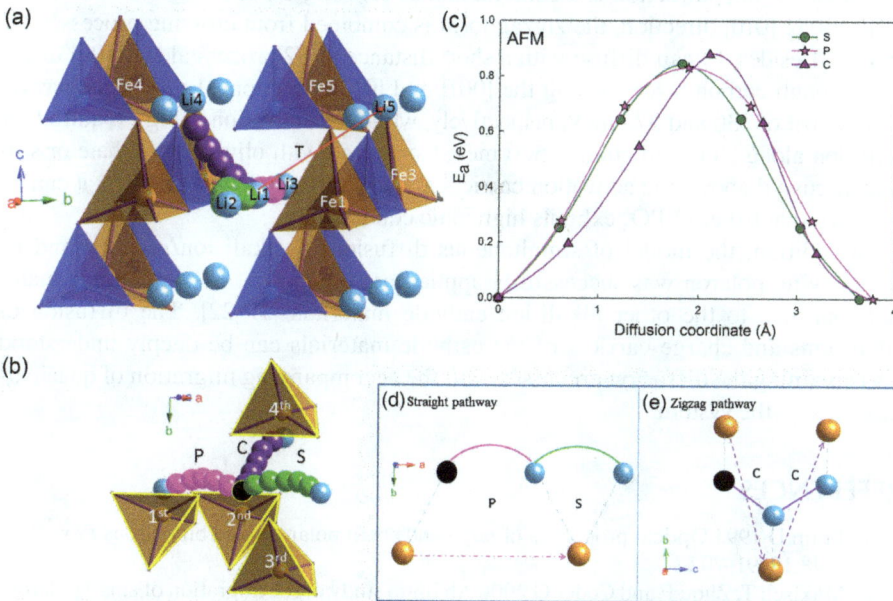

FIGURE 2.6 Four elementary diffusion processes (a) through layer process T; (b) single (green S), crossing (violet C), and parallel (pink P) processes; (c) activation energy of polaron–Li vacancy diffusion in Li_2FeSiO_4; (d) straight pathway; and (e) zigzag pathway [5]. (Copyright (2012) The Japan Society of Applied Physics.)

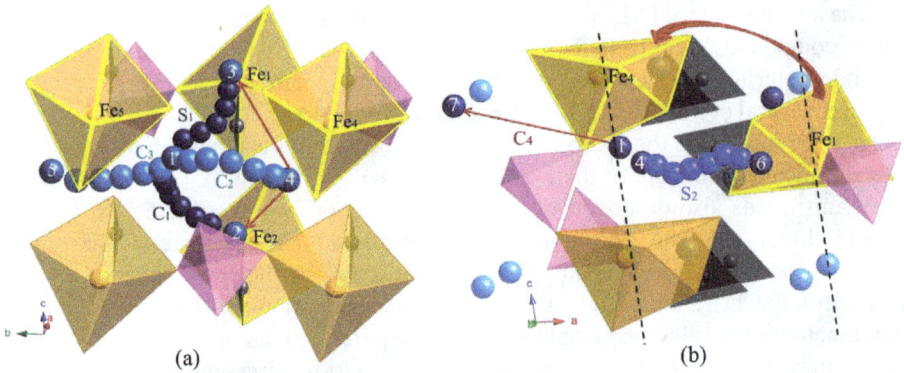

FIGURE 2.7 (a) Six intralayer EDPs and (b) interlayer EDPs in Li$_3$FeCO$_3$PO$_4$ [6]. (Copyright (2013) The Japan Society of Applied Physics.)

the density of states when a Li atom is removed. Seven diffusion processes were investigated including six intralayer EDPs and one interlayer EDP as shown in Figure 2.7 [6]. It is noted that four intralayer EDPs and one interlayer EDP are the most preferable ones. Among these processes, the single one has the lowest activation energy, implying that the effect of small polaron migration on Li-ion diffusion is substantial so that this effect lowers the diffusion of Li ion. Li ion can diffuse in three dimensions. The diffusion along [001] direction is a combination of crossing *C1* and single *S1* processes, while along [010] direction, the zigzag form is combined from crossing processes *C2* and *C3*. Besides, Li can diffuse with a short distance in *S2* process along [100] direction through carbon layers. Along the [001] and [010] directions, Li ion must pay an energy cost of 330 and 370 meV, respectively, while the activation energy required for diffusion along [100] direction is 500 meV. Compared with olivine phosphate or silicate discussed above, the activation barriers are considerably low. Therefore, it can be expected that LiFeCO$_3$PO$_4$ exhibits high ionic conductivity.

In addition, the model of simultaneous diffusion of alkali ion/vacancy and its companying polaron was successfully applied not only to the Li-ion cathode materials but also to the other alkali-ion cathode materials [17–22]. The diffusion of alkali ions and charge carriers of the cathode materials can be deeply understood via explaining the diffusion processes with the accompanying migration of quasiparticles inside the system.

REFERENCES

1. Emin D 1993 Optical properties of large and small polarons and bipolarons *Phys. Rev. B* **48** 13691–702.
2. Maxisch T, Zhou F and Ceder G 2006 Ab initio study of the migration of small polarons in olivine Li$_x$FePO$_4$ and their association with lithium ions and vacancies *Phys. Rev. B* **73** 104301.
3. Ong S P, Chevrier V L and Ceder G 2011 Comparison of small polaron migration and phase separation in olivine LiMnPO$_4$ and LiFePO$_4$ using hybrid density functional theory *Phys. Rev. B* **83** 075112.

4. Kong F, Longo R C, Park M-S, Yoon J, Yeon D-H, Park J-H, Wang W-H, Kc S, Doo S-G and Cho K 2015 Ab initio study of doping effects on $LiMnO_2$ and Li_2MnO_3 cathode materials for Li-ion batteries *J. Mater. Chem. A* **3** 8489–500.

5. Bui K M, Dinh V A and Ohno T 2012 Diffusion mechanism of polaron–Li vacancy complex in cathode material Li_2FeSiO_4 *Appl. Phys. Express* **5** 125802.

6. Duong D M, Dinh V A and Ohno T 2013 Quasi-three-dimensional diffusion of Li ions in $Li_3FePO_4CO_3$: First-principles calculations for cathode materials of Li-ion batteries *Appl. Phys. Express* **6** 115801.

7. Emin D 2012 *Polarons* (Cambridge: Cambridge University Press).

8. Hoang K and Johannes M 2018 Defect physics in complex energy materials *J. Phys. Condens. Matter* **30** 293001.

9. Watthaisong P, Jungthawan S, Hirunsit P and Suthirakun S 2019 Transport properties of electron small polarons in a V_2O_5 cathode of Li-ion batteries: a computational study *RSC Adv.* **9** 19483–94.

10. Saadoune I and Delmas C 1998 On the $Li_xNi_{0.8}Co_{0.2}O_2$ System *J. Solid State Chem.* **136** 8–15.

11. Fisher C A J, Hart Prieto V M and Islam M S 2008 Lithium battery materials $LiMPO_4$ (M = Mn, Fe, Co, and Ni): Insights into defect association, transport mechanisms, and doping behavior *Chem. Materials* **20** 5907–15.

12. Dinh V A, Nara J and Ohno T 2012 A new insight into the polaron–Li complex diffusion in cathode material $LiFe_{1-y}Mn_yPO_4$ for Li ion batteries *Appl. Phys. Express* **5** 045801.

13. Chung S-Y, Bloking J T and Chiang Y-M 2002 Electronically conductive phospho-olivines as lithium storage electrodes *Nat. Mater.* **1** 123–8.

14. Molenda J, Ojczyk W, Świerczek K, Zając W, Krok F, Dygas J and Liu R-S 2006 Diffusional mechanism of deintercalation in $LiFe_{1-y}Mn_yPO_4$ cathode material *Solid State Ionics* **177** 2617–24.

15. Nytén A, Abouimrane A, Armand M, Gustafsson T and Thomas J O 2005 Electrochemical performance of Li_2FeSiO_4 as a new Li-battery cathode material *Electrochem. Commun.* **7** 156–60.

16. Hautier G, Jain A, Chen H, Moore C, Ong S P and Ceder G 2011 Novel mixed poly-anions lithium-ion battery cathode materials predicted by high-throughput *ab initio* computations *J. Mater. Chem.* **21** 17147–53.

17. Bui K M, Dinh V A, Okada S and Ohno T 2015 Hybrid functional study of the NASICON-type $Na_3V_2(PO_4)_3$: Crystal and electronic structures and polaron-Na vacancy complex diffusion *Phys. Chem. Chem. Phys.* **17** 30433–30439.

18. Bui K M, Dinh V A, Okada S and Ohno T 2016 Na-ion diffusion in a NASICON-type solid electrolyte: A density functional study *Phys. Chem. Chem. Phys.* **18** 27226–17231.

19. Debbichi M, Debichi N, Dinh V A and Lebegue S 2016 First principles study of the crystal, electronic structure, and diffusion mechanism of polaron-Na vacancy of $Na_3MnPO_4CO_3$ for Na-ion battery applications *J. Phys. D: Appl. Phys.* **50** 045502.

20. Luong H D, Pham T D, Morikawa Y, Shibutani Y and Dinh V A 2018 Diffusion mechanism of Na ion – Polaron complex in potential cathode materials $NaVOPO_4$ and $VOPO_4$ for rechargeable sodium-ion batteries *Phys. Chem. Chem. Phys.* **20** 23625–23634.

21. Tran T L, Luong H D, Duong D M, Dinh N T and Dinh V A, Hybrid functional study on small polaron formation and ion diffusion in the cathode material $Na_2Mn_3(SO_4)_4$ *ACS Omega* **5** 10 5429–5435.

22. Luong H D, Dinh V A, Momida H and Oguchi T 2020 Insight into the diffusion mechanism of sodium ion – Polaron complexes in orthorhombic P2 layered cathode oxide Na_xMnO_2 *Phys. Chem. Chem. Phys.* **22** 18219–18228.

3 Enrichment of Optical Excitations of LiFeO$_2$

Vo Khuong Dien, Thi Han Nguyen,
and Ming-Fa Lin
National Cheng Kung University

CONTENTS

3.1 INTRODUCTION

Increasing demands for storing energy from wind and solar energy, mobile electronic equipment, and electronic transportations promote the development of reliable and cost-effective batteries [1–4]. Compared with other energy storage systems, lithium-ion batteries (LIBs) have been received a great deal of attention since they process desirable features, such as lightweight, long life cycle, fast charging time, and can provide a sizable electronic current for electronic devices [5–8]. A typical LIB system is a combination of the electrolyte sandwiched between the positive (anode) and the negative (cathode) electrodes. During the charging/discharging process, the Li$^+$ will migrate from the cathode $\longrightarrow$ electrolyte $\longrightarrow$ anode/anode $\longrightarrow$ electrolyte $\longrightarrow$ cathode materials [9]. The energy will be released/stored from/in this system in terms of chemical energy. Apparently, the physical/chemical/material pictures of these components are rather complicated and strongly related to the energy storage ability of the battery system [10–12].

Generally, commercial LIBs using solid or liquid electrolytes, such as Li$_2$SiO$_3$ or Li$_3$OCl ternary compounds, can provide a wide energy window [13,14], while the graphite layers or the ternary Li$_2$GeO$_3$ and Li$_4$Ti$_5$O$_{12}$ compounds were usually used for the anode ones [15,16]. On the other hand, ternary compounds such as LiCoO$_2$, LiNiO$_2$, and LiMn$_2$O$_4$ [17–19] are adopted in the traditional cathodes. However, these cathode materials have economic and environmental problems that limit their use in large-scale LIBs for electronic vehicles and electric storage devices. Recently, a lot of investment in research has been done to search for new cathode materials that satisfy the market demand and then reduce the pollution increasing. Among them, Li-Fe-O

DOI: 10.1201/9781003263807-3

compound has been paid more attention due to most abundance and nontoxicity of Fe. As a matter of fact, various types of lithium iron oxides have been investigated, for example, corundum-type α-Fe_2O_3, β-$NaMnO_2$-type $LiFeO_2$, inverse spinel-type Fe_3O_4, β-FeOOH-type $LiFeO_2$, β-$NaFeO_2$-type $LiFeO_2$, and layered α-$NaFeO_2$-type $LiFeO_2$. Among these materials, α-$NaFeO_2$-type $LiFeO_2$ has a relatively high theoretical capacity of 282 mAh g^{-1} in a one-electron reaction [20], simple and eco-friendly synthesis [21].

On the theoretical aspects, the various density functional theory (DFT) and molecule dynamic calculations were adopted to investigate the electronic and magnetic properties, the Li^+ transport mechanism, and the stability of electrode and electrolyte materials. Although the fundamental properties of the geometric, electronic, and magnetic properties of condensed materials have been investigated by both these methods [22,23], the systematic investigations on essential properties of these kind materials are absent. For example, the critical chemical/physical/material pictures, which relate to specific orbital hybridization in different chemical bonds, are very important since they are related to the ion transport mechanism and magnetic configuration of material, but the investigations to find out such phenomena are lacking up to now. The optical excitation, one of the most important properties, has not been completely comprehended. Furthermore, the close connection of the quasiparticle charge, spin density distributions, and the specific orbital hybridization with the energy-dependent optical excitation of such ternary compound has not been achieved so far.

Fortunately, previous theoretical simulations based on first-principles calculations can depict clearly the essential rich and unique features of emergence materials, especially for 3D ternary anode/cathode/electrolyte materials [24–26] and emergent 2D layered [27–29]. For example, systematic studies have been conducted on the features of ternary $Li_4Ti_5O_{12}$ anode, Li_2SiO_3, and Li_2GeO_3 electrolyte material [30–32], the essential properties of graphene/silicone-related materials [27–29,33–36]. Such investigations clearly illustrate that the quasiparticle charges, orbitals, and spins dominate all the fundamental properties. The high accurate simulation results and delicate analyses are capable of proposing significant pictures/mechanisms to fully understand the geometric, electronic, and magnetic properties, and optical excitations. The important single-/multi-orbital hybridizations in various chemical bonds are obtained from the optimization geometric, the atom-dominated band structures, the spatial charge densities and their variations after chemical modifications. The magnetic configurations in the host/guest material could be comprehended through spin-split/spin-degenerate energy bands, the spin density distributions, the net magnetic moments, and the spin-projected van Hove singularities. Furthermore, the energy-dependent optical excitations and the influence of electron–hole interaction on optical properties are thoroughly investigated by the dielectric function, absorption, reflectance coefficients, and energy loss function [37–39]. This developed framework, which is successfully conducted on silicone/graphene-related systems [40,41], the anode/cathode/electrolyte compounds [30,32,42] and could be generalized to other emergent materials or needs to be thoroughly tested in further investigations. It is thus expected to be very suitable for studying the rather complicated geometric, electronic, magnetic, and optical properties of mainstream Li^+-based batteries.

In this study, the theoretical framework is developed to comprehend the quasi-particle charges, spin orientations, and orbital hybridizations in the chemical bonding of the ternary LiFeO$_2$ compound. This strategy is based on the first-principles calculations on an optimized structure with position-dependent chemical bonding, the spin-dependent energy band structure with atom dominated at different ranges, the spatial spin and charge densities due to various orbitals, and the atom- and orbital-projected density of state related to spin directions and orbital overlaps. The energy-decomposed single-/multi-orbital hybridizations and spin polarizations will be utilized to account for the optical onset frequency, various prominent absorption structures, a tremendous plasmon response in terms of the dielectric functions, energy loss functions, reflectance spectra, and absorption coefficients under the distinct electric polarizations. The current study is of paramount importance not only for fundamental physics but also for technical applications. Most predicted results in this work require high-resolution experimental examinations.

3.2 COMPUTATIONAL DETAILS

The DFT method [43] via the Vienna Ab initio Simulation Package (VASP) [44] was utilized to optimize the geometric structure and calculation of the electronic and magnetic properties and optical excitations. The Perdew–Burke–Ernzerhof (PBE) [45] generalized gradient approximation was used for the exchange-correlation functional. We employed the projected augmented wave (PAW) pseudopotentials to describe the electronic wave functions in the core region [46]. The cutoff energy for the expansion of the plane wave basis was set to 500 eV. The Brillouin zone was integrated with a special k-point mesh of $15 \times 15 \times 15$ in the Monkhorst–Pack sampling technique for geometric optimization. The convergence condition of the ground state is set to be 10^{-8} eV between two consecutive simulation steps, and all atoms could allow to fully relax during geometric optimization until the Hellmann–Feynman force acting on each atom was smaller than 0.01 eV.

When a material is disturbed by an electromagnetic wave, all charge carriers in this condensed-matter system will screen the external field and create charge density fluctuations and thus generate the induce-charge currents. From the quantum mechanics point of view, such phenomena could be considered as the scattering between the photons–quanta of electromagnetic wave and electrons in condensed matters. Apparently, these scattering mush be satisfied the conservation of momentum, energy as well as Pauli exclusive principle. The picture for single-particle excitation can be described by the Kubo formula [47].

$$\varepsilon_2^{\uparrow(\downarrow)}(\omega) = \frac{8\pi^2 e^2}{\omega^2} \sum_{vck} \left| e.v k^{\uparrow(\downarrow)} |v| c k^{\uparrow(\downarrow)} \right|^2 \delta\left(\omega - E_{ck}^{\uparrow(\downarrow)} - E_{vk}^{\uparrow(\downarrow)}\right),$$

where the available transition channels and strength of each excitation peak are directly related to the joined density of state, $\delta\left(\omega - E_{ck}^{\uparrow(\downarrow)} - E_{vk}^{\uparrow(\downarrow)}\right)$, and the velocity matrix element, $\left| e.v k^{\uparrow(\downarrow)} |v| c k^{\uparrow(\downarrow)} \right|^2$, respectively. To evaluate the optical response of $LiFeO_2$ from first-principles calculations, we adopted the strategy of Michael Rohlfing and Steven G. Louie [48]. In which, the quasiparticle Green's function and the screened Coulomb interactions (G_0W_0) approach using 250 eV energy cutoff for the response functions and the Brillouin zone was integrated with a special k-points mesh of $12 \times 12 \times 12$ in the Gamma sampling technique to obtain the corrected density of states and electronic band structure and quasiparticle excitations of $LiFeO_2$.

Generally, the excited electrons and holes could be strongly combined through a suitable condition such as a large electronic band gap for low screening ability. The excitons, the new composite quasiparticle due to the mutual interactions of electrons and holes, are expected to dominate the main features of optical properties; for example, the redshift of the optical gap, the modifications of shape, energy, and intensity of each adsorption peak. Using the k-point sampling, energy cutoff, and number of bands setting the same as in the GW calculation, those exciton states can be determined by solving the standard Bethe–Salpeter equation (BSE) [49]:

$$\varepsilon_2^{\uparrow(\downarrow)}(\omega) = \frac{8\pi^2 e^2}{\omega^2} \sum_{vck} \left| e.0^{\uparrow(\downarrow)} |v| S^{\uparrow(\downarrow)} \right|^2 \delta\left(\omega - \Omega_s^{\uparrow(\downarrow)}\right).$$

Due to the dielectric functions describing the causal response, the real part of the dielectric functions, $\varepsilon_1(\omega)$, can be evaluated from the imaginary one $\varepsilon_2(\omega)$ through the Kramers–Kronig relations,

$$\varepsilon_1(\omega) = 1 + \frac{2}{\pi} P \int_0^\infty \frac{\varepsilon_2(\omega')\omega'}{\omega^2 - \omega'^2} d\omega'$$

Other optical properties, such as the energy loss functions $\left(\text{Im}\left[-1/\varepsilon(\omega)\right]\right)$, reflectivity $\left(R(\omega)\right)$, and absorption coefficients $\left(\alpha(\omega)\right)$, can be obtained from the imaginary part $\left(\varepsilon_2(\omega)\right)$ and real part $\left(\varepsilon_1(\omega)\right)$ of dielectric functions:

$$\text{Im}\left[\frac{-1}{\varepsilon}\right] = \frac{\varepsilon_2(\omega)}{\varepsilon_1^2(\omega) + \varepsilon_2^2(\omega)},$$

$$R(\omega) = \left| \frac{\sqrt{\varepsilon_1(\omega) + \varepsilon_2(\omega)} - 1}{\sqrt{\varepsilon_1(\omega) + \varepsilon_2(\omega)} + 1} \right|^2,$$

$$\alpha(\omega) = 1/\sqrt{2}\left[\sqrt{\varepsilon_1^2(\omega) + \varepsilon_2^2(\omega)} + \varepsilon_1(\omega)\right]^{1/2}.$$

The close connection of the quasiparticle charges, spin polarizations, and orbital hybridization and the effect of the coupled quasiparticle on optical excitations will be discussed in detail in the current work.

3.3 RESULTS AND DISCUSSION

The LiFeO$_2$ compound crystallizes in the trigonal structure with R-3m space group (Figure 3.1a and b). The calculated lattice constants of this structure are 2.88, 2.88, and 14.31 Å for x-, y-, and z-directions, respectively, which are very good in agreement with previous theoretical and experimental values [50–53]. The basic building block of LiFeO$_2$ is stacked alternately of the FeO$_2$ layer with the Li layer along the z-axis, in which each Fe/Li atom occupies the center of an octahedron of O atoms (Figure 3.1c). As a result, there are a total of 18 equivalent Fe-O bonds (1.97 Å) and 18 identical Li-O bonds (2.13 Å). The highly ordered arrangement of the atoms and the anisotropy of the geometric structure, which are originated from the complex orbital hybridization, will support Li$^+$ migration and responsible for the anisotropic behavior of optical excitation. In addition to the lattice constant of 3D ternary compound, other quantities, such as the morphology or material particle size, could be measured by SEM, the top view of nanoscale materials could be clarified by TEM, while STM measurement is usually used to provide the side view information. Such experiments have been successfully applied for multi-walled carbon nanotube [54], the stacking configuration in multi-layer graphene [55], or the geometric profile of graphene nanoribbons [56] and are therefore very suitable for the ternary LiFeO$_2$ compound. The predicted results in the strongly anisotropic atomic ordering, the layer structure of LiFeO$_2$ compound, the unit cell, and the bond length combined with experimental results are very useful in confirming the complicated orbital hybridizations in the chemical bonds.

The energy band structure along the high symmetry point of the LiFeO$_2$ compound was calculated and shown in Figure 3.2a. The reference point also regarded as Fermi energy was set at the middle of the conduction band and valence band. They present various energy sub-bands with various dispersion characteristics such as oscillatory,

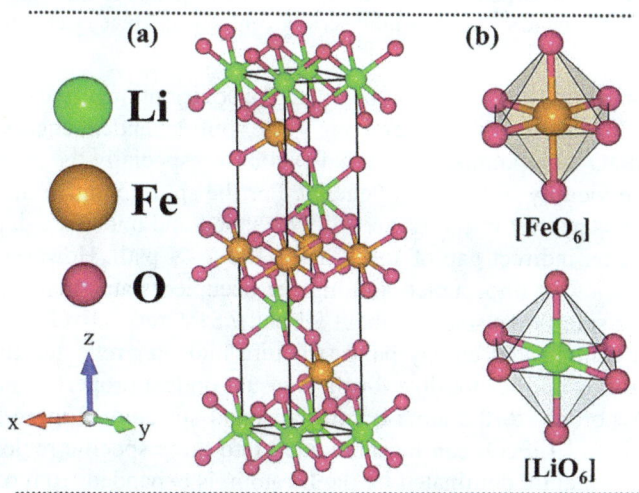

FIGURE 3.1 The optimal geometric structure of LiFeO$_2$ compound with (a) side view, (b) top view, respectively, and (c) the octahedron structure of [LiO$_6$] and [FeO$_6$].

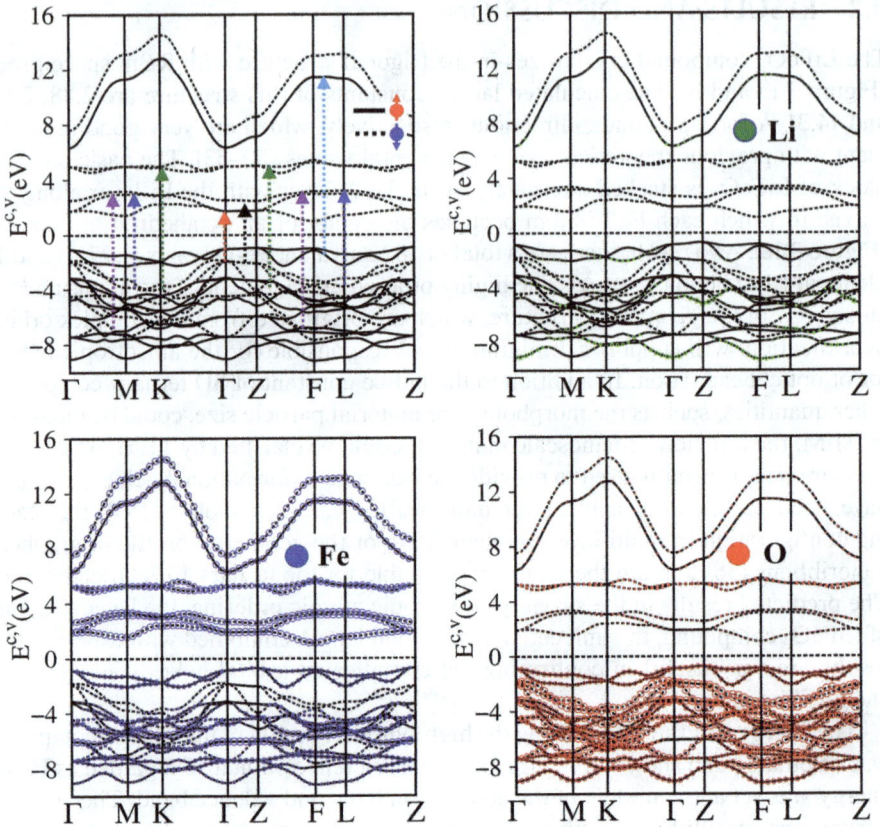

FIGURE 3.2 (a) Band structure along the high symmetry points in the wave-vector space, with (b) Li-, (c) Fe-, and (d) O-atom dominances (green, red, and blue circles, respectively).

camel's back shape, or dispersionless owing to the contribution of the certain orbitals and atoms in the unit cell. Very interestingly, the spin-dependent energy band structure of the LiFeO$_2$ compound is expressed explicitly, especially the remarkable spin splitting in the vicinity of the Fermi energy. For the spin-up and spin-down states, the energy bands nearest to the E$_F$ are fully occupied and unoccupied, respectively, which leads to an indirect gap of 1.9 eV along the Γ-X path. However, the energy spacing for the lowest unoccupied and highest occupied states that are considered for spin-up and spin-down states is about 3.1 and 7.5 eV, respectively. The large spin splitting with complicated energy band structure indicates reflected the ferromagnetic configuration and responsible for the unusual optical properties in the LiFeO$_2$ compound. According to the atom-dominated band structure (Figure 3.2b–d), the energy spectrum of LiFeO$_2$ can be subdivided into three specific regions. The first region, which is mainly dominated by the Fe atom, is expanded from 6.5 to 14.7 eV. The second one with the energy range between 1 and 5 eV is co-dominated by Fe and O atoms. Similarly, the last region with the energy around −1 to −9 eV is mainly derived from Fe and O atoms. For the entire energy spectrum, the contribution of the Li atom is a minority. However, some salient features of LiFeO$_2$ will disappear in the absence of the Li dominations. The band edge states related to the prominence optical excitation peaks could be clarified through the close combination of the current-dominated energy spectrum with spatial spin density distribution/charge density distribution, and the spin-polarized density of states.

The magnetic properties of the anode/cathode/electrolyte compounds could be comprehended through the spin density distribution and the magnetic moment (Table 3.1), in which the net magnetic moment in a unit cell will be determined by the competition of spin-up and spin-down components. The LiFeO$_2$ compound presents the ferromagnetic configuration with the total magnetic moment equal to 4.624 μ_B as expected since the Fe atom introduces four spin-up electrons. In fact, the spin-up density is the most dominant part and relies on the Fe atom (Figure 3.3a) with the typical magnetic moment equal to 4.06 μ_B, which is directly reflected by the fully occupied state in strongly dispersive energy bands below the Fermi level (the third region). Furthermore, O and Li atoms, which are nonmagnetic before combination, also present the partial minor magnetic contributions (Figure 3.3b) with these values lower than the Fe atom under one and two orders, respectively. As a

TABLE 3.1

The Magnetic Moment of the Specific Orbital and Atom of LiFeO$_2$

	Magnetic Moment (μ_B)			
Ion	s	p	d	tot
Li	0.009	0	0	0.009
Fe	0.023	0.026	4.011	4.060
O	0.018	0.260	0	0.277
O	0.018	0.260	0	0.277
Total	0.061	0.551	4.011	4.624

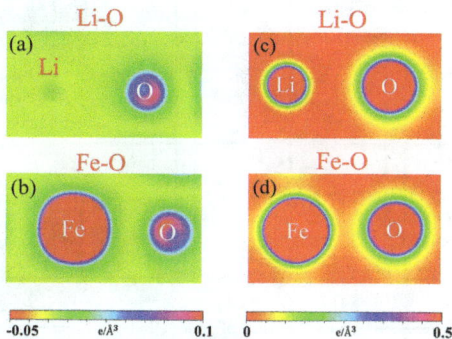

FIGURE 3.3 (a and b) Spin density distribution. (c and d) Charge density distributions related to the significant orbital hybridizations in the Fe-O bonds and Li-O ones, respectively.

result, the net magnetic moments will be expected to be sensitive to change during Li^+ transportation.

The orbital character in chemical bonds can be realized by considering the spatial charge density of Li, Fe, and O atoms in the chemical bonds. As for the O atom, the red-blue (inner) and the green-yellow (outer) regions arise from the (2s) and $(2p_x, 2p_y, 2p_z)$ orbitals, respectively. Similarly, the inner and outer parts of the Fe atom correspond to the $(3d_{xy}, 3d_{yz}, 3d_{xz}, 3d_{x^2-y^2}, 3d_{z^2})$ and 4s orbitals. When they combine together to form the Fe-O chemical bond, the serious distortion of these spherical charges (Figure 3.3c) is clearly evident in the appearance of the important multi-hybridization of Fe-$(3d_{xy}, 3d_{yz}, 3d_{xz}, 3d_{x^2-y^2}, 3d_{z^2})$ and O-$(2p_x, 2p_y, 2p_z)$ orbitals and the hybridizations of Fe-(4s) & O-$(2p_x, 2p_y, 2p_z)$ orbitals. For the Li-O chemical bond, the charge density around the Li atom can also be divided into two parts, the inner part corresponding to the 1s orbitals, while the outer ones belonging to 2s one. As shown in Figure 3d, the slight deformation of Li and O spatial charge densities between Li-O chemical bonds presents multi-Li-2s and O-$(2p_x, 2p_y, 2p_z)$ orbitals combination. Very interestingly, the O-(2s) orbital negligibly dominates at the low-lying energy and therefore almost does not participate in the chemical bondings (the spherical sharp of oxygen in Li-O and Fe-O bonds).

The orbital hybridizations and spin polarizations in Li-O and Fe-O bonds could be clarified by the atom and orbital density of states. As revealed in Figure 3.4a–c, the density of states of 3D ternary $LiFeO_2$ compound mostly presents the shoulder and asymmetric van Hove singularities since they mainly originated from the oscillatory, local minimum/maximum, or almost dispersionless relation energy sub-bands. The merge well of various atoms and orbitals evident the complicated orbital hybridizations in Li-O and Fe-O bonds. Very interestingly, the domination and energy of the spin-up and spin-down are extremely different; this indicated that $LiFeO_2$ possesses very strong ferromagnetic behavior, being good in agreement with spin-splitting band structure and spin density distribution. According to the distribution of atoms

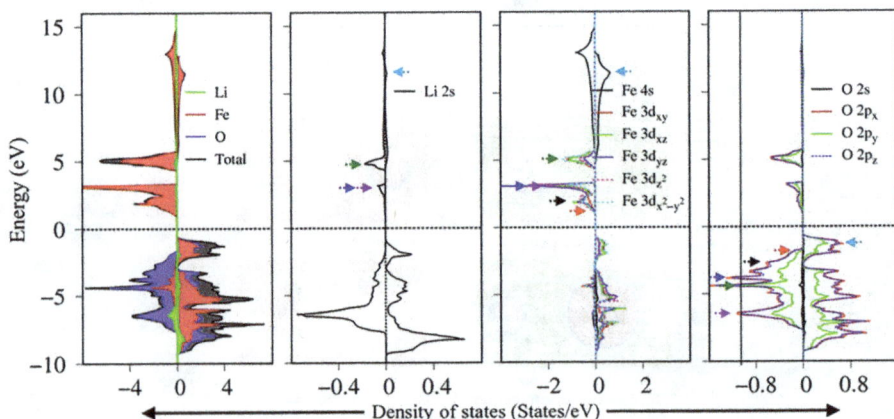

FIGURE 3.4 Density of states under the different components: (a) Li-2s, (b) Fe-$(4s, 3d_{xy}, 3d_{yz}, 3d_{xz}, 3d_{x^2-y^2}, 3d_{z^2})$, and (c) O-$(2s, 2p_x, 2p_y, 2p_z)$ orbitals.

and orbitals, the orbital hybridization of the three mentioned energy regions can be in turn identified: The first region is governed by Li-2s and O-($2p_x$, $2p_y$, $2p_z$) and Fe-4s and O-($2p_x$, $2p_y$, $2p_z$) orbital hybridizations. The second region is a mixture of Li-2s and O-($2p_x$, $2p_y$, $2p_z$) and Fe-(4s, $3d_{xy}$, $3d_{xz}$, $3d_{yz}$, $3d_{x^2-y^2}$, $3d_{z^2}$) orbitals, and the last one is related to Li-2s and O-($2p_x$, $2p_y$, $2p_z$) and Fe-(4s, $3d_{xy}$, $3d_{xz}$, $3d_{yz}$, $3d_{x^2-y^2}$, $3d_{z^2}$) and O-($2p_x$, $2p_y$, $2p_z$) orbital hybridizations. This delicate analysis will elucidate the spin polarization and orbital hybridization that are related to the prominence of optical excitations. In addition to the geometric structure, the examination of rich and unique electronic features required high-resolution equipment. For instance, the electronic band structure including various energy dispersions can be depicted by using ARPESS measurement [57,58], the STS examination [59,60] can provide the asymmetric or shoulder van Hove singularities around the Fermi level, and furthermore, spin-polarized STS also provides the magnetic information of materials. To date, such experimental measurements have been successfully used to confirm occupied electronic states and dimension-diversified van Hove singularities of monolayer graphene [61], few-layer graphene systems [62], adatom absorption graphene [63,64], 1D graphene nanoribbons [65], and other 2D materials [66,67]. Apparently, these measurements are available for verifying the asymmetric of the occupied and unoccupied states, the presence of various van Hove singularities, and the spin-split/degeneracy of the LiFeO$_2$ compound.

As a direct consequence of electromagnetic wave absorption, the dielectric function is complex and expresses the main electronic and magnetic features. The optical excitation of the ferromagnetic LiFeO$_2$ compound within the absence of the coherent charge screening is described by the imaginary part $[\varepsilon_2(\omega)]$ of the dielectric function. As presented in Figure 3.5a, the calculated $\varepsilon_2(\omega)$ of the x- and y-polarizations is the same, but these curves for the z-polarization are totally different owing to the different atomic arrangements. This indicates that the trigonal LiFeO$_2$ has strongly optical anisotropy. The threshold frequency, or so-called the optical gap, is situated at 3.1 and 3.2 eV for x-, y-, and z-polarizations, respectively. However, it is reduced by 0.2 eV and the excitation strength is enhanced as a consequence of electron–hole coupling (Figure 3.5c). The slightly redshift of the optical gap however with the remarkable change of these curves evident a weak but very significant impact of the excitonic effects. Beyond the threshold frequency, it also possesses a lot of special structures with a variety of shapes. The shoulder structure, symmetric/asymmetric arises from the transition of the electron with the same spin orientation in band edge states/the strong van Hove singularities. For instance, the transition from the spin-down of O-($2p_x$, $2p_y$, $2p_z$) to Fe ($3d_{xy}$, $3d_{yz}$, $3d_{xz}$, $d_{x^2-y^2}$, $3d_{z^2}$) with the same spin orientation will lead to the threshold structure. The promotion of electrons from the highest occupied to the lowest unoccupied is forbidden due to different spin ordering. To explain the microscopic origin characters of all individual electronic transitions, we performed a delicate connection of the concise physical/chemical picture with the prominence optical excitations response, in which the excitation peaks were denoted by the colored triangles in Figure 3.5a, the vertical transition was described by the arrowhead in Figure 3.2a, and the orbital properties of each channel were presented in the density of states (Figure 3.4a–c) and shown in Table 3.2. Obviously, owing to the difference in band structures, the electron with different spin orientations has dramatically different dielectric responses.

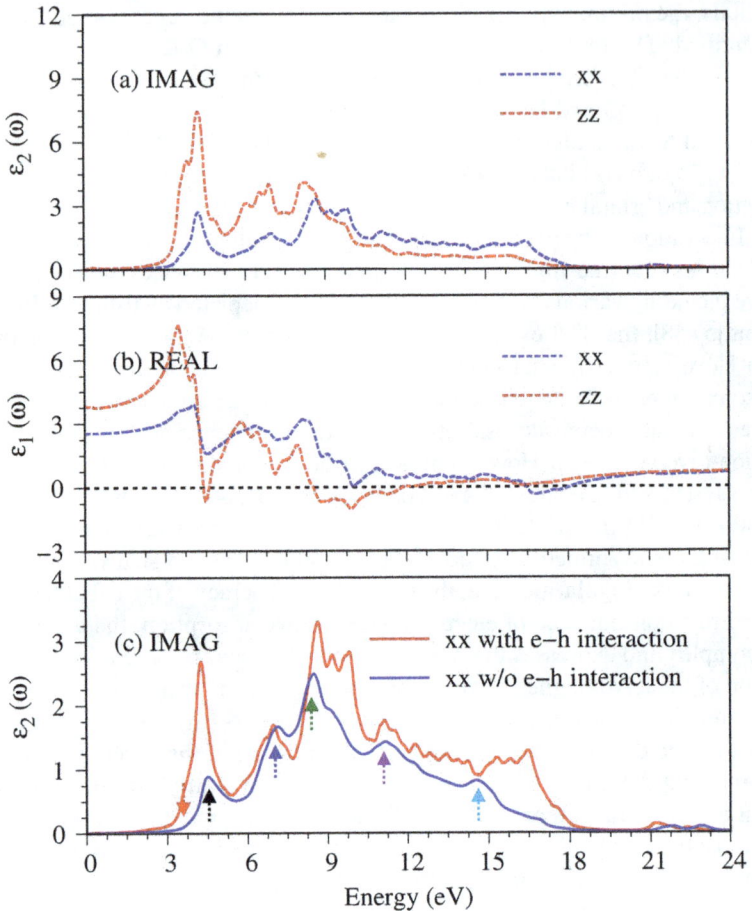

FIGURE 3.5 (a) The imaginary $\varepsilon_2(\omega)$ and (b) real $\varepsilon_1(\omega)$ parts of dielectric functions with the excitonic effects. (c) Comparison of $\varepsilon_2(\omega)$ of x-direction with and without excitonic effects.

Considering the real part of the dielectric function $[\varepsilon_1(\omega)]$, as shown in Figure 3.5b, its special structure involves various asymmetry or logarithm-relation peaks because it is generated by the imaginary part one (Figure 3.5a) through the Kramers–Kronig relations. Its characteristic is nearly independent of the frequency in the low-energy region. For instance, the values of background dielectric constant are equal to 4 and 2.7 eV for x-, y-, and z-directions, respectively, and almost remain the same up until the energy crosses the threshold frequency. However, beyond the optical gap, its relation becomes more complicated with the explicitly sensitive change to the frequency increase. Very interestingly, at some specific frequencies, $\varepsilon_1(\omega)$ crosses to zero, and these positions will correspond to the plasmon modes with the weak Landau damping if they present with insignificant interband transitions. The zero points at 18 or 20 eV for x-, y-, or z-direction, respectively, as a typical example in this case.

TABLE 3.2

Calculated Prominent Absorption Peaks and the Leading Interband Transitions of Each Peak

Energy (eV)	Colors of Arrows	Spin	Orbital Hybridizations-Leading Excitations (Fe-O)
3.1	Red	↓	Fe $(3d_{xy}, 3d_{yz}, 3d_{xz}, 3d_{x^2-y^2}, 3d_{z^2})$-O $(2p_x, 2p_y, 2p_z)$
4.7	Black	↓	Fe $(3d_{xy}, 3d_{yz}, 3d_{xz}, 3d_{x^2-y^2}, 3d_{z^2})$-O $(2p_x, 2p_y, 2p_z)$
7	Blue	↓	Fe $(3d_{xy}, 3d_{yz}, 3d_{xz}, 3d_{x^2-y^2}, 3d_{z^2})$-O $(2p_x, 2p_y, 2p_z)$
8.6	Green	↓	Fe $(3d_{xy}, 3d_{yz}, 3d_{xz}, 3d_{x^2-y^2}, 3d_{z^2})$-O $(2p_x, 2p_y, 2p_z)$
11.3	Pink	↓	Fe $(3d_{xy}, 3d_{yz}, 3d_{xz}, 3d_{x^2-y^2}, 3d_{z^2})$-O $(2p_x, 2p_y, 2p_z)$
14.5	Cyan	↑	Fe (4s)-O $(2p_x, 2p_y, 2p_z)$

However, the dip curve of $\varepsilon_1(\omega)$ at 4.5 eV and 9 eV will become meaningless since it accompanies by very strong single-particle excitation.

The energy loss function, being defined as $\mathrm{Im}[-1/\varepsilon(\omega)]$ (Figure 3.6a), is useful to elucidate the coherent oscillation of valence charges and can be detected by electron energy loss spectrum or X-ray inelastic scattering, in which each prominence peak is referred to as plasmon, and the collective excitation of free charges appears at the frequencies where $\varepsilon_1(\omega)$ passes through zero and $\varepsilon_2(\omega)$ is nearly vanishing. For the opposite cases, the finite $\varepsilon_2(\omega)$ implies that the plasmon modes might decay into single-particle excitations owing to the Landau damping. The most notable peak that can be mentioned existent at 17 and 18 eV for x-, y-, and z-directions, respectively, corresponds to these zero points of $\varepsilon_1(\omega)$ at the same energy and is originated from the excitation of Li-2s, O-$(2p_x, 2p_y, 2p_z)$ and Fe-$(4s, 3d_{xy}, 3d_{yz}, 3d_{xz}, 3d_{x^2-y^2}, 3d_{z^2})$ orbitals, and the contribution of 2s orbital of oxygen to these peaks is ignored since it is an unremarkable contribution in the active regions. Furthermore, the energy loss function also presents the several relatively weak plasmon modes since they appear with disappearance of zero points or accompany with an extremely Landau damping.

When a photon beam is arriving at the boundary of material, some of them will throw back from the front surface while another will transmit and incorporate with the valence electrons. The reflection and absorption are two usual optical phenomena and reflect the main features of the coherent and single optical excitations. The reflection coefficient in the low-energy region is weakly dependent on energy, with typical values in the range of 0.1 and 0.06 for x-, y-, and z-directions, respectively (Figure 3.6b), while the absorption coefficient is zero due to the absence of optical excitations. However, during the excitation events, the reflection changes sensitively and exhibits a large fluctuation, especially explicitly variation (~30%) for z-direction at the strongest plasmon mode owing to the contribution of coherent excitation. Similarly, for the absorption coefficient, the presence of large variation with various prominence channels as a consequence of the various orbital and spin contributions.

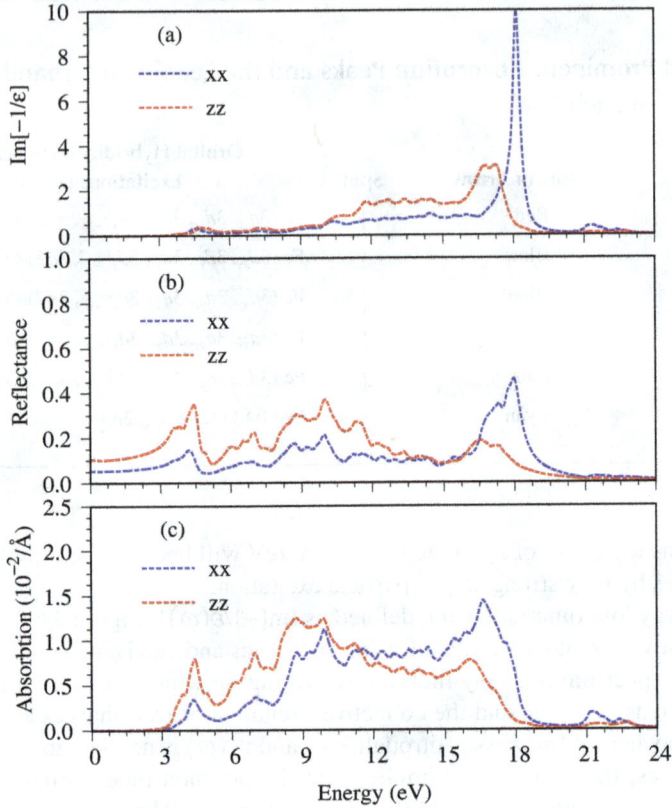

FIGURE 3.6 The various optical properties: (a) energy loss functions, (b) reflectance spectra, and (c) absorption coefficients under x-, y-, and z-electric-polarization directions.

The inverse values of the absorption coefficient (Figure 3.6c) mostly lie in the range of 67–200 Å, meaning that the electromagnetic wave that interacts with the $LiFeO_2$ compound will easily be absorbed through the rich optical excitations.

Our calculations on the optical properties of the $LiFeO_2$ compound can be detected by the reflections, absorption, and transmission measurements [68–71], from which other useful information, such as diversified optical excitation peaks, the slight redshift of the optical gap due to the excitonic effect, and the strong anisotropy of the optical responses, can be carried out, for example, based on the high-resolution reflection spectrum with a wide frequency range $\omega < 26$ eV [72]. Taft and Philipp can point out the main features of the optical excitations of Bernal graphite, such features associated with the free conduction electrons, the valence π electrons [the C-$2p_z$-orbitals], and the deeper-energy σ ones [the C-$(2s, 2p_x, 2p_y)$-orbitals]. These quasiparticle behaviors directly reflect the energy band structure. Apparently, our analysis of spin polarizations and orbital hybridizations that are related to optical properties of the $LiFeO_2$ compound is very difficult to achieve in experiments.

3.4 CONCLUSIONS

In the current work, DFT was used to investigate the geometric, electronic, and optical properties of LiFeO$_2$. As a result, the relation between spin polarizations and orbital hybridizations with electronic and magnetic properties and optical response can be successfully achieved. The state-of-the-art analysis is very useful for fully comprehending the diversified properties in anode/cathode/electrolyte and other emerging materials.

The 3D LiFeO$_2$ compound presents unique features. The Moiré superlattice with high atomic ordering and anisotropy geometric, the spin-polarization band structure with various atom dominations, the spatial spin and charge redistribution, and a lot of van Hove singularities due to the spin-polarizations and extreme point dispersions. As a consequence, the band edge states with orbital character and spin-orientation that assign the optical excitations could be well characteristics. Regarded optical properties, the weak but important impart of excitonic effect exhibit by slightly reduced of the optical gap and extremely enhance the single-particle excitations. The optical response also involves low reflectance/high transmission coefficients at an energy lower than the optical gap. The presence of various prominent excitation peaks, a large prominent plasmon mode due to the contribution of certain valence electrons at beyond 17 eV, the sensitivity of the absorption and reflectance spectrum upon the excitation events. The current developed framework could be adopted for the anode-/cathode-/electrolyte-related systems or other novel materials.

ACKNOWLEDGMENTS

This work is supported by the Hi-GEM Research Center and the Taiwan Ministry of Science and Technology under grant number MOST 108–2212-M-006–022-MY3, MOST 109–2811-M-006–505, and MOST 108–3017-F-006-003.

REFERENCES

1. Devabhaktuni V, Alam M, Depuru S S S R, Green II R C, Nims D and Near C 2013 Solar energy: Trends and enabling technologies *Renewable and Sustainable Energy Reviews* **19** 555–64.
2. Whittingham M S 2012 History, evolution, and future status of energy storage *Proceedings of the IEEE* **100** 1518–34.
3. Yilmaz M and Krein P T 2012 Review of battery charger topologies, charging power levels, and infrastructure for plug-in electric and hybrid vehicles *ITPE* **28** 2151–69.
4. Deng D 2015 Li-ion batteries: Basics, progress, and challenges *Energy Science & Engineering* **3** 385–418.
5. Boyer M J and Hwang G S 2016 Recent progress in first-principles simulations of anode materials and interfaces for lithium ion batteries *Current Opinion in Chemical Engineering* **13** 75–81.
6. Sygletou M, Petridis C, Kymakis E and Stratakis E 2017 Advanced photonic processes for photovoltaic and energy storage systems *Advanced Materials* **29** 1700335.
7. Baxter J, Bian Z, Chen G, Danielson D, Dresselhaus M S, Fedorov A G, Fisher T S, Jones C W, Maginn E and Kortshagen U 2009 Nanoscale design to enable the revolution in renewable energy *Energy & Environmental Science* **2** 559–88.

8. Singh A and Kalra V 2019 Electrospun nanostructures for conversion type cathode (S, Se) based lithium and sodium batteries *Journal of Materials Chemistry A* **7** 11613–50.

9. Lu X, Yu M, Zhai T, Wang G, Xie S, Liu T, Liang C, Tong Y and Li Y 2013 High energy density asymmetric quasi-solid-state supercapacitor based on porous vanadium nitride nanowire anode *Nano Letters* **13** 2628–33.

10. Hall P J, Mirzaeian M, Fletcher S I, Sillars F B, Rennie A J, Shitta-Bey G O, Wilson G, Cruden A and Carter R 2010 Energy storage in electrochemical capacitors: Designing functional materials to improve performance *Energy & Environmental Science* **3** 1238–51.

11. Von der Kammer F, Ferguson P L, Holden P A, Masion A, Rogers K R, Klaine S J, Koelmans A A, Horne N and Unrine J M 2012 Analysis of engineered nanomaterials in complex matrices (environment and biota): General considerations and conceptual case studies *Environmental Toxicology and Chemistry* **31** 32–49.

12. Stambouli A B and Traversa E 2002 Solid oxide fuel cells (SOFCs): A review of an environmentally clean and efficient source of energy *Renewable and Sustainable Energy Reviews* **6** 433–55.

13. Gao Z, Sun H, Fu L, Ye F, Zhang Y, Luo W and Huang Y 2018 Promises, challenges, and recent progress of inorganic solid-state electrolytes for all-solid-state lithium batteries *Advanced Materials* **30** 1705702.

14. Zhang Z, Shao Y, Lotsch B, Hu Y-S, Li H, Janek J, Nazar L F, Nan C-W, Maier J and Armand M 2018 New horizons for inorganic solid state ion conductors *Energy & Environmental Science* **11** 1945–76.

15. Yi T-F, Xie Y, Zhu Y-R, Zhu R-S and Shen H 2013 Structural and thermodynamic stability of Li4Ti5O12 anode material for lithium-ion battery *Journal of Power Sources* **222** 448–54.

16. Hochgatterer N S, Schweiger M R, Koller S, Raimann P R, Wöhrle T, Wurm C and Winter M 2008 Silicon/graphite composite electrodes for high-capacity anodes: Influence of binder chemistry on cycling stability *Electrochemical and Solid State Letters* **11** A76.

17. Lyu Y, Wu X, Wang K, Feng Z, Cheng T, Liu Y, Wang M, Chen R, Xu L and Zhou J 2020 An overview on the advances of $LiCoO_2$ cathodes for lithium-ion batteries *Advanced Energy Materials* **11** 2000982.

18. Kim D, Shim H C, Yun T G, Hyun S and Han S M 2016 High throughput combinatorial analysis of mechanical and electrochemical properties of Li [NixCoyMnz] O_2 cathode *Extreme Mechanics Letters* **9** 439–48.

19. Mandal S, Amarilla J M, Ibanez J and Rojo J M 2001 The role of carbon black in LiMn2O4-based composites as cathodes for rechargeable lithium batteries *Journal of the Electrochemical Society* **148** A24.

20. Rahman M M, Wang J-Z, Hassan M F, Chou S, Chen Z and Liu H K 2011 Nanocrystalline porous α-$LiFeO_2$–C composite—An environmentally friendly cathode for the lithium-ion battery *Energy & Environmental Science* **4** 952–7.

21. Wang Y, Wang J, Liao H, Qian X, Wang M, Song G and Cheng S 2014 Facile synthesis of electrochemically active α-$LiFeO_2$ nanoparticles in absolute ethanol at ambient temperature *RSC Advances* **4** 3753–7.

22. Guan J, Jia C, Li Y, Liu Z, Wang J, Yang Z, Gu C, Su D, Houk K N and Zhang D 2018 Direct single-molecule dynamic detection of chemical reactions *Science Advances* **4** eaar2177.

23. Petersen M, Hafner J and Marsman M 2006 Structural, electronic and magnetic properties of Gd investigated by DFT+ U methods: Bulk, clean and H-covered (0001) surfaces *Journal of Physics: Condensed Matter* **18** 7021.

24. Giorgi G, Fujisawa J-I, Segawa H and Yamashita K 2014 Cation role in structural and electronic properties of 3D organic–inorganic halide perovskites: A DFT analysis *The Journal of Physical Chemistry C* **118** 12176–83.

25. da Silveira Lacerda L H and de Lazaro S R 2018 Multiferroism and magnetic ordering in new NiBO$_3$ (B= Ti, Ge, Zr, Sn, Hf and Pb) materials: A DFT study *Journal of Magnetism and Magnetic Materials* **465** 412–20.

26. Mirtamizdoust B, Ghaedi M, Hanifehpour Y, Mague J T and Joo S W 2016 Synthesis, structural characterization, thermal analysis, and DFT calculation of a novel zinc (II)-trifluoro-β-diketonate 3D supramolecular nano organic-inorganic compound with 1, 3, 5-triazine derivative *Materials Chemistry and Physics* **182** 101–9.

27. Tran N T T, Nguyen D K, Glukhova O E and Lin M-F 2017 Coverage-dependent essential properties of halogenated graphene: A DFT study *Scientific Reports* **7** 1–13.

28. Nguyen D K, Tran N T T, Chiu Y-H and Lin M-F 2019 Concentration-diversified magnetic and electronic properties of halogen-adsorbed silicene *Scientific Reports* **9** 1–15.

29. Lin S-Y, Tran N T T, Chang S-L, Su W-P and Lin M-F 2018 *Structure-and Adatom-Enriched Essential Properties of Graphene Nanoribbons*: CRC Press..

30. Nguyen T D H, Pham H D, Lin S-Y and Lin M-F 2020 Featured properties of Li+-based battery anode: Li$_4$Ti$_5$O$_{12}$ *RSC Advances* **10** 14071–9.

31. Han N T, Dien V K, Thuy Tran N T, Nguyen D K, Su W-P and Lin M-F 2020 First-principles studies of electronic properties in lithium metasilicate (Li$_2$SiO$_3$) *RSC Advances* **10** 24721–9.

32. Khuong Dien V, Thi Han N, Nguyen T D H, Huynh T M D, Pham H D and Lin M-F 2020 Geometric and electronic properties of Li$_2$GeO$_3$ *Frontiers in Materials* **7**.

33. Tran N T T, Lin S-Y, Glukhova O E and Lin M-F 2015 Configuration-induced rich electronic properties of bilayer graphene *The Journal of Physical Chemistry C* **119** 10623–30.

34. Tran N T T, Lin S-Y, Lin C-Y and Lin M-F 2017 *Geometric and Electronic Properties of Graphene-Related Systems: Chemical Bonding Schemes*: CRC Press.

35. Tran N T T, Gumbs G, Nguyen D K and Lin M-F 2020 Fundamental properties of metal-adsorbed silicene: A DFT study *ACS Omega* **5** 13760–9.

36. Lin S-Y, Liu H-Y, Nguyen D K, Tran N T T, Pham H D, Chang S-L, Lin C-Y and Lin M-F 2020 *Silicene-Based Layered Materials*: IOP Publishing Limited.

37. Chantler C and Bourke J 2019 Low-energy electron properties: Electron inelastic mean free path, energy loss function and the dielectric function. Recent measurements, applications, and the plasmon-coupling theory *Ultramicroscopy* **201** 38–48.

38. Onida G, Reining L and Rubio A 2002 Electronic excitations: Density-functional versus many-body Green's-function approaches *Reviews of Modern Physics* **74** 601.

39. Senthilkumar P, Dhanuskodi S, Thomas A R and Philip R 2017 Enhancement of non-linear optical and temperature dependent dielectric properties of Ce: BaTiO$_3$ nano and submicron particles *Materials Research Express* **4** 085027.

40. Nguyen D K, Tran N T T, Nguyen T T and Lin M-F 2018 Diverse electronic and magnetic properties of chlorination-related graphene nanoribbons *Scientific Reports* **8** 1–12.

41. Lin S-Y, Chang S-L, Shyu F-L, Lu J-M and Lin M-F 2015 Feature-rich electronic properties in graphene ripples *Carbon* **86** 207–16.

42. Lin M-F, Hsu W-D and Huang J-L *Lithium-Ion Batteries and Solar Cells: Physical, Chemical, and Materials Properties*: CRC Press.

43. Gross E K and Dreizler R M 2013 *Density Functional Theory* vol 337: Springer Science & Business Media.

44. Hafner J 2008 Ab-initio simulations of materials using VASP: Density-functional theory and beyond *Journal of Computational Chemistry* **29** 2044–78.

45. Peng H and Perdew J P 2017 Rehabilitation of the Perdew-Burke-Ernzerhof generalized gradient approximation for layered materials *PhRvB* **95** 081105.

46. Blöchl P E 1994 Projector augmented-wave method *PhRvB* **50** 17953.

47. Crépieux A and Bruno P 2001 Theory of the anomalous hall effect from the Kubo formula and the Dirac equation *PhRvB* **64** 014416.

48. Rohlfing M and Louie S G 2000 Electron-hole excitations and optical spectra from first principles *PhRvB* **62** 4927–44.

49. Nakanishi N 1969 A general survey of the theory of the Bethe-Salpeter equation *PThPS* **43** 1–81.

50. Boufelfel A 2013 Electronic structure and magnetism in the layered $LiFeO_2$: DFT+ U calculations *Journal of Magnetism and Magnetic Materials* **343** 92–8.

51. Kanno R, Shirane T, Kawamoto Y, Takeda Y, Takano M, Ohashi M and Yamaguchi Y 1996 Synthesis, structure, and electrochemical properties of a new lithium iron oxide, $LiFeO_2$, with a corrugated layer structure *Journal of the Electrochemical Society* **143** 2435.

52. Carlier D, Ménétrier M, Grey C P, Delmas C and Ceder G 2003 Understanding the NMR shifts in paramagnetic transition metal oxides using density functional theory calculations *PhRvB* **67** 174103.

53. Shirane T, Kanno R, Kawamoto Y, Takeda Y, Takano M, Kamiyama T and Izumi F 1995 Structure and physical properties of lithium iron oxide, $LiFeO_2$, synthesized by ionic exchange reaction *Solid State Ionics* **79** 227–33.

54. Kim Y, Hayashi T, Osawa K, Dresselhaus M and Endo M 2003 Annealing effect on disordered multi-wall carbon nanotubes *Chemical Physics Letters* **380** 319–24.

55. Shen Y and Wu H 2012 Interlayer shear effect on multilayer graphene subjected to bending *Applied Physics Letters* **100** 101909.

56. Sun Y, Zheng Z, Cheng J, Liu J, Liu J and Li S 2013 The un-symmetric hybridization of graphene surface plasmons incorporating graphene sheets and nano-ribbons *Applied Physics Letters* **103** 241116.

57. Ichihashi F, Dong X, Inoue A, Kawaguchi T, Kuwahara M, Ito T, Harada S, Tagawa M and Ujihara T 2018 Development of angle-resolved spectroscopy system of electrons emitted from a surface with negative electron affinity state *Review of Scientific Instruments* **89** 073103.

58. Ichinokura S 2017 *Observation of Superconductivity in Epitaxially Grown Atomic Layers: In Situ Electrical Transport Measurements*: Springer.

59. Wolf E L 2012 *Principles of Electron Tunneling Spectroscopy* vol 152: Oxford University Press.

60. Brar V W, Zhang Y, Yayon Y, Ohta T, McChesney J L, Bostwick A, Rotenberg E, Horn K and Crommie M F 2007 Scanning tunneling spectroscopy of inhomogeneous electronic structure in monolayer and bilayer graphene on SiC *Applied Physics Letters* **91** 122102.

61. Wehling T, Black-Schaffer A M and Balatsky A V 2014 Dirac materials *Advances in Physics* **63** 1–76.

62. Kim S, Ihm J, Choi H J and Son Y-W 2013 Minimal single-particle Hamiltonian for charge carriers in epitaxial graphene on 4H-SiC (0001): Broken-symmetry states at Dirac points *Solid State Communications* **175** 83–9.

63. Haberer D, Petaccia L, Wang Y, Quian H, Farjam M, Jafari S, Sachdev H, Federov A, Usachov D and Vyalikh D 2011 Electronic properties of hydrogenated quasi-free-standing graphene *Physica Status Solidi (B)* **248** 2639–43.

64. Haberer D, Petaccia L, Farjam M, Taioli S, Jafari S, Nefedov A, Zhang W, Calliari L, Scarduelli G and Dora B 2011 Direct observation of a dispersionless impurity band in hydrogenated graphene *PhRvB* **83** 165433.

65. Ruffieux P, Cai J, Plumb N C, Patthey L, Prezzi D, Ferretti A, Molinari E, Feng X, Müllen K and Pignedoli C A 2012 Electronic structure of atomically precise graphene nanoribbons *ACS Nano* **6** 6930–5.

66. Lin C-L, Arafune R, Kawahara K, Kanno M, Tsukahara N, Minamitani E, Kim Y, Kawai M and Takagi N 2013 Substrate-induced symmetry breaking in silicene *Physical Review Letters* **110** 076801.

67. Du Y, Zhuang J, Liu H, Xu X, Eilers S, Wu K, Cheng P, Zhao J, Pi X and See K W 2014 Tuning the band gap in silicene by oxidation *ACS nano* **8** 10019–25.
68. BV D C and HYDRAULICS D 1962 Absorption spectroscopy.
69. Mirabella F M 1992 *Internal Reflection Spectroscopy: Theory and Applications* vol 15: CRC Press.
70. Brydson R 2020 *Electron Energy Loss Spectroscopy*: Garland Science.
71. Ibach H and Mills D L 2013 *Electron Energy Loss Spectroscopy and Surface Vibrations*: Academic Press.
72. Taft E and Philipp H 1965 Optical properties of graphite *PhRv* **138** A197.

4 Positive Electrode Stability in Higher Voltage Region

Akira Yano, Masahiro Shikano, and Hikari Sakaebe
National Institute of Advanced Industrial
Science and Technology (AIST)

CONTENTS

4.1 INTRODUCTION

With the steadily increasing demand for high-energy-density Li-ion batteries, it has become necessary to develop positive electrode materials that exhibit excellent performance. Layered rock-salt-structured active materials of $LiMO_2$ type (M is a metal), such as $LiCoO_2$, $LiNi(CoAl)O_2$, and $LiNi_{1/3}Co_{1/3}Mn_{1/3}O_2$, exhibit a theoretical capacity as high as 270–280 mAh g^{-1} and an average discharging voltage of approximately 4 V (vs. Li$^+$/Li) [1–3]. However, their actual capacity in commercial Li-ion batteries is typically limited to approximately 60% of their theoretical capacity; further, their charging voltage does not exceed 4.4 V. This is because the electrodes rapidly degrade at higher charging voltages.

In addition to the stability of the bulk of the active material, the stability of the interface between the electrode and the electrolyte is also of great importance. Studies report that the surface structure of $LiCoO_2$ changes from layered rock-salt-like to spinel-like during cycling [4]. Moreover, a rock-salt-like phase with a NaCl structure is typically formed near the surfaces of the active materials: $LiNi_{0.8}Co_{0.2}O_2$, $LiNi_{0.8}Co_{0.15}Al_{0.05}O_2$, and $LiNi_{1/3}Co_{1/3}Mn_{1/3}O_2$ [5–7]. Thus, it is believed that the degradation of this electrode/electrolyte interface is the primary cause of capacity fading.

DOI: 10.1201/9781003263807-4

The application of a surface coating on $LiMO_2$ active materials is an effective technique for stabilizing the electrode/electrolyte interface, especially for high-voltage charge/discharge. Various kinds of coating materials, such as oxides, phosphates, and fluorides, have been studied for this purpose [8–12]. Of these, Al- and Zr-based oxides are the most effective coating materials. Because the solubility of Al cations in $LiMO_2$ is high and that of Zr cations is low, each coating forms a different surface structure. However, because the coating phases were formed in a thin-film region with a thickness of several nanometers, their structures were not elucidated in detail. To address this issue, we investigated the surface structures of Al- and Z-oxide-coated $LiNi_{1/3}Co_{1/3}Mn_{1/3}O_2$ and $LiCoO_2$, which are finely prepared using an original solgel method via high-resolution scanning transmission electron microscopy (HR-STEM). Moreover, the effects of the coatings on the charging/discharging performance have been studied at a charging voltage of 4.5 V or higher [13–16].

Li-ion batteries, especially for applications in electric vehicles, are required to exhibit fast charge/discharge capabilities and long cycle lives. To improve these attributes, it is essential to understand the kinetics of the Li-ion transfer reaction at the electrode/electrolyte interface. Studies on oxide coatings report an improvement in the current-rate and cycle performance, suggesting that the surface coatings not only stabilize the electrode/electrolyte interface but also affect the Li-ion transfer kinetics at the interface [17–24]. However, it remains unclear why these oxides with their poor electronic and ionic conductivities could improve the current-rate performance. Therefore, the kinetics of Li-ion transfer at the electrode/electrolyte interface of the $LiCoO_2$ surface coated with Zr oxide or Al oxide was examined in detail by an electrochemical impedance analysis [16].

Epitaxial thin-film electrodes, which can function as surface-enhanced electrodes, are used to analyze interfacial reactions. These electrodes have very flat surfaces and include no additives, thus simplifying the surface reactions. Furthermore, the lattice orientation is controlled by epitaxial growth; consequently, the reaction field can be restricted. We also investigated the Li-ion transfer kinetics at the interface using surface-coated/uncoated epitaxial $LiCoO_2$ films [25]. Li_2ZrO_3 was used as the coating material, because Zr-oxide coatings have been reported to improve current-rate performances [14,18,20].

In this study, the surface structures of Al- and Zr-oxide-coated $LiNi_{1/3}Co_{1/3}Mn_{1/3}O_2$ and $LiCoO_2$ electrodes were described first; subsequently, the effect of surface coating on the charge/discharge characteristics in the high-voltage region was shown. Next, kinetic analysis using an epitaxial thin-film electrode was performed. The underlying mechanism of interface stabilization by surface coating was investigated along with the kinetics of Li-ion transfer at the interface.

4.2 SURFACE STRUCTURE OF THE COATED ACTIVE MATERIALS

$LiNi_{1/3}Co_{1/3}Mn_{1/3}O_2$ and $LiCoO_2$ powders with secondary particle diameters of 10 and 7 μm, respectively, were used as the active materials. Thin and uniform coatings of Zr oxide and Al oxide were formed via the solgel method [13,14], using zirconium tetrapropoxide and aluminum isopropoxide as precursors, respectively. Dipropylene glycol was used as the chelating agent for the coatings. By substituting the ligand of the precursor with dipropylene glycol, a uniform gel net was formed on the surface of the active materials. The solgel-coated samples were calcined at 500°C for 2 hours [13–15].

4.2.1 $LiNi_{1/3}Co_{1/3}Mn_{1/3}O_2$ AND $LiCoO_2$ COATED WITH AL OXIDE

Figure 4.1a shows the scanning electron microscopy (SEM) images of bare and Al-oxide-coated $LiNi_{1/3}Co_{1/3}Mn_{1/3}O_2$ with an Al-oxide content of 0.5 wt%. No significant difference was discernible between these samples, which were clearly composed of primary particles no bigger than approximately 1 μm in diameter. Deposited materials were also scarcely observed on the coated $LiNi_{1/3}Co_{1/3}Mn_{1/3}O_2$ particles. Figure 4.1b shows energy-dispersive X-ray spectroscopy (EDX) elemental maps of the Al-oxide-coated $LiNi_{1/3}Co_{1/3}Mn_{1/3}O_2$, thereby demonstrating that Al was uniformly distributed over the $LiNi_{1/3}Co_{1/3}Mn_{1/3}O_2$ particles.

Figure 4.2a and b shows cross-sectional STEM-EDX maps and a concentration depth profile of the Al-oxide-coated $LiNi_{1/3}Co_{1/3}Mn_{1/3}O_2$, respectively. A continuous layer with a high concentration of Al was formed from the surface to a depth of several nanometers. The concentration of Al was approximately 45 at% at the outermost surface, with Ni, Co, and Mn all coexistent with Al; however, this Al concentration was continuously reduced with an increase in depth; it was almost zero at a depth of 10 nm. These SEM and STEM-EDX results indicate that a thin coating layer was uniformly formed over the whole surface of the Al-oxide-coated $LiNi_{1/3}Co_{1/3}Mn_{1/3}O_2$.

(a)　　　　　　　　　　　　　　(b)

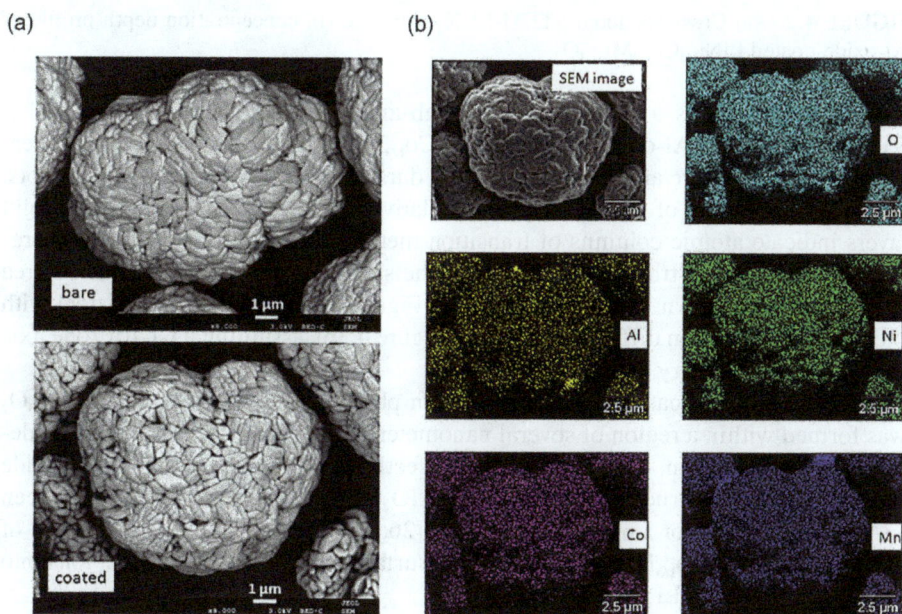

FIGURE 4.1 (a) SEM images of bare and Al-oxide-coated $LiNi_{1/3}Co_{1/3}Mn_{1/3}O_2$. (b) EDX elemental maps of Al-oxide-coated $LiNi_{1/3}Co_{1/3}Mn_{1/3}O_2$.

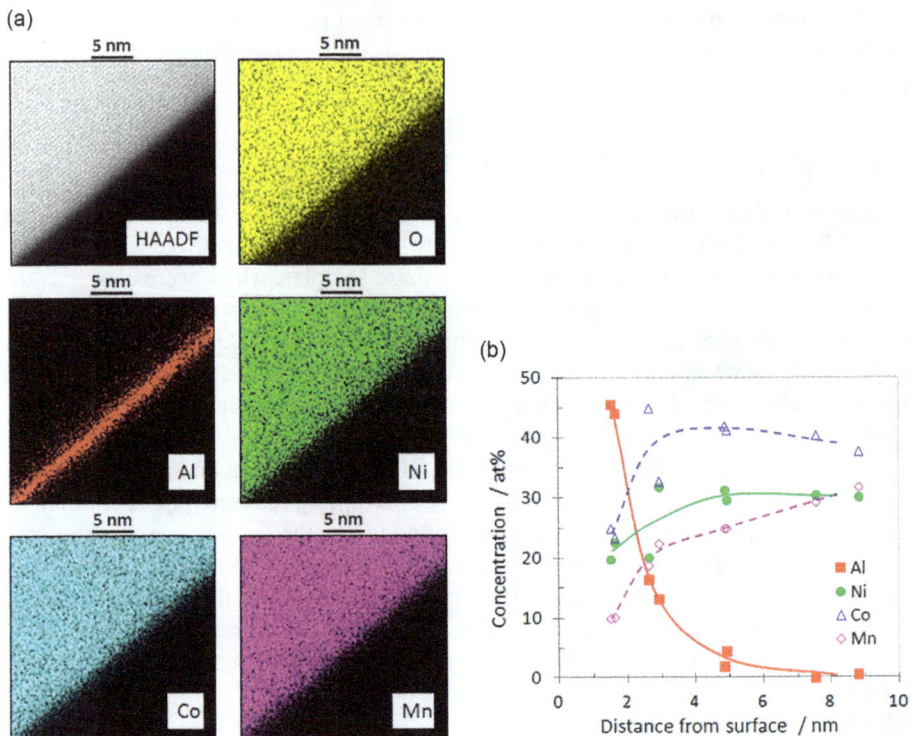

FIGURE 4.2 (a) Cross-sectional STEM-EDX maps and (b) concentration depth profile of Al-oxide-coated $LiNi_{1/3}Co_{1/3}Mn_{1/3}O_2$.

Figure 4.3a shows a high-resolution, high-angle annular dark-field (HAADF)-STEM image of the Al-oxide-coated $LiNi_{1/3}Co_{1/3}Mn_{1/3}O_2$. It was observed that layers with bright spots were arranged linearly and dark layers did not comprise any spots. Since the brightness of HAADF spots is relative to the atomic number, the bright layers indicate atomic columns of transition metals, whereas the dark layers represent Li. Fast Fourier transform patterns of the surface area (FFT1) and inside area (FFT2) are also shown in the inset image in Figure 4.3a. They were consistent with the calculated electron diffraction pattern (Figure 4.3b), assuming a [-1-10] zone axis for the $R\bar{3}m$ symmetry.

These results indicate that a solid-solution phase of $LiAlO_2$-$LiNi_{1/3}Co_{1/3}Mn_{1/3}O_2$ was formed within a region of several nanometers from the surface of the Al-oxide-coated $LiNi_{1/3}Co_{1/3}Mn_{1/3}O_2$. This seems perfectly plausible given that Li-Al-oxide has the same $R\bar{3}m$ structure as that of α-$LiAlO_2$ and forms a solid solution between $LiAlO_2$ and $LiNiO_2$ or $LiAlO_2$ and $LiCoO_2$ [26,27]. Furthermore, the formation of the $LiAlO_2$-$LiNi_{1/3}Co_{1/3}Mn_{1/3}O_2$ phase at the surface ensures diffusion of Li ions into the bulk through the Li sites of this phase.

Figure 4.4a shows the STEM-EDX mapping of Al-oxide-coated $LiCoO_2$ with 0.5 wt% Al-oxide content, and Figure 4.4b shows the concentration depth profile that corresponds to the green line in Figure 4.4a. A continuous coating layer with high

(a)

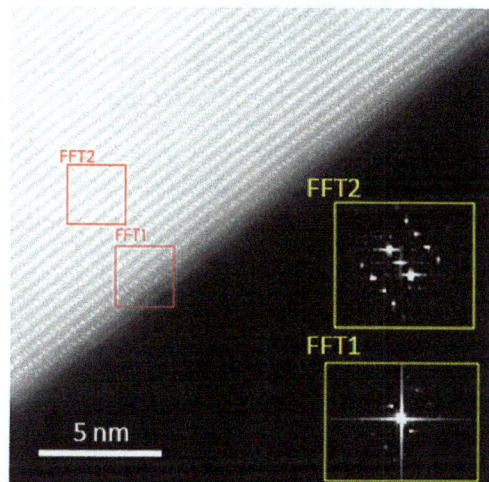

(b)

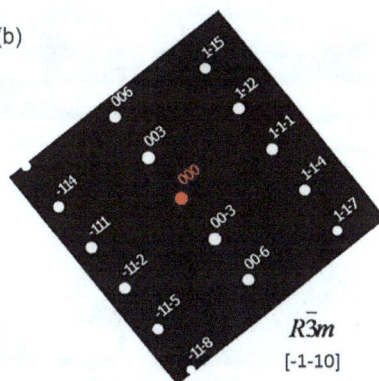

FIGURE 4.3 (a) High-resolution STEM image of Al-oxide-coated $LiNi_{1/3}Co_{1/3}Mn_{1/3}O_2$. The insets show fast Fourier transform patterns of a surface area (FFT1) and an inside area (FFT2). (b) Calculated electron diffraction pattern assuming a [-1-10] zone axis for the $R\bar{3}m$ symmetry.

Al concentration was formed on the surface with a depth of several nanometers, and the atomic concentration ratio (Al:Co) was approximately 1:2. In the high-resolution STEM-HAADF image of Figure 4.4c, atomic columns of the layered rock-salt arrangement were observed inside and on the surface of the coated $LiCoO_2$. These results indicate that a solid-solution phase of $LiAlO_2$-$LiCoO_2$, with a thickness of several nanometers, was formed on the surface after coating.

4.2.2 $LiNi_{1/3}Co_{1/3}Mn_{1/3}O_2$ COATED WITH ZR OXIDE

Figure 4.5a shows surface SEM images of the Zr-oxide-coated $LiNi_{1/3}Co_{1/3}Mn_{1/3}O_2$ samples with Zr-oxide contents of 0.125, 0.25, 0.5, and 1.0 wt%, respectively. Primary particles of coated $LiNi_{1/3}Co_{1/3}Mn_{1/3}O_2$ with diameters in the submicron range were observed. The deposited material was distributed uniformly over the primary particles for all the samples. Figure 4.5b shows the high-magnification SEM images of these samples. The deposited coatings consisted of island-like grains that were 10–20 nm in diameter; furthermore, uncoated areas (i.e., the sea-like zones) were also observed. With a decrease in the Zr-oxide content, the amount of the island-like grains decreased, while the sea-like zones increased in size. The results of X-ray diffraction and STEM analyses indicate that these island-like grains were crystals of ZrO_2 that did not contain Li [14]. Since ZrO_2 crystals exhibit very poor Li-ion conductivity, it can be assumed that these sea-like zones were the electrode/electrolyte interface where the transfer of Li ions took place.

(a)

(b)

(c)

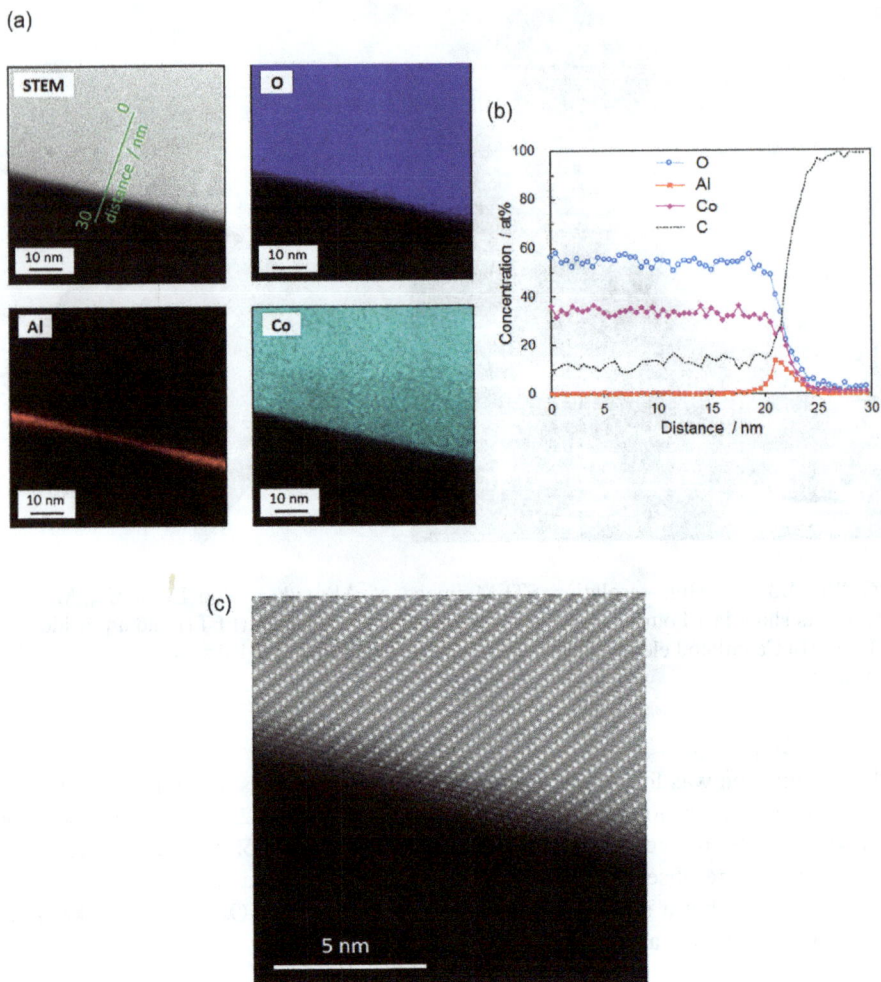

FIGURE 4.4 (a) STEM-EDX mapping, (b) concentration depth profile, and (c) STEM-HAADF image of Al-oxide-coated $LiCoO_2$.

Figure 4.6a shows a HAADF image of the Zr-oxide-coated $LiNi_{1/3}Co_{1/3}Mn_{1/3}O_2$ sample with a 0.125 wt% Zr-oxide content. In addition to the island-like grains, sea-like zones were also observed, wherein the grains were not deposited. Figure 4.6b and c show the concentration depth profiles (0.5 nm pitch in depth) for the sample; these correspond to the line that runs across the sea-like zone, as shown in the inset of Figure 4.6a. The vertical axes in Figure 4.6b and c represent the intensity and atomic concentration, respectively. Zr was detected in the region extending from the surface with a depth of approximately 2 nm, along with Ni, Co, Mn, and O. This indicates that a compound oxide phase containing Zr, Ni, Co, and Mn was formed in the surface region of the sea-like zones. The Zr concentration at the outmost surface of this region was comparable to that of the other metal elements.

(a)

500nm

(b)

100nm

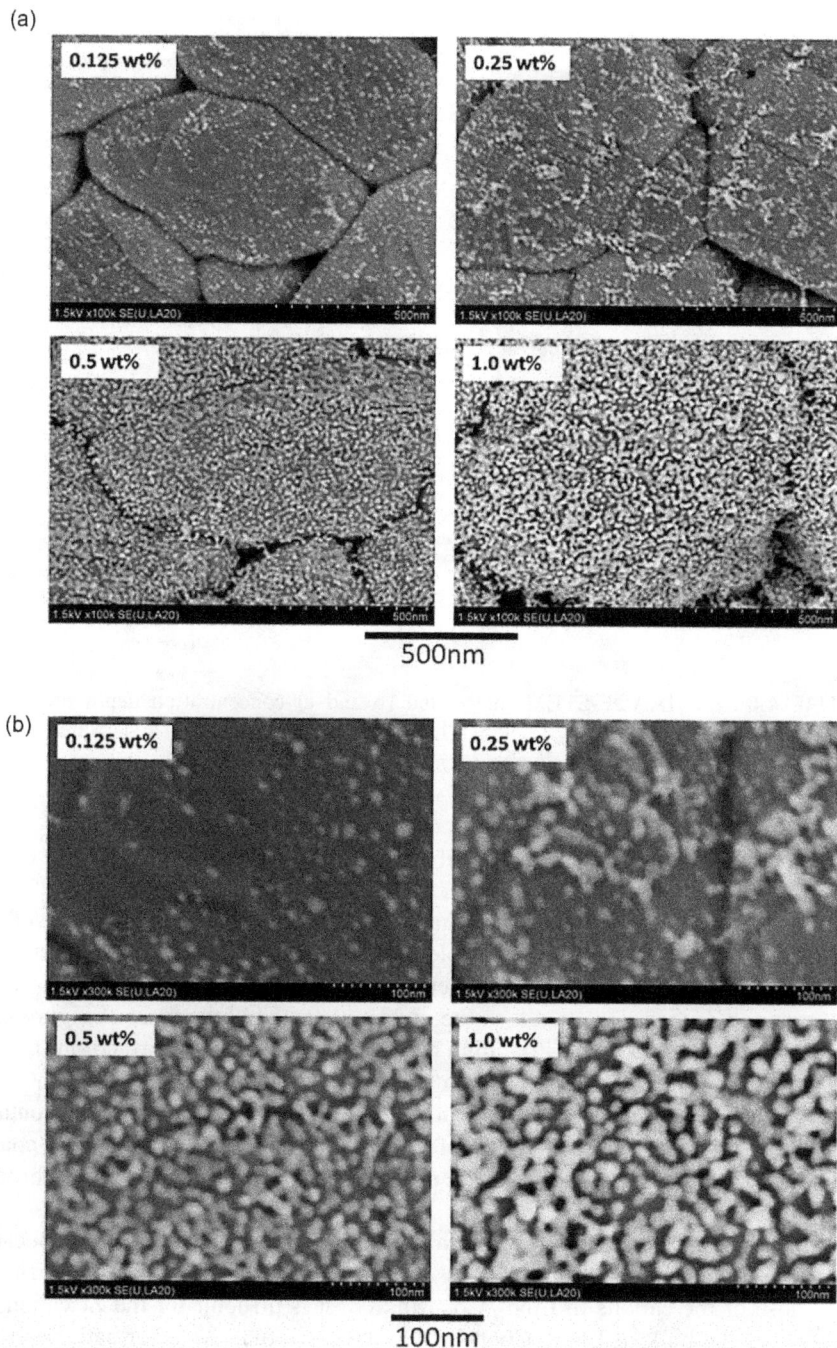

FIGURE 4.5 SEM images of Zr-oxide-coated $LiNi_{1/3}Co_{1/3}Mn_{1/3}O_2$ with Zr-oxide content of 0.125, 0.25, 0.5, and 1.0 wt.%. (a) Low magnification (100k) and (b) high magnification (300k) images.

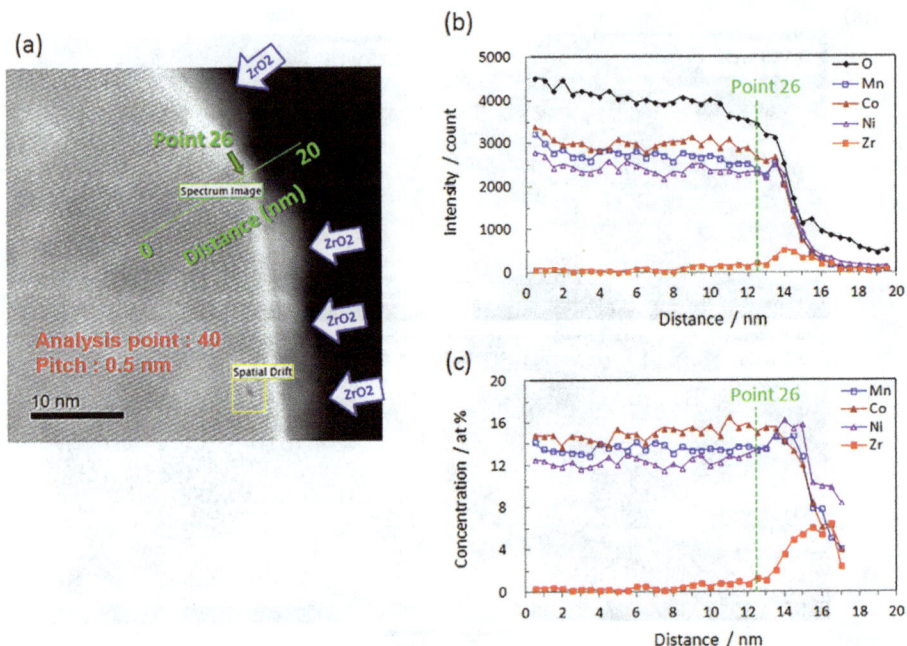

FIGURE 4.6 (a) HAADF-STEM image and (b and c) concentration depth profiles of Zr-oxide-coated $LiNi_{1/3}Co_{1/3}Mn_{1/3}O_2$ with 0.125 wt.%. Vertical axes of (b) and (c) are intensity and atomic concentration, respectively.

Figure 4.7a shows a cross-sectional annular bright-field (ABF) image of the Zr-oxide-coated $LiNi_{1/3}Co_{1/3}Mn_{1/3}O_2$ sample with a Zr-oxide content of 0.125 wt%. In this image, the island-like grains and sea-like zones can also be observed. In this case, the $LiNi_{1/3}Co_{1/3}Mn_{1/3}O_2$ sample exhibits a surface step in the direction of the electron beam, as shown in the schematic in Figure 4.7c. Figure 4.7b shows the EDX map for Zr that corresponds to the area represented in the inset of Figure 4.7a. Further, a continuous layer with a high concentration of Zr was formed on the surface of the sea-like zone with a depth of approximately 2 nm. It is evident from these STEM-EDX results that a uniform thin coating of a compound oxide containing Zr was formed over the surface of the sea-like zones of the Zr-oxide-coated $LiNi_{1/3}Co_{1/3}Mn_{1/3}O_2$. It can be considered that Li ions diffuse into the bulk through the composite oxide phase.

The formation mechanism of this layer can be considered as follows. Because the ionic radius (r_i) and valency of the Zr cation (Zr^{4+}, $r_i = 0.072$ nm) are different from those of the cations in $LiNi_{1/3}Co_{1/3}Mn_{1/3}O_2$, it is difficult for the Zr cations to diffuse into the bulk of $LiNi_{1/3}Co_{1/3}Mn_{1/3}O_2$ and substitute the other cations. If the discharge-state valences are Li^+, Ni^{2+}, Co^{3+}, and Mn^{4+}, the values of r_i are 0.076, 0.069, 0.055, and 0.053 nm, respectively. However, the degree of freedom of the atoms in the surface region is higher than that in the bulk, and the equilibrium conditions are different from those in the bulk. Therefore, a certain amount of Zr cations

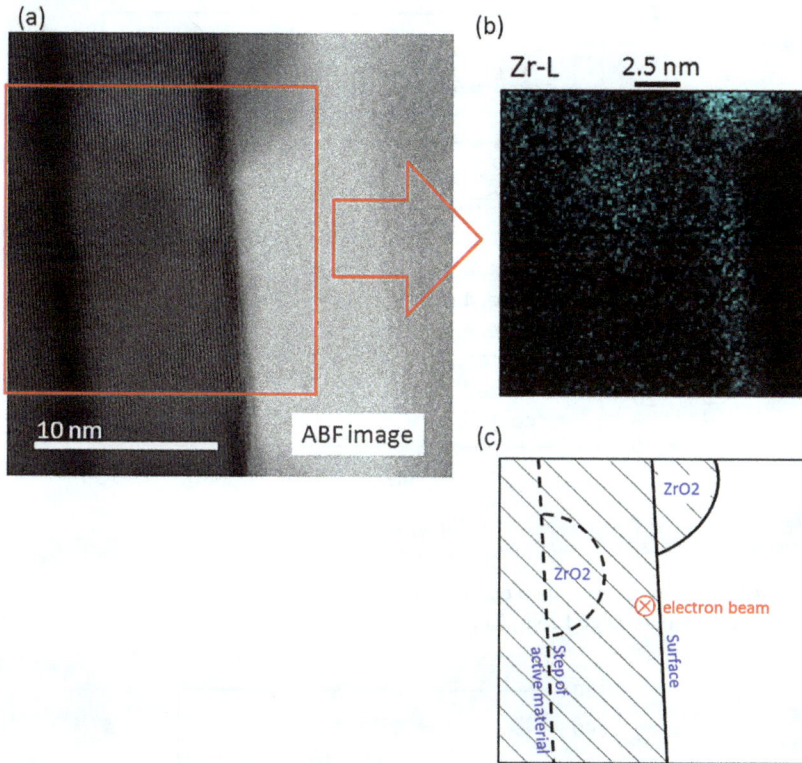

FIGURE 4.7 (a) ABF-STEM image of Zr-oxide-coated $LiNi_{1/3}Co_{1/3}Mn_{1/3}O_2$ with 0.125 wt.%, (b) EDX map of Zr corresponding to the selected area, and (c) schematic diagram of the ABF-STEM image.

can diffuse into the surface region. However, at the same time, excess Zr cations in the gel net that cannot mix into the surface region are deposited on the surfaces of the $LiNi_{1/3}Co_{1/3}Mn_{1/3}O_2$ as ZrO_2 particles.

4.3 EFFECT OF SURFACE COATING ON CHARGE/DISCHARGE CHARACTERISTICS IN HIGH-VOLTAGE REGION

The charge/discharge characteristics of the surface-coated $LiNi_{1/3}Co_{1/3}Mn_{1/3}O_2$ and $LiCoO_2$ were evaluated at a charging voltage of 4.5 V or higher. The composite positive electrodes were fabricated using a mixture of 90 wt% active material, 5 wt% acetylene black, and 5 wt% polyvinylidene fluoride on an Al current collector. Further, a Li metal foil, a polypropylene membrane, and 1.0 mol dm^{-3} of $LiPF_6$ dissolved in a 1:2 (v/v) solution of ethylene carbonate (EC) and diethyl carbonate (DEC) were used as the negative electrode, separator, and electrolyte, respectively [13,14,16].

Figure 4.8 shows the cycle performance of bare and Al-oxide-coated $LiNi_{1/3}Co_{1/3}Mn_{1/3}O_2$ (0.5 wt%) at a charge cutoff voltages (V_c) of 4.5, 4.6, and 4.7 V (discharge cutoff = 2.5 V, current rate = 1 C), respectively. This demonstrates that the

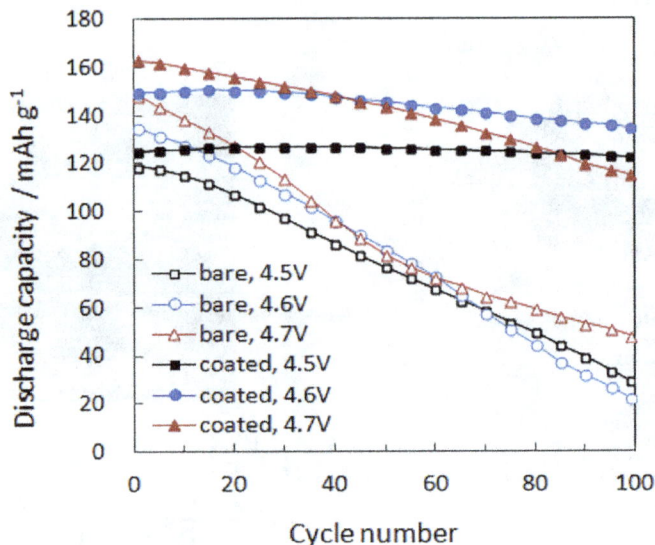

FIGURE 4.8 Cycle performance examined at charge cutoff voltages of 4.5, 4.6, and 4.7 V for bare and Al-oxide-coated $LiNi_{1/3}Co_{1/3}Mn_{1/3}O_2$ (discharge cutoff = 2.5 V, current rate = 1 C).

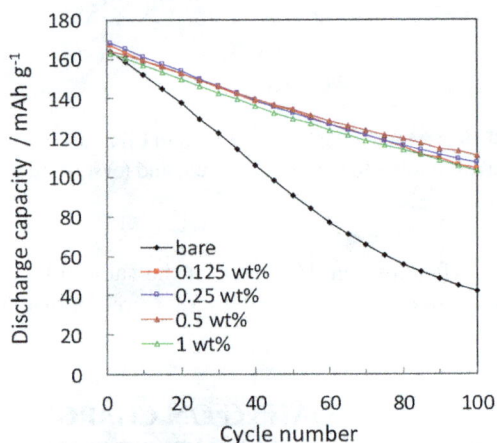

FIGURE 4.9 Discharge capacities during cycling of bare and Zr-oxide-coated $LiNi_{1/3}Co_{1/3}Mn_{1/3}O_2$ at voltage range of 2.5–4.6 V and current rate of 1 C.

cycle performance was effectively improved by the Al oxide coating, regardless of V_c. Figure 4.9 shows the cycle performance of the $LiNi_{1/3}Co_{1/3}Mn_{1/3}O_2$ coated with 0.125–0.5 wt% Zr-oxide in the voltage range of 2.5–4.6 V and a current rate of 1 C, along with that of the bare $LiNi_{1/3}Co_{1/3}Mn_{1/3}O_2$. The cyclability of $LiNi_{1/3}Co_{1/3}Mn_{1/3}O_2$ was improved markedly by the Zr-oxide coating; this was true regardless of the coating content. Figure 4.10 shows the charge transfer resistance (R_{ct}) at different open-circuit voltages (OCVs), which was evaluated using electrochemical impedance

(a)

(b)

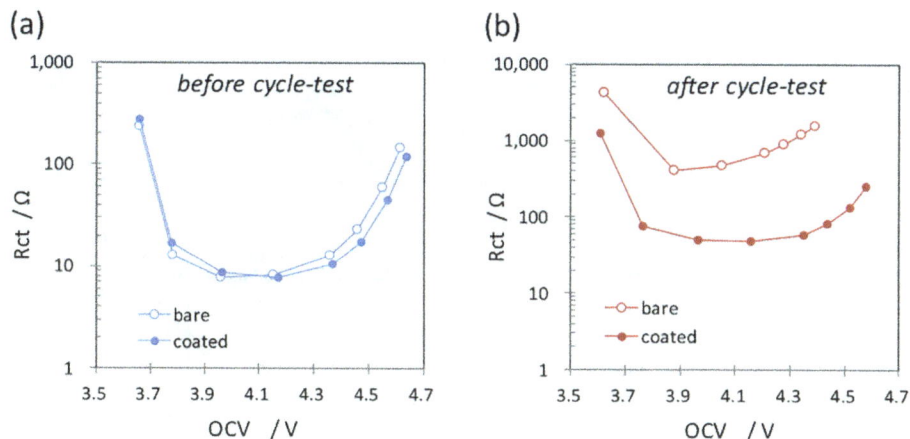

FIGURE 4.10 Charge transfer resistances (R_{ct}) analyzed at different open-circuit voltages (OCVs) for bare and Al-oxide-coated $LiNi_{1/3}Co_{1/3}Mn_{1/3}O_2$. (a) Before and (b) after cycle tests (2.5–4.7 V, 100 cycles).

spectroscopy, before and after the cycle test (2.5–4.7 V, 100 cycles) of bare and Al-oxide-coated $LiNi_{1/3}Co_{1/3}Mn_{1/3}O_2$. The increase in R_{ct} with the cycle was significantly suppressed by the coating. Since R_{ct} is the resistance of Li-ion transfer at the electrode/electrolyte interface, this result indicates that the interface was stabilized by the coating.

Figure 4.11 shows the discharge capacities and average discharge voltages of bare, Zr-oxide-coated (0.18 wt%), and Al-oxide-coated (0.11 wt%) $LiCoO_2$ during cycle tests in the voltage range of 2.5–4.5 V and a current rate of 1 C. The capacity and voltage reversibility were significantly increased by both Zr-oxide and Al-oxide coatings. The improvement in the high-voltage charge/discharge characteristics of the coatings was also confirmed in $LiCoO_2$.

A major cause of cycling-related degradation of the $LiMO_2$ electrode is the formation of rock-salt-like or spinel-like phases at the surface region during charging [4–7,13,15]. When $LiMO_2$ is charged at high voltages, most of the M cations are oxidized to tetravalent MO_2. Then, rock-salt-like (such as NiO) or spinel-like (such as Co_3O_4) phases form because of the change in the valency of M and the detachment of O:

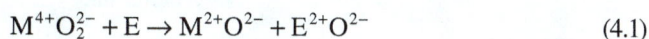

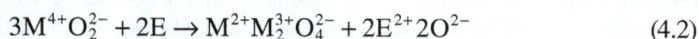

$$M^{4+}O_2^{2-} + E \rightarrow M^{2+}O^{2-} + E^{2+}O^{2-} \tag{4.1}$$

$$3M^{4+}O_2^{2-} + 2E \rightarrow M^{2+}M_2^{3+}O_4^{2-} + 2E^{2+}2O^{2-} \tag{4.2}$$

where E is the electrolyte or impurity that functions as a reducer. Because trivalent Al and tetravalent Zr cations do not readily change their valency, the detachment of O is suppressed by the sharing of O by Al or Zr with Ni, Co, or Mn in the compound oxide phase in the form of octahedral MO_6. Therefore, it can be considered that the compound oxide phase containing Al or Zr suppresses the formation of the rock-salt-like or spinel-like phases and improves the cycling performance.

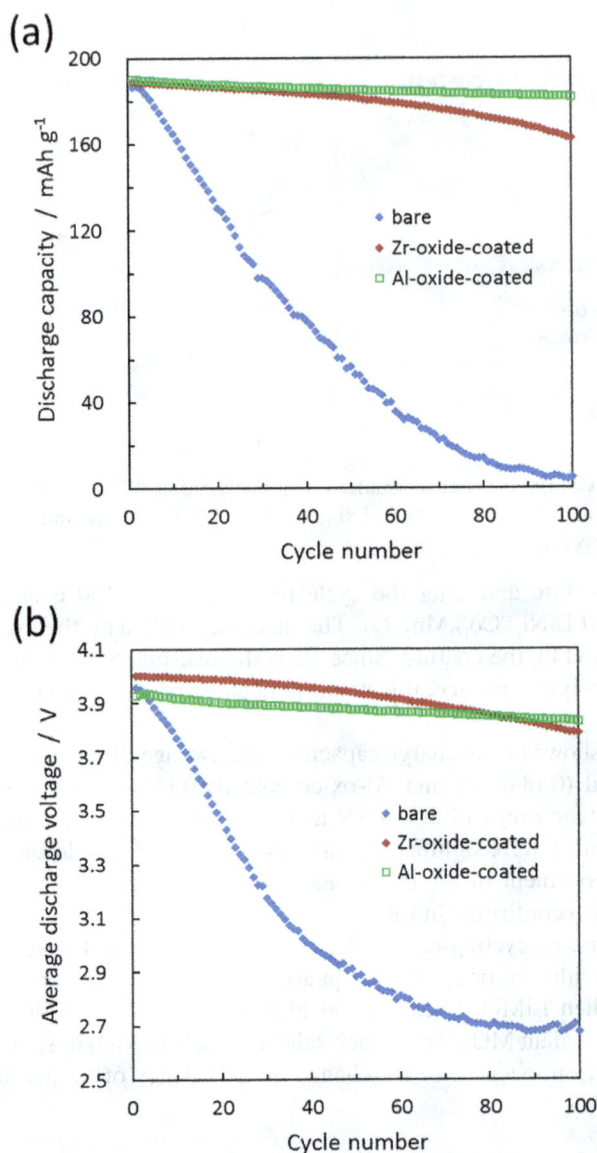

FIGURE 4.11 (a) Discharge capacities and (b) average discharge voltages during cycle test at current rate of 1C for bare, Zr-oxide-coated, and Al-oxide-coated LiCoO$_2$.

Figure 4.12a–c show charge/discharge curves at the discharge current rates of 0.2, 1, and 3 C for bare, Zr-oxide-coated, and Al-oxide-coated LiCoO$_2$, respectively. The polarization increased with an increasing current rate for all types of LiCoO$_2$; however, this increase was smaller for Zr-oxide-coated LiCoO$_2$ as compared to that observed for the other two. The average discharge voltages at 3 C for bare, Zr-oxide-coated, and Al-oxide-coated LiCoO$_2$ were 3.79, 3.89, and 3.77 V, respectively. This

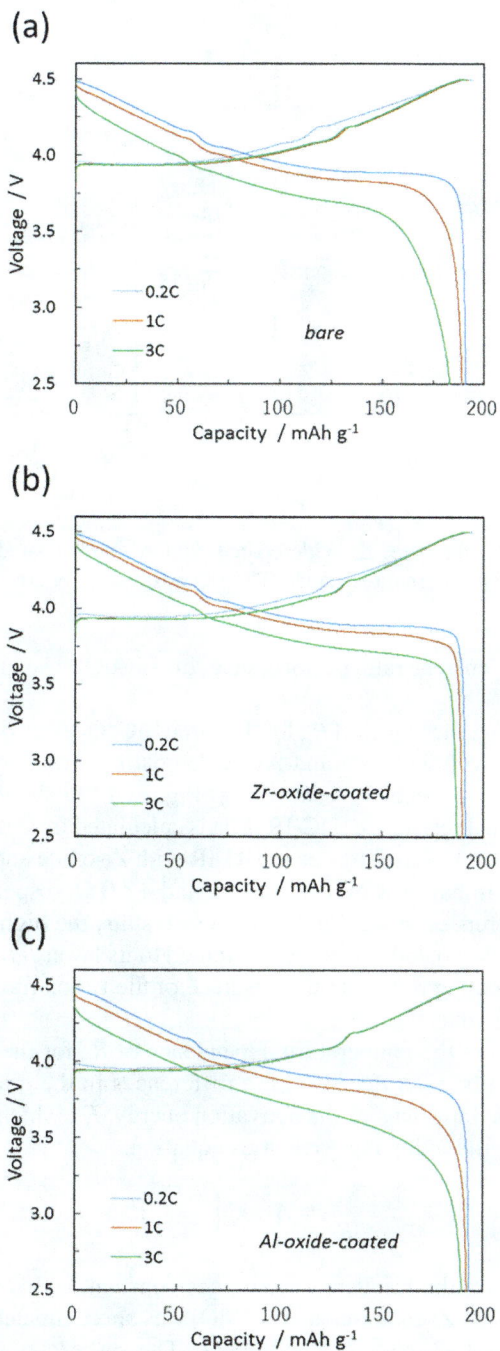

FIGURE 4.12 Charge/discharge curves at discharge current rate of 0.2, 1, and 3 C. (a) Bare LiCoO$_2$, (b) Zr-oxide-coated LiCoO$_2$, and (c) Al-oxide-coated LiCoO$_2$.

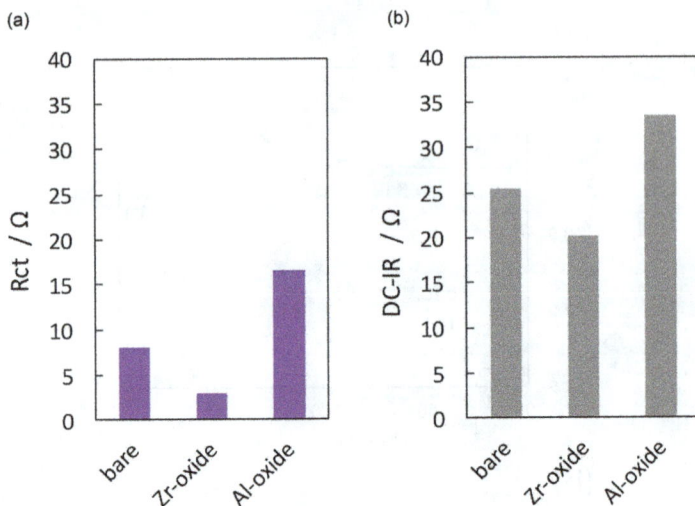

FIGURE 4.13 (a) R_{ct} for bare, Zr-oxide-coated, and Al-oxide-coated LiCoO$_2$ at OCV of 4.5 V. (b) Direct current-internal resistance (DC-IR) estimated from the rate performance for the samples.

indicates that the current-rate performance of LiCoO$_2$ was improved with the Zr-oxide coating.

Figure 4.13a shows the values of R_{ct} for the three LiCoO$_2$ electrodes at OCV = 4.5 V. The changes in R_{ct} with Zr-oxide and Al-oxide coatings compared to those in bare LiCoO$_2$ were −5.1 and +8.5 Ω, respectively. Figure 4.13b shows the values of the direct current-internal resistance (DC-IR) that is calculated from the discharge curves shown in Figure 4.12. The differences in DC-IR with Zr-oxide and Al-oxide coatings compared to those in bare LiCoO$_2$ were −5.2 and +8.0 Ω, respectively, which were similar to the differences in R_{ct}. Based on these results, the higher current-rate performance of Zr-oxide-coated LiCoO$_2$ is attributed to its lower R_{ct}. This also indicates that the Zr-oxide coating reduced the resistance of the Li-ion transfer reaction at the electrode/electrolyte interface.

Figure 4.14a shows the temperature dependence of R_{ct} for the three electrodes at OCV = 4.2 V. It can be seen that there are differences in the gradients of the lines. Figure 4.14b shows the values of the activation energy, E_a, which is calculated from the gradients using the following Arrhenius equation:

$$\frac{1}{R_{ct}} = A \cdot \exp\left(\frac{-E_a}{RT}\right) \tag{4.3}$$

where A, R, and T are the frequency factor, gas constant, and absolute temperature, respectively. The E_a of Zr-oxide-coated LiCoO$_2$ was approximately 10 kJ mol^{-1} lower than those of bare and Al-oxide-coated LiCoO$_2$. The lower R_{ct} of the Zr-oxide-coated LiCoO$_2$ can be attributed to the lower E_a. These results indicate that the surface coating not only stabilizes the electrode/electrolyte interface but also affects the kinetics of Li-ion transfer at the interface.

(a)

(b)

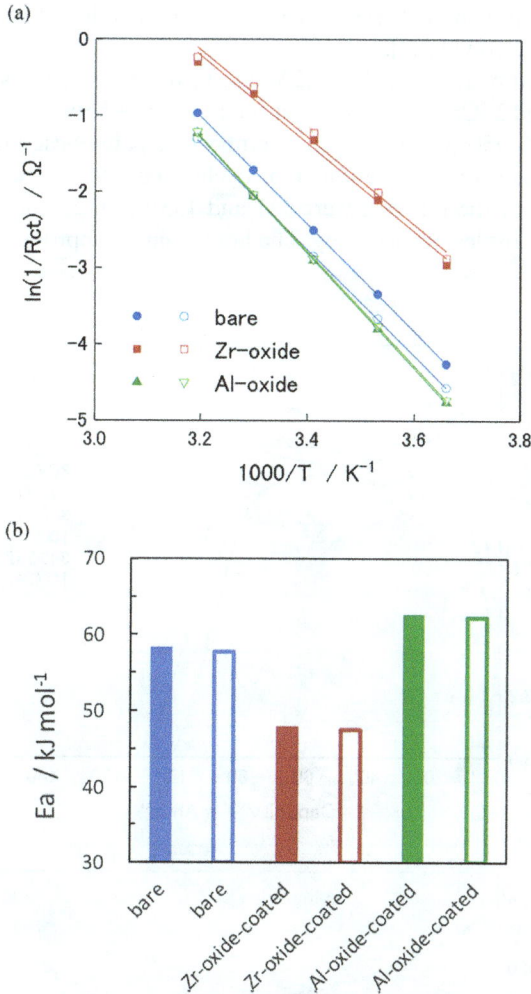

FIGURE 4.14 (a) The temperature dependence of R_{ct} at OCV = 4.2 V and (b) the activation energy (E_a) for bare, Zr-oxide-coated, and Al-oxide-coated $LiCoO_2$.

4.4 ANALYSIS USING EPITAXIAL THIN-FILM ELECTRODE

The Li-ion transfer kinetics at the electrode/electrolyte interface using surface-coated/uncoated epitaxial films was investigated. $LiCoO_2$ and Li_2ZrO_3 were used as the positive electrode and coating material, respectively. To investigate the Li-ion transfer reaction on a single-crystal plane, epitaxial $LiCoO_2$ films with intercalation-active (104) planes were fabricated on $SrRuO_3$(100)/Nb-doped $SrTiO_3$(100) substrates using pulsed laser deposition. A single orientation of (104), film thickness of approximately 18 nm, and flat surface with a roughness of less than 1 nm were confirmed by X-ray diffractometry and X-ray reflectometry. The Li_2ZrO_3 film, which has an island-like structure or a phase with a density gradient, was formed with a thickness of approximately 1 nm on the surface. A 1.0 mol dm^{-3} solution of LiPF$_6$ in EC + DEC was used as the electrolyte [25].

Figure 4.15a and b show the 3.0–4.2 V charge/discharge curves of the uncoated and Li$_2$ZrO$_3$-coated LiCoO$_2$ electrodes at a current rate between 1 and 300 C (current density of 2.5–750 µA cm^{-2}). The numbers in parentheses in the legends of these figures indicate the total number of cycles. The discharge capacities at 1 C at the initial stage of the rate test were 111 and 133 mA h g^{-1} for the uncoated and coated LiCoO$_2$ electrodes, respectively. The latter value is approximately equal to the

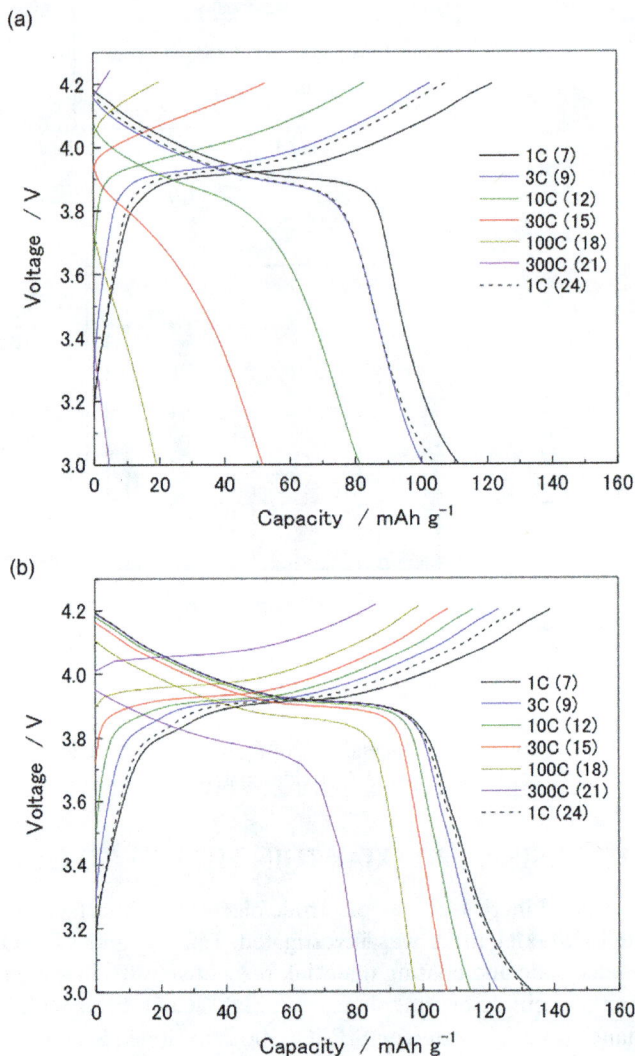

FIGURE 4.15 Charge/discharge curves during rate test with 3.0–4.2 V. (a) Uncoated and (b) Li$_2$ZrO$_3$-coated LiCoO$_2$. The numbers in parentheses in the legends of these figures indicate the total number of cycles.

standard discharge capacity of the $LiCoO_2$ powder material (approximately 140 mA h g^{-1} at 4.2 V). The uncoated $LiCoO_2$ showed a significant increase in polarization and capacity reduction at a rate higher than 30 C. On the other hand, the coated $LiCoO_2$ sample showed decreased polarization and retained a considerable capacity up to 300 C. The discharge capacity retentions at 300 C compared to the initial capacity at 1 C were 5% and 65% for the uncoated and coated $LiCoO_2$ samples, respectively. The rate capability was significantly improved by the surface coating.

Figure 4.16a and b show the 3.0–4.5 V charge/discharge curves of uncoated and Li_2ZrO_3-coated $LiCoO_2$ at current rates between 1.7 and 170 C (4.2–420 µA cm^{-2}). The

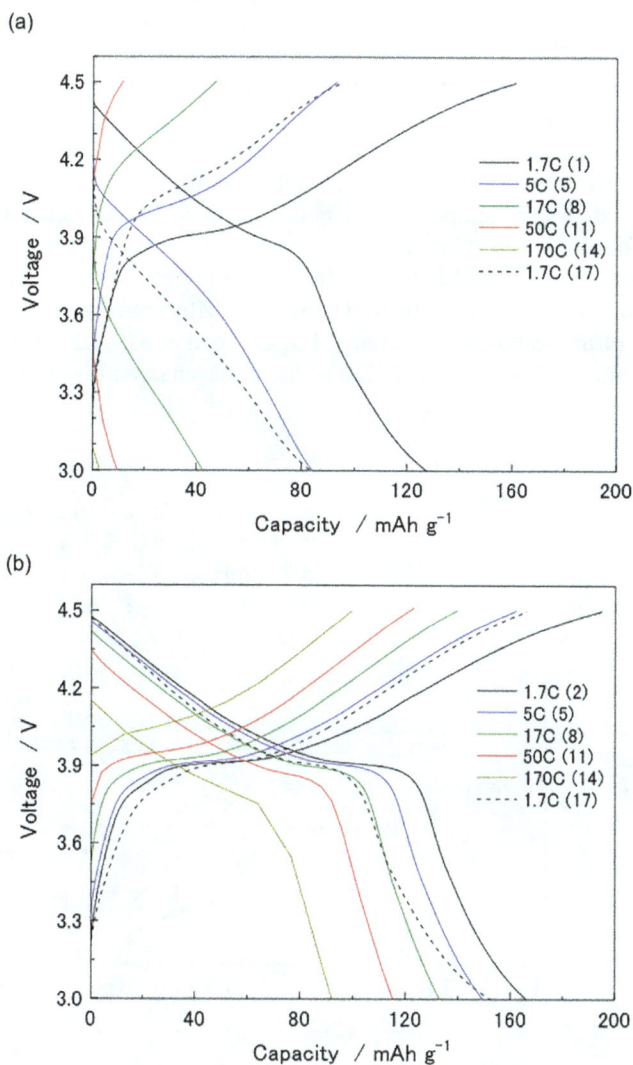

FIGURE 4.16 Charge/discharge curves during rate test with 3.0–4.5 V. (a) Uncoated and (b) Li_2ZrO_3-coated $LiCoO_2$.

discharge capacities at 1.7 C at the initial stage of the rate test were 128 and 166 mA h g^{-1} for the uncoated and coated LiCoO$_2$ electrodes, respectively. The capacity enhancement due to the high-voltage charging to 4.5 V was more effectively demonstrated in the case of the coated LiCoO$_2$. The uncoated LiCoO$_2$ hardly worked at a rate of 50 C or more, while the coated LiCoO$_2$ could be charged/discharged up to 170 C. The discharge capacity retentions at 170 C, as compared to the initial capacity at 1.7 C, were 2% and 55% for the uncoated and coated LiCoO$_2$ electrodes, respectively. The rate capability at high voltage charge/discharge was also markedly improved by the surface coating. Moreover, in the uncoated LiCoO$_2$ sample, the 1.7 C curve at the later stage of the rate test (black broken line) showed considerably greater polarization when compared to the 1.7 C curve at the initial stages of the test (black solid line). This indicates that severe degradation of the electrode/electrolyte interface had occurred by the high-voltage charge/discharge of 3.0–4.5 V. In contrast, the differences between the initial and subsequent curves at 1.7 C were relatively small in the case of the coated LiCoO$_2$ sample. Thus, the surface coating significantly suppressed the degradation of the electrode/electrolyte interface even under high-voltage charge/discharge conditions.

Figure 4.17 shows the temperature dependence of R_{ct} at an electrode potential of 4.0 V in the fourth charge. The samples were uncoated LiCoO$_2$ electrodes charged/discharged at 3.0–4.2 V and Li$_2$ZrO$_3$-coated LiCoO$_2$ electrodes charged/discharged at 3.0–4.2 and 3.0–4.5 V, respectively. There were differences in the gradients of the lines. The E_a of uncoated LiCoO$_2$ charged/discharged at 3.0–4.2 V was 55 kJ mol^{-1}, and those of the Li$_2$ZrO$_3$-coated LiCoO$_2$ electrode charged/discharged at 3.0–4.2 and 3.0–4.5 V were 42 and 47 kJ mol^{-1}, respectively. The Li$_2$ZrO$_3$ coating not only

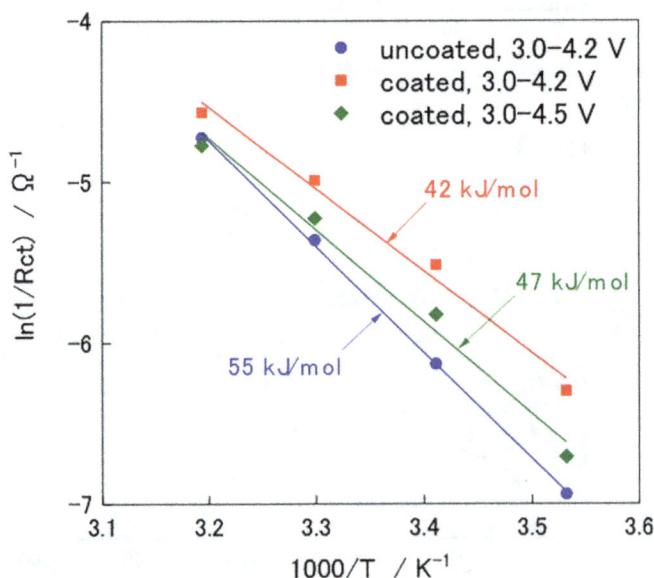

FIGURE 4.17 The temperature dependence of R_{ct} at electrode potential of 4.0 V in the fourth charge for uncoated LiCoO$_2$ charged/discharged with 3.0–4.2 V and Li$_2$ZrO$_3$-coated LiCoO$_2$ charged/discharged with 3.0–4.2 and 3.0–4.5 V.

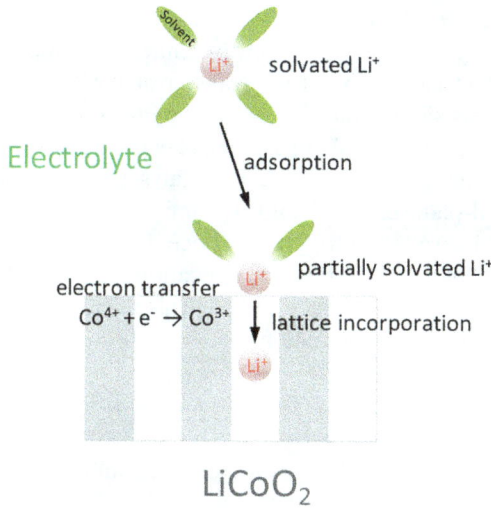

Electrolyte

solvated Li⁺

adsorption

partially solvated Li⁺

electron transfer
$Co^{4+} + e^- \rightarrow Co^{3+}$

lattice incorporation

$LiCoO_2$

FIGURE 4.18 Schematic diagram of Li-ion transfer at the electrode/electrolyte interface.

stabilized the electrode/electrolyte interface but also reduced the E_a value for Li-ion transfer at the interface. Similar to the results of our study, previous studies using composite electrodes and a polycrystalline thin-film electrode revealed that oxide coatings decrease E_a for charge-transfer resistance [14,16,17,20]. A decrease in E_a was also observed in this study using an electrode with a single-crystal plane, confirming that surface modification by coating intrinsically affects the Li-ion transfer mechanism at the interface.

An adatom model has been proposed as a mechanism for the Li-ion transfer reaction at the intercalation electrode and electrolyte interface [28–30]. Recently, computational studies of interface reactions in intercalation materials have also been reported [31,32]. A schematic diagram of the Li-ion transfer reaction based on these studies is shown in Figure 4.18. The reaction from the solution phase (electrolyte) to the solid phase ($LiCoO_2$) is thought to proceed as follows:

1. The solvated Li⁺ in the electrolyte is adsorbed onto the electrode surface.
2. The solvation shell breaks down, which leads to partially solvated Li⁺.
3. The electron transfer reaction ($Co^{4+} + e^-$) occurs, and the remaining solvent molecules desorb.
4. Desolvated Li⁺ is incorporated into the crystal lattice of the electrode.

The low E_a for the Li_2ZrO_3-coated $LiCoO_2$ electrode suggests that it exhibits a low activation barrier for the lattice incorporation of Li ions and/or electron transfer.

4.5 CONCLUSIONS

The solid-solution phase of $LiAlO_2$-$LiMO_2$ was formed uniformly at the surface of the Al-oxide-coated $LiNi_{1/3}Co_{1/3}Mn_{1/3}O_2$ and $LiCoO_2$. The oxide phase containing Zr, Ni, Co, and Mn was formed continuously in the surface region of the sea-like zones for Zr-oxide-coated $LiNi_{1/3}Co_{1/3}Mn_{1/3}O_2$. The cycle performance in the high-voltage region was significantly improved by either Zr-oxide or Al-oxide coating. The increase in the resistance of Li-ion transfer at the electrode/electrolyte interface with the cycle was significantly suppressed by the coating. It can be considered that the oxide phases containing Al or Zr suppress the formation of rock-salt-like or spinel-like phases at the surface region. In the analyses using the composite electrode and the single-crystal-plane electrode, it was confirmed that the Zr-based coating not only stabilized the electrode/electrolyte interface, but also reduced the E_a value for Li-ion transfer at the interface. Surface modification is an essential factor that stabilizes the electrode/electrolyte interface and controls the Li-ion transfer kinetics at the interface, even in the high-voltage region.

REFERENCES

1. Ohzuku T and Ueda A 1994 Why transition metal (di)oxides are the most attractive materials for batteries *Solid State Ionics* **69** 201–11.
2. Ohzuku T, Ueda A and Nagayama M 1993 Electrochemistry and structural chemistry of $LiNiO_2$ (R-3m) for 4 volt secondary lithium cells *J. Electrochem. Soc.* **140** 1862–70.
3. Mizushima K, Jones P C, Wiseman P J and Goodenough J B 1980 $LixCoO_2$ ($0<x\leq1$): A new cathode material for batteries of high-energy density *Mater. Res. Bull.* **15** 783–9.
4. Wang H F, Jang Y I, Huang B Y, Sadoway D R and Chiang Y T 1999 TEM study of electrochemical cycling-induced damage and disorder in $LiCoO_2$ cathodes for rechargeable lithium batteries *J. Electrochem. Soc.* **146** 473–80.
5. Abraham D P, Twesten R D, Balasubramanian M, Kropf J, Fischer D, McBreen J, Petrov I and Amine K 2003 Microscopy and spectroscopy of lithium nickel oxide-based particles used in high power lithium-ion cells *J. Electrochem. Soc.* **150** A1450–A6.
6. Muto S, Sasano Y, Tatsumi K, Sasaki T, Horibuchi K, Takeuchi Y and Ukyo Y 2009 Capacity-fading mechanisms of $LiNiO_2$-based lithium-ion batteries II. Diagnostic analysis by electron microscopy and spectroscopy *J. Electrochem. Soc.* **156** A371–A7.
7. Shikano M, Kobayashi H, Koike S, Sakaebe H, Saito Y, Hori H, Kageyama H and Tatsumi K 2011 X-ray absorption near-edge structure study on positive electrodes of degraded lithium-ion battery *J. Power Sources* **196** 6881–3.
8. Cho J, Kim Y J and Park B 2000 Novel $LiCoO_2$ cathode material with Al_2O_3 coating for a li ion cell *Chem. Mater.* **12** 3788–91.
9. Chen Z H and Dahn J R 2004 Methods to obtain excellent capacity retention in $LiCoO_2$ cycled to 4.5 V *Electrochimica Acta* **49** 1079–90.
10. Li C, Zhang H P, Fu L J, Liu H, Wu Y P, Ram E, Holze R and Wu H Q 2006 Cathode materials modified by surface coating for lithium ion batteries *Electrochimica Acta* **51** 3872–83.
11. Sun Y K, Han J M, Myung S T, Lee S W and Amine K 2006 Significant improvement of high voltage cycling behavior AlF_3-coated $LiCoO_2$ cathode *Electrochem. Commun.* **8** 821–6.
12. Appapillai A T, Mansour A N, Cho J and Shao-Horn Y 2007 Microstructure of $LiCoO_2$ with and without "AlPO4" nanoparticle coating: Combined STEM and XPS studies *Chem. Mater.* **19** 5748–57.
13. Yano A, Aoyama S, Shikano M, Sakaebe H, Tatsumi K and Ogumi Z 2015 Surface structure and high-voltage charge/discharge characteristics of Al-oxide coated $LiNi_{1/3}Co_{1/3}Mn_{1/3}O_2$ cathodes *J. Electrochem. Soc.* **162** A3137–A44.

14. Yano A, Ueda A, Shikano M, Sakaebe H and Ogumi Z 2016 Surface structure and high-voltage charging/discharging performance of low-content Zr-oxide-coated $LiNi_{1/3}Co_{1/3}Mn_{1/3}O_2$ *J. Electrochem. Soc.* **163** A75–A82.
15. Yano A, Shikano M, Ueda A, Sakaebe H and Ogumi Z 2017 $LiCoO_2$ degradation behavior in the high-voltage phase transition region and improved reversibility with surface coating *J. Electrochem. Soc.* **164** A6116–A22.
16. Yano A, Hirayama M and Kanno R 2019 Kinetics of Li-ion transfer at the electrode/electrolyte interface and current rate performance of $LiCoO_2$ surface-coated with zirconium oxide and aluminum oxide *Electrochemistry* **87** 234–41.
17. Iriyama Y, Kurita H, Yamada I, Abe T and Ogumi Z 2004 Effects of surface modification by MgO on interfacial reactions of lithium cobalt oxide thin film electrode *J. Power Sources* **137** 111–6.
18. Ni J, Zhou H, Chen J and Zhang X 2008 Improved electrochemical performance of layered $LiNi_{0.4}Co_{0.2}Mn_{0.4}O_2$ via Li_2ZrO_3 coating *Electrochimica Acta* **53** 3075–83.
19. Park B C, Kim H B, Myung S T, Amine K, Belharouak I, Lee S M and Sun Y K 2008 Improvement of structural and electrochemical properties of AlF3-coated Li $Ni_{1/3}Co_{1/3}Mn_{1/3}$ O-2 cathode materials on high voltage region *J. Power Sources* **178** 826–31.
20. Huang Y Y, Chen J T, Ni J F, Zhou H H and Zhang X X 2009 A modified ZrO_2-coating process to improve electrochemical performance of $Li(Ni_{1/3}Co_{1/3}Mn_{1/3})O_2$ *J. Power Sources* **188** 538–45.
21. Dou J Q, Kang X Y, Wumaier T, Yu H W, Hua N, Han Y and Xu G Q 2012 Effect of lithium boron oxide glass coating on the electrochemical performance of $LiNi_{1/3}Co_{1/3}Mn_{1/3}O_2$ *J. Solid State Electrochem.* **16** 1481–6.
22. Dai X Y, Wang L P, Xu J, Wang Y, Zhou A J and Li J Z 2014 Improved electrochemical performance of $LiCoO_2$ electrodes with ZnO coating by radio frequency magnetron sputtering *Acs Appl. Mater. Interfaces* **6** 15853–9.
23. Liu K, Yang G-L, Dong Y, Shi T and Chen L 2015 Enhanced cycling stability and rate performance of $Li[Ni_{0.5}Co_{0.2}Mn_{0.3}]O_2$ by CeO_2 coating at high cut-off voltage *J. Power Sources* **281** 370–7.
24. Teranishi T, Yoshikawa Y, Sakuma R, Okamura H, Hashimoto H, Hayashi H, Fujii T, Kishimoto A and Takeda Y 2015 High-rate capabilities of ferroelectric $BaTiO_3$–$LiCoO_2$ composites with optimized $BaTiO_3$ loading for Li-ion batteries *ECS Electrochem. Lett.* **4** A137–A40.
25. Yano A, Hikima K, Hata J, Suzuki K, Hirayama M and Kanno R 2018 Kinetics and stability of Li-ion transfer at the $LiCoO_2$ (104) plane and electrolyte interface *J. Electrochem. Soc.* **165** A3221–A9.
26. Ohzuku T, Ohzuku T and Kouguchi M 1995 Synthesis and characterization of $LiAl_{1/4}Ni_{3/4}O_2$ (R-3m) for lithium-ion (shuttlecock) batteries *J. Electrochem. Soc.* **142** 4033–9.
27. Jang Y I, Huang B Y, Wang H F, Sadoway D R, Ceder G, Chiang Y M, Liu H and Tamura H 1999 $LiAlyCol-yO_2$ (R-3m) intercalation cathode for rechargeable lithium batteries *J. Electrochem. Soc.* **146** 862–8.
28. Bruce P G and Saidi M Y 1992 The mechanism of electrointercalation *J. Electroanal. Chem.* **322** 93–105.
29. Nakayama M, Ikuta H, Uchimoto Y and Wakihara M 2003 Study on the AC impedance spectroscopy for the Li insertion reaction of $LixLa_{1/3}NbO_3$ at the electrode–electrolyte interface *J. Phys. Chem. B* **107** 10603–7.
30. Kobayashi S and Uchimoto Y 2005 Lithium ion phase-transfer reaction at the interface between the lithium manganese oxide electrode and the nonaqueous electrolyte *J. Phys. Chem. B* **109** 13322–6.

31. Haruyama J, Ikeshoji T and Otani M 2018 Analysis of lithium insertion/desorption reaction at interfaces between graphite electrodes and electrolyte solution using density functional + implicit solvation theory *J. Phys. Chem. C* **122** 9804–10.

32. Stauffer S K and Vilciauskas L 2018 Computational study of chemical and electrochemical intercalation of Li into $Li_{1+x}Ti_2O_4$ spinel structures *J. Phys. Chem. C* **122** 7779–89.

5 Layered Cathode Materials for Sodium-Ion Batteries (SIBs)

Synthesis, Structure, and Characterization

Hoang V. Nguyen, Minh L. Nguyen, Kha M. Le, Thinh G. Phung, Huong T. D. Nguyen, Man V. Tran, and Phung M.L. Le
Viet Nam National University - Ho Chi Minh City (VNU HCM)
Ho Chi Minh University of Science (HCMUS)

CONTENTS

DOI: 10.1201/9781003263807-5

The latest advanced energy storage system, the Li-ion battery, has enabled revolutionary technological advances in electric/electronic fields and those related to them over the past decades. The intercalation mechanism of Li-ion batteries made them different from the previous types of rechargeable batteries. In that mechanism, the intercalation Li^+ ion incorporates/removes to/from the host lattice of the electrode material under the electric force. The alkaline, earth alkaline metal ions and small sized ions such as Li^+, Na^+, Ca^{2+}, Mg^{2+}, Al^{3+}, Zn^{2+}... are all capable of exchanging from the appropriated host structures that paved the bloom of research in post-Li-ion batteries. Among them, Na-ion batteries are the mature technologies that are evaluated at the industrial level somewhere. The research on Na-ion batteries is almost suspended due to the ultimate features of Li-ion batteries such as high energy density, voltage, and lightweight. From the early 21st century, Na-ion batteries have been growing up to offer us efficient energy storage solutions with reduced cost and sustainability because of the cheap and abundant sodium sources.

Theoretically, the intercalation mechanism allows preservation of the host lattice, but it is not always the truth in the case of intercalation of big ion size like Na^+ ions. Therefore, many challenges are encountering in the development of the electrode material for Na-ion batteries. In other words, the research on the electrode materials, especially the cathode materials bring the success of Na-ion batteries in the commercial market.

In this chapter, we briefly introduce the reason for the revisit of Na-ion rechargeable batteries, their structure, and their working principle. Then, we present the structure of layered transition metal oxide and the properties of typical types P2- and O3-type layered structure materials. In the next section, we will summarize the synthesized pathways of layered structure cathode materials. Next, the phase transition of the P2- and O3-type layered structure was summarized. Finally, the structure-electrochemical properties of single/multiple transition metal oxides with layered structure were reviewed.

5.1 THE RECHARGEABLE NA-ION BATTERIES

In our day, energy storage systems (EESs) are essential to support intelligent equipments and electric vehicles. Currently, lithium-ion batteries have dominated the market of portable power sources and automobiles; however, for large-scale storage applications, the lithium supply becomes a big challenge.

An alternative to lithium-ion batteries, research breakthrough on the cheap and abundant material components for EESs, has been opening up of the new century to other alkali-metal-based technology. Owing to the development of carbon materials and layered metal oxides, sodium-ion batteries (SIBs) have drawn a lot of attention because of their high feasibility, high energy density, and a huge available resource from the ocean (sodium chloride). In recent years, SIBs have been developed towards the priority of minimizing the cobalt content in the electrode material to reduce costs and toxicity to meet the new regulation of green, safe, and low-cost future batteries.

In general, SIB contains two main electrodes: a positive electrode and a negative electrode. A separator is placed between them to prevent the short-circuit. A

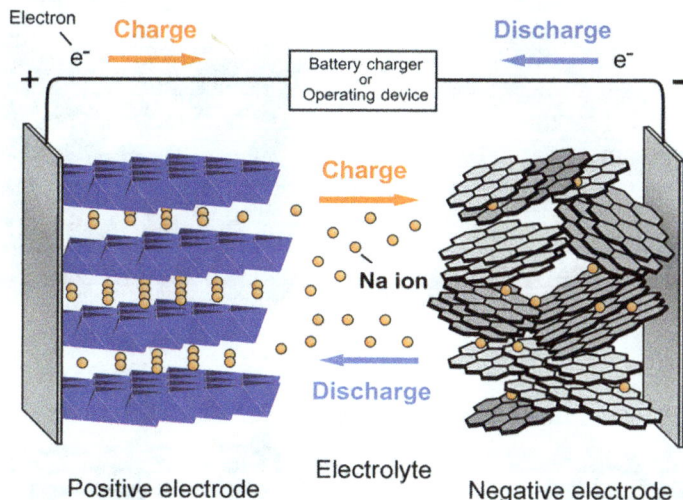

FIGURE 5.1 Structure and principle of Na-ion. (Reproduced with permission [1].)

non-aqueous electrolyte is responsible to ensure the reversible diffusion of Na$^+$ ions from/to cathode or anode (Figure 5.1). During the process of charging, an oxidation reaction occurs on the positive electrode, and a reduction reaction occurs on the negative one. Thus, it is important to use a true Na-ion system where Na ions are exchanged between cathodes and anodes in a "rocking-chair" manner.

The principle of SIB is described by the following reaction equations, with an example of using NaCoO$_2$ as the positive electrode and hard carbon (HC) as the negative one:

Positive electrode	$Na_{1-x}CoO_2 + xNa^+ + xe \leftrightarrow NaCoO_2$	(5.1)
	$Na_{1-x}CoO_2 + xNa^+ + xe \leftrightarrow NaCoO_2$	
Negative electrode	$Na_xC \leftrightarrow C + xNa^+ + xe^-$	(5.2)
Total	$Na_{1-x}CoO_2 + Na_xC \leftrightarrow NaCoO_2 + Na_xC$	(5.3)

Like LIBs, the energy and power density of SIBs depend on the selection of cathode and anode; therefore, it is important to use materials that have a high theoretical capacity, the high potential of the positive electrode, and the low potential of the negative one. Generally, the capacity of the positive electrode is much lower than the negative electrode. In this way, the positive electrode materials determine the energy and power of the batteries.

Various cathode materials for SIBs have been reported; for instance, layer and tunnel type transition metal oxides, transition metal sulfides and fluorides, oxyanion compounds, Prussian blue analogs [2]. Similar to LIBs, highly reversible cathode materials based on the intercalation reaction, which involves the interstitial introduction of a guest species (Na$^+$ in the present context), are needed for high capacity and

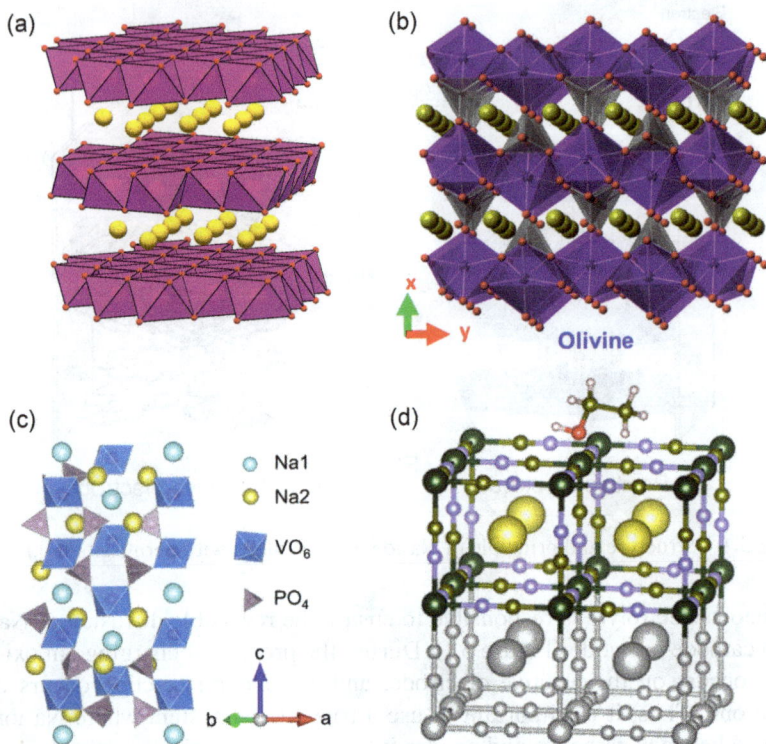

FIGURE 5.2 Typical structures of the positive electrodes in Na-ion batteries: (a) Layered structure of Na_xMO_2, (b) Olivine $NaFePO_4$ [8], (c) NASICON $Na_3V_2(PO_4)_3$ [9], (d) Prussian blue [10]. (Reproduced with permission [8–10].)

good cyclability of SIBs. The ion storage ability inside the structure of these materials is different for Li^+ and Na^+ as the Na^+ ion has a large atomic radius. Consequently, the common number of Na coordination is about 6 or 8 and the spinel structure can't be formed for residing a sodium-ion.

These electrode materials are mainly categorized into oxides, polyanions such as phosphates, pyrophosphates, fluorosulfates, oxychlorides, and Na superionic conductor (NASICON) types, and metal-organic compounds (Figure 5.2).

The layer structure such as $NaFe_{0.5}Co_{0.5}O_2$, $NaNi_{1/3}Mn_{2/3}O_2$, $Na_xFe_{0.5}Mn_{0.5}O_2$, etc... [3–7] showed a good ability of Na-ion storage, its capacity over 120 mAh/g, and good stability for long-cycling. Especially, common and friendly environmental metals such as Fe, Mn have been replaced with toxic Co to improve the specific capacity.

Regardless of the negative electrodes, different electrode reaction mechanisms have been suggested (Figure 5.3) to explain the high capability of Na-ion storage. HC and $Na_2Ti_3O_7$ exhibit high discharge capacity for SIB application and are suitable for the commercial market [11]. The layer structure such as P2-$Na_{2/3}Ni_{1/6}Mg_{1/6}Ti_{2/3}O_2$ [12] using redox of titanium is also a promising anode.

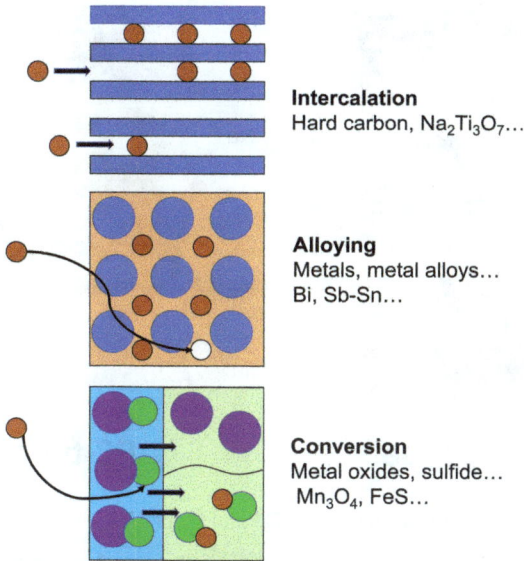

FIGURE 5.3 Some intercalation mechanisms representing the negative electrode materials in Na-ion batteries.

5.2 STRUCTURAL DESCRIPTION OF THE LAYERED STRUCTURE

5.2.1 CRYSTALLINE STRUCTURE

Sodium layered transition metal oxides (Na_xMO_2) typically deliver a high reversible capacity and high volumetric energy density with low cost; thus they have been intensively studied for grid-scale application. Layered oxides have formula A_xMO_2 ($0.5 \leq x \leq 1$) where A stands for alkaline metal ion such as Na^+ ion for sodium layered compounds and M represents a transition metal with +3 oxidation states or a mixture of trivalent and tetravalent (or eventually divalent) elements. In the structure, Na^+ ions layers are sandwiched between edge-sharing octahedra MO_6 sheets stacking along the c-axis. Delmas et al. [13] divided layered structure into several groups depending on the environment of sodium ions between $(MO_2)_n$ layers: octahedra (O), prism (P), and tetrahedra (T), followed by a figure indicating the number of repeated $(MO_2)_n$ sheet in the lattice. Sometimes, the symbol can be used to express the distortion in-plane of the lattice. Because of the small volume of the tetrahedral voids, the (T) phase is not common at Na_xMO_2 material, where Na^+ ion has a large ionic radius (1.02 Å), which hardly inserts into tetrahedral sites of (T) structure reversibly and leads to the large volume expansion during cycling.

The common structures of sodium layer metal oxides are O3 type and P2 type as shown in Figure 5.4. In the O3 structure, the oxygen stacking type is ABCABC and the Na^+ ions are located in the octahedral interstitials of the cubic close-packed oxygen array. O3-type layered phase has a very small vacancy content with cation-ordered rock-salt superstructure oxides [14]. O3 structures belong to the space group *R-3m* (166).

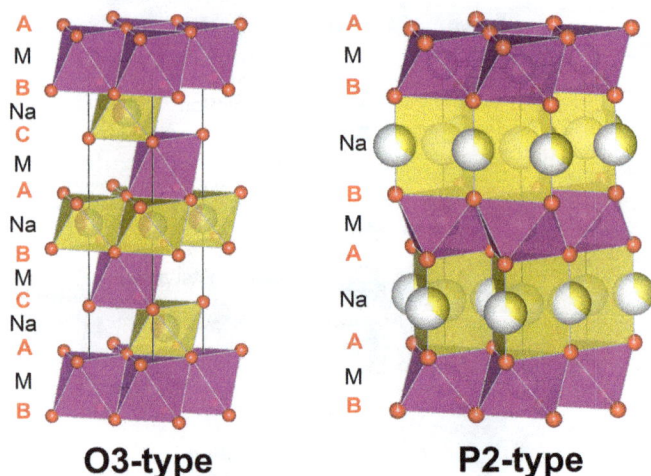

O3-type   **P2-type**

FIGURE 5.4   The crystalline structures of O3 and P2 phases.

P2 structure is with the oxygen stacking sequence ABBA. Na^+ ions occupy one-half the edge-shared prisms (Na_e) while the rest of ions lie in the face-shared prisms (Na_f). P2 type can be classified into the space group $P6_3/mmc$ (194). P2 structure allows more Na^+ ions to diffuse into the pristine structure due to the higher spacious content. As a result, P2-type layered oxides exhibit more specific capacity than O3 phases.

5.2.2   Phase Transition and Thermodynamic Behavior

Phase transition upon sodium ion extraction/insertion in/out of the structure is the primary issue that encourages a lot of strategies for improving cycling stability, rate capability of the layered cathode for sodium-ion batteries have been investigated. Phase transition is more likely to occur in sodium layered compounds because big Na^+ ion size induces strong interaction between the ion and the structure.

5.2.2.1   O3-Type Sodium Transition Metal Oxides

For lithium-ion batteries, the O3-type lithium transition metal oxides ($LiMO_2$) have been intensively investigated as the positive electrode materials such as $LiCoO_2$ (LCO), $Li[Ni_{0.8}Co_{0.15}Al_{0.05}]O_2$ (NCA), and $Li[Ni_{1/3}Mn_{1/3}Co_{1/3}]O_2$ (NMC) are commonly used for commercial Li-ion batteries. On the other hand, O3-type (including O′3-type) sodium compounds, $NaMeO_2$ (Me = 3d transition metal such as Fe [15], Cr [16], V [17], Ni [18], Mn [19], and Co [20] and their sodium intercalation/de-intercalation behavior have been reported. Moreover, multiple transition metal oxides such as $Na_x[Ni_{1/2}Mn_{1/2}]O_2$, $Na_x[Ni_{0.6}Co_{0.4}]O_2$, $Na[Fe_{1/2}Co_{1/2}]O_2$, $Na[Fe_{1/2}Ni_{1/2}]O_2$, $Na[Ni_{1/3}Co_{1/3}Fe_{1/3}]O_2$, $Na[Fe_x(Ni_{1/2}Mn_{1/2})_{1-x}]O_2$, $Na[Ni_xMn_yCo_z]O_2$ and sodium-excess O3-type compounds have also been studied as the positive electrode materials for sodium-ion batteries.

The O3-type cathode materials are considered fascinating candidates for SIBs due to the high-energy-density, uncomplicated synthesis process, and sufficient sodium-ion reservoirs, which serve as a significant success factor in the full-cell system. However, O3-phase commonly suffers from a series of phase transformations and structural evolution during the electrochemical cycling, which induces complex electrochemical behaviors and inevitably causes large expansion and shrinkage of the lattice volume. The frequent phase transformations associated with multiple voltage steps, as well as sluggish kinetics, are inclined to result in poor rate performance, which is unfavorable to the structural stability. Therefore, suppression or reduction of phase transitions and extension of solid solution zone through chemical element substitution is crucial to the improvement in electrochemistry properties of layered oxides in SIBs.

Komaba et al. successfully synthesized O3-type $NaNi_{0.5}Mn_{0.5}O_2$ layered oxide by a solid-state method [21]. The material delivered the capacity of 105–125 mAh g^{-1} in the voltage range of 2.2–3.8 V. When the electrode was changed to 4.5 V, a large capacity of 185 mAh g^{-1} was even obtained but with poor reversibility caused by the significant expansion of interslab space due to the O3 phase in the $NaNi_{0.5}Mn_{0.5}O_2$ that continuously changes to the O′3, P3, P′3, and then P″3 phase during the sodium-ion extraction process (Figure 5.5). Inside, the P″3 phase

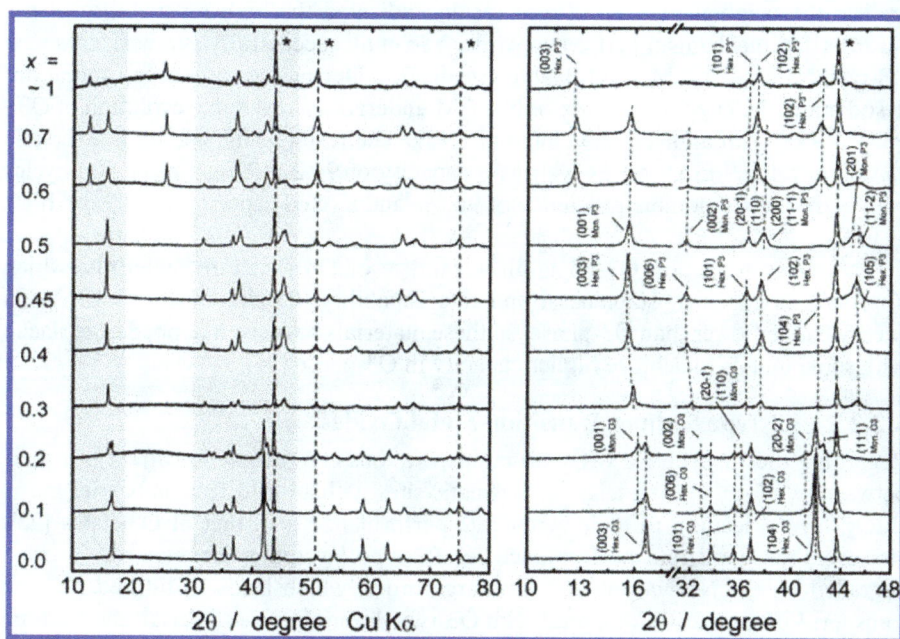

FIGURE 5.5 (Left) *Ex situ* XRD patterns of the $Na_{1-x}Ni_{0.5}Mn_{0.5}O_2$ composite electrodes on a nickel current collector, which were electrochemically prepared in the sodium cells. (Right) Highlighted XRD patterns. Asterisks show a nickel mesh used as a current collector, Reproduced with permission [21].

is assumed to incorporate the solvent molecules in the interstitial space between the $[Ni_{0.5}Mn_{0.5}]O_2$ slabs because of the anomalously large interslab distances (~7.0 Å) such as those of the layered oxides are observed. Therefore, $NaNi_{0.5}Mn_{0.5}O_2$ is a robust and durable positive electrode material.

Phase transition of O3 to O'3 is not common in single and binary systems but seems likely occurring in tertiary systems. Zhao et al. successfully synthesized a high-entropy O3-type $NaNi_{0.12}Cu_{0.12}Mg_{0.12}Fe_{0.15}Co_{0.15}Mn_{0.1}Ti_{0.1}Sn_{0.1}Sb_{0.04}O_2$ layered cathode material and revealed the unprecedented new concept of high-entropy chemistry [22]. In the initial stage of charge, the high-entropy material induced a phase transition from O3 to O3' structure, where the newly formed O3' phase shares the same rhombohedral structure with a space group of R-$3m$ with the pristine O3 structure; however, the cell parameters changed owing to the different Na content. The highly reversible O3-P3 phase transformation is delivered with more than 60% of the total capacity stored in the O3-type region. The large facilitation of the layered O3-type structure is caused by the entropy stabilization on the host matrix, which helps for superior rate performance and very good cycling stability.

In addition, Yao et al. reported the structural evolution of $NaFe_{0.3}Ni_{0.35}Mn_{0.35}O_2$ compounds during sodiation/desodiation [23]. The behavior of complicated phase transitions of O3hex–O'3mon–P3hex–P'3mon–P3'hex comes from the gliding of $[MO_2]$ sheets and structure distortion. When the electrode is discharged to 2.0 V, an opposite evolution is exhibited and a set of well-defined O3 phase peaks are recovered to the original position. These results indicated the high reversibility of the Na-insertion mechanism [21]. Meanwhile, Yao et al. successfully synthesized a new O3-type $NaFe_{0.45}Co_{0.5}Mg_{0.05}O_2$ layered oxide [23]. During the sodium-ion sodiation/desodiation, the crystal structure of NaFCM undergoes a reversible evolution of O3–P3–P'3–O3' with an average voltage of 3.1 V. The results demonstrate a reversible capacity of 139.9 mAh g^{-1} as well as a capacity of 94.2 mAh g^{-1} after 500 cycles with an initial Coulombic efficiency of 96.6% and a rated capacity of 73.9 mAh g^{-1} at 10 C.

The phase transition O3-P3 is an advantage of O3 structure materials, which improves its capacity, stableness, and rate capability. P3-phase has the diffusion energy barrier lower than O3-phase, so these materials have been dopped or replaced with some metals to achieve higher capacity in O3-phase [24].

5.2.2.2 P2-Type Sodium Transition Metal Oxides

Regarding the P2-type Na_xMO_2 structures, Na ions can diffuse for direct transport between two face-sharing trigonal prismatic sites with a low diffusion barrier in the $[MO_2]$ layers, leading to their better rate performance than that of O3-types [25]. However, the good rate performance for P2-type layered oxide material is often degraded by the Na^+-vacancies and charge order, which leads to limited Na^+ ion transport kinetics [26]. Compared with O3-type layered oxide material, the P2-type cathode usually suffers from a short cycling lifespan during Na-ion intercalation and deintercalation due to the inevitable phase transformation at high voltage [26]. In addition, the P2-phase contains vacancies in the Na layers and is only stable at low Na concentration (sodium deficiency) in the pristine state, which limits its charge capacity leading to the first cycle. Coulombic efficiency is higher than 100% when

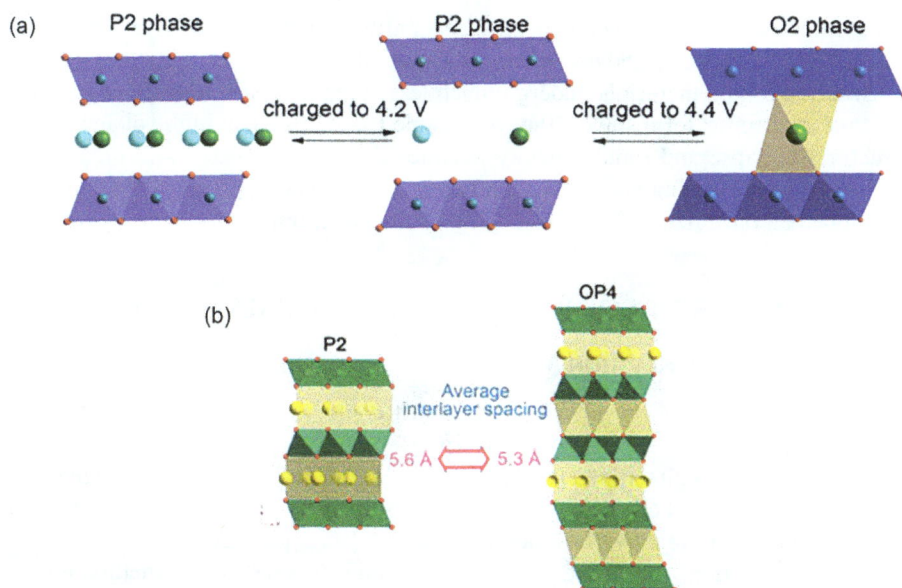

FIGURE 5.6 The phase transition: (a) P2-O2 [29] and (b) P2-OP4 [30]. (Reproduced with permission [29,30].)

cycling in a sodium half-cell. Indeed, the $[MO_2]$ sheets are prompted to glide into a–b plane as a consequence of repulsion force increase between successive O layers, leading to the transformation from P2 (ABBA) to O2 (ABCB) type (the P2-O2 phase transition takes place when $x < 0.3$) (Figure 5.6). This is an important stabilization mechanism against the degradation induced by phase transformation for high-performance SIBs [27].

Cao et al. successfully investigated P2-type $Na_{0.67}Ni_{0.3-x}Cu_xMn_{0.7}O_2$ material by a simple sol-gel method [28]. Copper substitution effectively suppresses Na^+/ vacancy ordering transition occurring at the high potential and the phase transformation compared to the copper-free electrode. When the electrode is firstly charged to 4.5 V and discharged to 2.0 V, XRD patterns of the copper-free $Na_{0.67}Ni_{0.3}Mn_{0.7}O_2$ electrode appeared at 20.1° referring to O2-characteristic peaks and disappeared (or was weak) at (004), (100), and (104) related-peaks. However, the copper-substituted electrodes $Na_{0.67}Ni_{0.3-x}Cu_xMn_{0.7}O_2$ ($x = 0.1, 0.2,$ and 0.3) didn't show these changes since P2-O2 phase transitions didn't occur in the charging process, although these peaks still shifted due to lattice distortions induced by Na^+ extraction.

In addition, Sun and co-workers successfully synthesized P2-type $Na_{0.55}[Ni_{0.1}Fe_{0.1}Mn_{0.8}]O_2$ cathode material for high-energy-density SIBs [31]. On charging, when voltage further increases in the range of 4.1–4.3 V, XRD result of material significantly changed due to the thermodynamical instability of prismatic sites without Na ions, which is ascribed to the OP4-type $Na_{0.55}[Ni_{0.1}Fe_{0.1}Mn_{0.8}]O_2$ with the P_6m2 space group. During the Na^+ ion intercalation process, OP4 phase has prismatic sites as well as octahedral sites, and the high reversibility of the P2 to OP4 phase transformation leads to lower structural instability, which accounts for a high discharge specific capacity and excellent cyclic stability.

Briefly, P2-type materials undergo much less complex phase transitions than the O3-type structures; for instance, the irreversible P2-O2 phase at high voltage causes structural collapse and rapid capacity decline. To limit the phase transition of the material type P2 during Na^+ ion intercalation/deintercalation process, many reports mentioned an optimized amount of doping metal substitution into the MO_2 layers.

5.3 SYNTHESIS OF SODIUM LAYERED MATERIALS

5.3.1 SOLID-STATE REACTION METHOD

The solid-state reaction is considered the most common method to synthesize the polycrystalline material for sodium-ion batteries from solid reagents. The reaction usually occurs at a high temperature. Critical factors affecting the solid-state reaction are morphological and chemical properties of the reagents such as surface area, free energy change, and reactivity. In addition, temperature, pressure, and the environment of the reaction need to be considered. The advantages of this method enable simplicity and large-scale production [32].

Hwang and co-workers comprehensively studied $Na[Ni_xCo_yMn_z]O_2$ ($x = 1/3$, 0.5, 0.6, and 0.8) cathodes to determine the optimal composition by varying the transition metal ratio and treatment temperature condition since the electrochemical, structural, and thermal properties in O3-type layered cathodes are strongly dependent on the transition metal composition [33]. When increasing Ni content, the material delivered capacity increased proportionally about to 15 mA g^{-1} with values of 187.1 mAh g^{-1} for Na-NCM 811, 150 mAh g^{-1} for NaNCM 622, 146.1 mAh g^{-1} for Na-NCM 523, and 141.1 mAh g^{-1} for Na-NCM 333. The capacity retention of $Na[Ni_xCo_yMn_z]O_2$ was exhibited with values of 80% for Na-NCM 333, 65.5% for Na-NCM 523, 57% for Na-NCM 622, and 50% for Na-NCM 811 after 100 cycles at 75 mA g^{-1}. These results show that the better capacity retentions displayed with low Ni compositions although a higher capacity was obtained with higher Ni compositions due to the electrochemical activity of $Ni^{2+/3+/4+}$.

The solid-state reactions method can be widely used to produce polycrystalline cathode materials with high throughput. Hollow and porous cathode nanostructures can be easily synthesized with short Na^+ diffusion path lengths for high rate and long-term cycling stability. However, this method is unbelievable to control the final size and regular shape of the material in terms of nanostructure synthesis. Moreover, it is also difficult to mix homogeneously all the solid precursors to obtain nanostructured materials with well-controlled morphology.

5.3.2 SOL-GEL METHOD

The sol-gel method is one of the well-established synthetic approaches to prepare cathode materials. The sol-gel method required low processing temperature and the liquid precursor solution is polymerized to form gelish structure, forming particles with high purity. This method has excellent control over the shapes and size of the materials

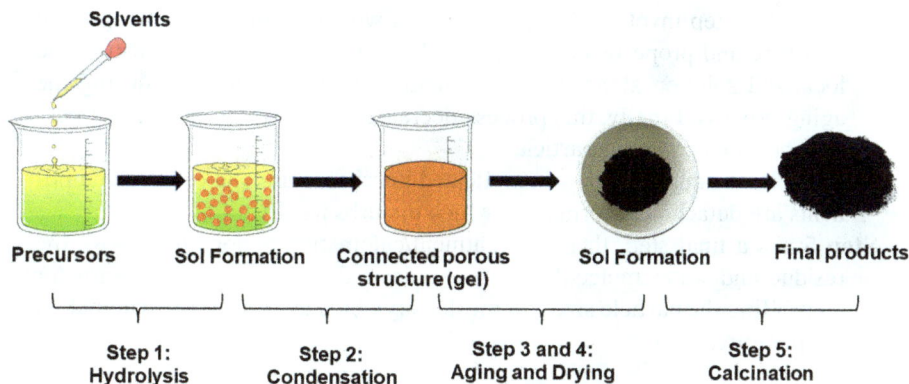

FIGURE 5.7 Process of the synthesis steps by the sol-gel method.

that directly affect material properties. In general, the sol-gel method can be performed in five key steps: hydrolysis, polycondensation, aging, drying, and thermal decomposition [34] as described in Figure 5.7.

> **Step 1**: The precursors such as metal alkoxides are hydrolyzed in water or organic solvent. For the synthesis of layered material, water or organic solvent supplied oxygen which is necessary for the formation of metal oxide. If water is used as a reaction medium, this method is known as the aqueous sol-gel method, and the use of organic solvent is termed the nonaqueous sol-gel route. When adding water or organic solvent into the precursor's solution, an acid or a base also helps in the hydrolysis of the precursors. The hydrolysis process is given below:

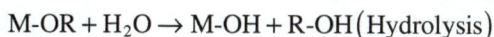

$$M\text{-}OR + H_2O \rightarrow M\text{-}OH + R\text{-}OH \,(\text{Hydrolysis})$$

> Where M = metal, R = alkyl group (C_nH_{2n+1})
>
> **Step 2**: Water or organic solvent is eliminated, and metal oxide linkages are formed when condensation of adjacent molecules and polymeric networks grow to form colloidal dimensions in the liquid state. This process takes place in two phases: olation and oxolation. The process of forming a hydroxyl (–OH–a) bridge between two metal centers (metal–hydroxy-metal bonds) is called olation and oxolation is a process in which an oxo (–O–) bridge is formed between two metal centers (metal–oxo-metal bonds). The condensation is described below:

$$M\text{-}OH + XO\text{-}M \rightarrow M\text{-}O\text{-}M + X\text{-}OH \,(\text{Condensation})$$

> The result of the condensation increases the viscosity of the solvent leading to the formation of a porous structure maintaining a liquid phase called gel. The alkoxide precursor and pH of the solution are the factors affecting the size and the cross-linking within the colloidal particles.

Step 3: This step involves the aging process which continuously changes the structure and properties of the gel. Polycondensation continues within the localized solution along with precipitation of the gel network during the aging process. Finally, this process decreases porosity and increases thickness between colloidal particles.

Step 4: The drying process is complicated because water and organic components are detached to form a gel which disturbs its structure.

Step 5: As a final step, thermal treatment/calcination is done to remove the residue and water molecules from the desired sample. The key factor for controlling the particle size and the density of the material is the calcination temperature.

Tang et al. synthesized the P2-type layered $Na_{0.67}Ni_{0.33}Mn_{0.67}O_2$ (NMO) and $Na_{0.67}Ni_{0.28}Mg_{0.05}Mn_{0.67}O_2$ (NMMO) samples were prepared via a typical sol-gel method with citric acid as a chelating agent followed by calcination at 900°C for 12 hours in the air to obtain final target samples [35]. The electrochemical properties of NMO and NMMO at different rates from 0.1 to 5 C and then back to 0.1 C in the voltage range of 2.5–4.3 V are shown in Figure 5.8b. NMO and NMMO electrodes present a discharge capacity of 33.6 and 88.2 mAh g^{-1} at 5 C, respectively. When the current density returns to 0.1 C, NMO and NMMO electrodes deliver a discharge capacity of 90.1 and 112.7 mAh g^{-1}, corresponding to 60.7% and 86.4% of the initial capacities, respectively. Figure 5.8c displays the cycling stability of NMO and NMMO electrodes at a rate of 0.2 C. The NMO electrode shows a discharge capacity of 55.6 mAh g^{-1} with a poor capacity retention of 41.5% after 100 cycles. By comparison, the NMMO electrode shows a discharge capacity of 82.3 mAh g^{-1} which corresponds to the capacity retention of 63.8%.

P-type $Na_{0.67}Co_{0.5}Mn_{0.5}O_2$ cathode material with hierarchical architectures was synthesized by Zhou et al. through a facile and simple sol-gel route, using metal acetate salts and citric acid as chelating agents. To obtain the colloidal sol, the solution was kept at 80°C under constant agitation, and then the achieved sol was dried overnight at 100°C in air. The resulting powders were firstly annealed at 500°C for 6 hours in air. The material phase transformed from P3 to P2 with P2/P3 coexisting in the intermediate temperature (800°C) when the calcinating temperature was increased from 700°C–950°C [36,37]. The electrode presented a discharge capacity of 147, 132, 110, and 98 mAh g^{-1} at 0.1, 1, 3, and 10 C, respectively. Even at 30 C (5.1 A g^{-1}), the reversible capacity still reached 88 mAh g^{-1}, which indicates that 60% capacity retention was achieved. By comparison, P2/P3- $Na_{0.67}Co_{0.5}Mn_{0.5}O_2$ cathode delivered discharge capacities of 156.1, 141.8, and 126.6 mAh g^{-1} with increased Coulombic efficiency from 1 to 5 C, respectively, as shown in refs [36,37]. Remarkably, the discharge capacity of P2/P3- $Na_{0.67}Co_{0.5}Mn_{0.5}O_2$ cathode is 67 mAh g^{-1} with a high-capacity retention of 84.5%. However, P2- $Na_{0.67}Co_{0.5}Mn_{0.5}O_2$ material can be charged and discharged at 30 C high rate and showed a discharge capacity of 88 mAh g^{-1}. In short, the calcinating temperature has an influence on the synthesis of materials by sol-gel method, resulting in the formation of different types of structures within the material.

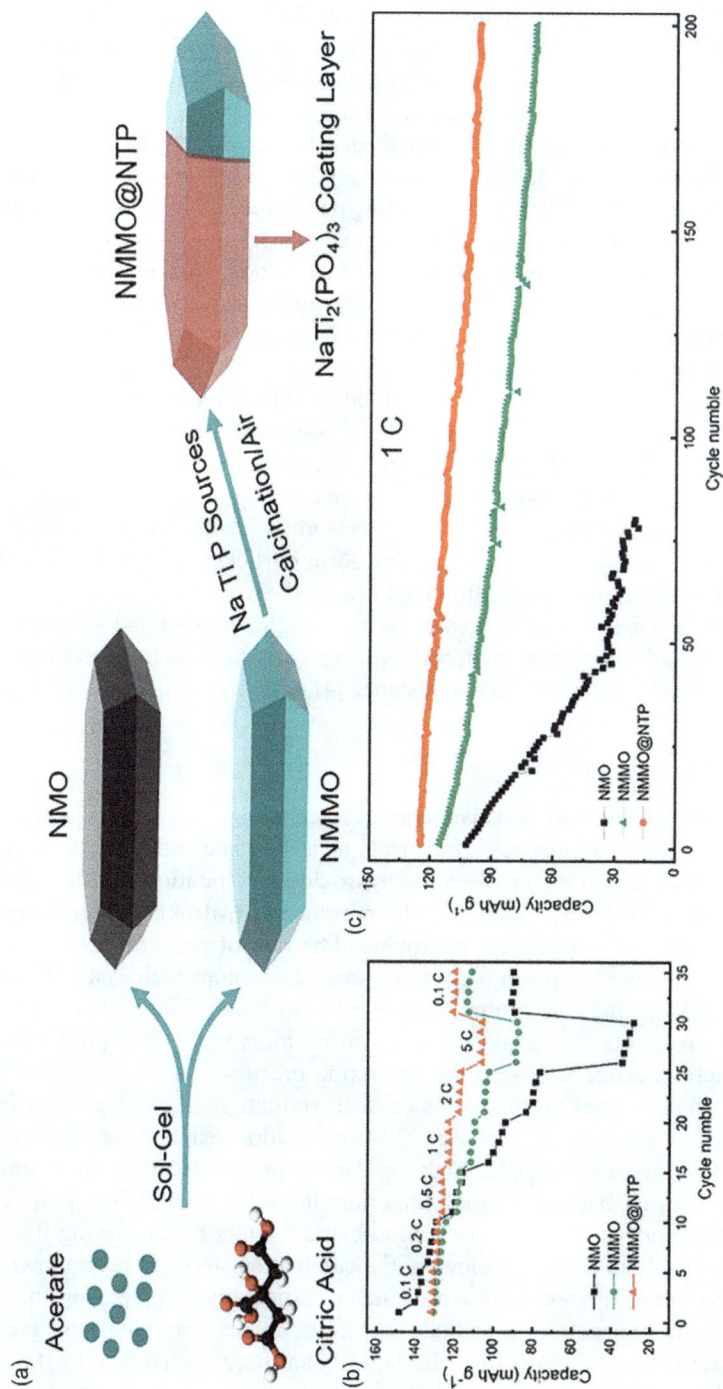

FIGURE 5.8 (a) Schematic illustration for synthesizing NMO, NMMO, and NMMO@NTP (b) Rate capability comparison of three electrodes at various current densities. (c) Cycling performance of three electrodes at 0.2 C. (Reproduced with permission [35].)

In recent years, many attempts have been devoted to synthesize the solid solutions of layered $Na[Ni_xMn_yCo_z]O_2$ to obtain low-cost and high-capacity Na-insertion cathodes. Yang et al. synthesized pure P2-phase $Na_{0.67}[Mn_{0.65}Ni_{0.15}Co_{0.2}]O_2$ micro flakes by a citric acid-assisted sol-gel method [38]. Sathiya et al. has prepared O3-type $NaNi_{1/3}Mn_{1/3}Co_{1/3}O_2$ material via sol-gel route using the acetate of Na, Co, Ni, and Mn in the molar ratio 1.05:0.33:0.33:0.33, and citric acid as a chelating agent [39]. The SEM image shown in Figure 5.9b indicates the agglomeration of flaky particles with sizes ranging between 1 and 10 μm. The cycling performance of $NaNi_{1/3}Mn_{1/3}Co_{1/3}O_2$ electrode delivered reversible intercalation of 0.5 Na, leading to a capacity of 120 mAh g^{-1} in the voltage range of 2.0–3.75 V. In addition, Sun et al. investigated micrometer-sized layered P2-type $Na_{2/3}[Mn_{0.54}Ni_{0.13}Co_{0.13}]O_2$ cathode with uniform size and distribution with a citric acid assisted sol-gel method at 850°C for 12 hours [40]. The citric acid acted as a chelating agent to reduce the particle agglomeration and enhance the structural integrity of the final product. Figure 5.9c displays well-defined particles with very smooth surfaces and edges with an average particle size of approximately 1.5 μm. P2-type $Na_{2/3}[Mn_{0.54}Ni_{0.13}Co_{0.13}]O_2$ electrode was studied within a potential range of 2.0–4.5 V and delivered the highest discharge capacity of 123 mAh g^{-1} at 1 C, as well as excellent cyclic stability.

In conclusion, the sol-gel method helps uniform particle size and distribution of particles, which results from the addition of citric acid as a chelating agent during the synthesis process. This method decreases particle agglomeration and aids in reducing the distance for ionic migration because the ion insertion into the host material is mainly dominated by the ion diffusion distance [41,42].

5.3.3 Co-Precipitation Method

The co-precipitation method is a wet synthesis technique that well-established to prepare the spherical nanoparticles materials at large-scale industrial simply and economically. This technique involves the hydroxide precipitation reaction of some metal cations in an aqueous solution in the presence of hydroxide sources such as sodium hydroxide and ammonium hydroxide. The rate of precipitation is hard to control, but by keeping the pH at a constant level, the nanoparticles can be eventually obtained. When there is more cation needed to homogenous coprecipitation, the precipitation agents like oxalate anion, citric anion, etc. are required to slow down the reaction. After washing, the hydroxide precursors were separated from the aqueous solution; they were calcinated with sodium sources to convert into a sodium transition metal oxide material. Co-precipitation reaction has been extensively taken to synthesize spherical high-tap density precursors for high volumetric energy density Li-ion battery cathode materials in commercialization [43]. These hydroxide precursors can be used for Na-ion battery synthesis by mixing the dried precursors with sodium sources following the calcination steps. In recent years, the core-shell structure has been intensively studied, especially in preparing the core-shell and gradient concentration cathode materials. In the core-shell and gradient concentration structures (Figure 5.10), the capacity has been determined by the core's composition while the shell's composition affects the rate capability, structural, and thermal stability. Therefore, the hydroxide precursors have high Ni content in the

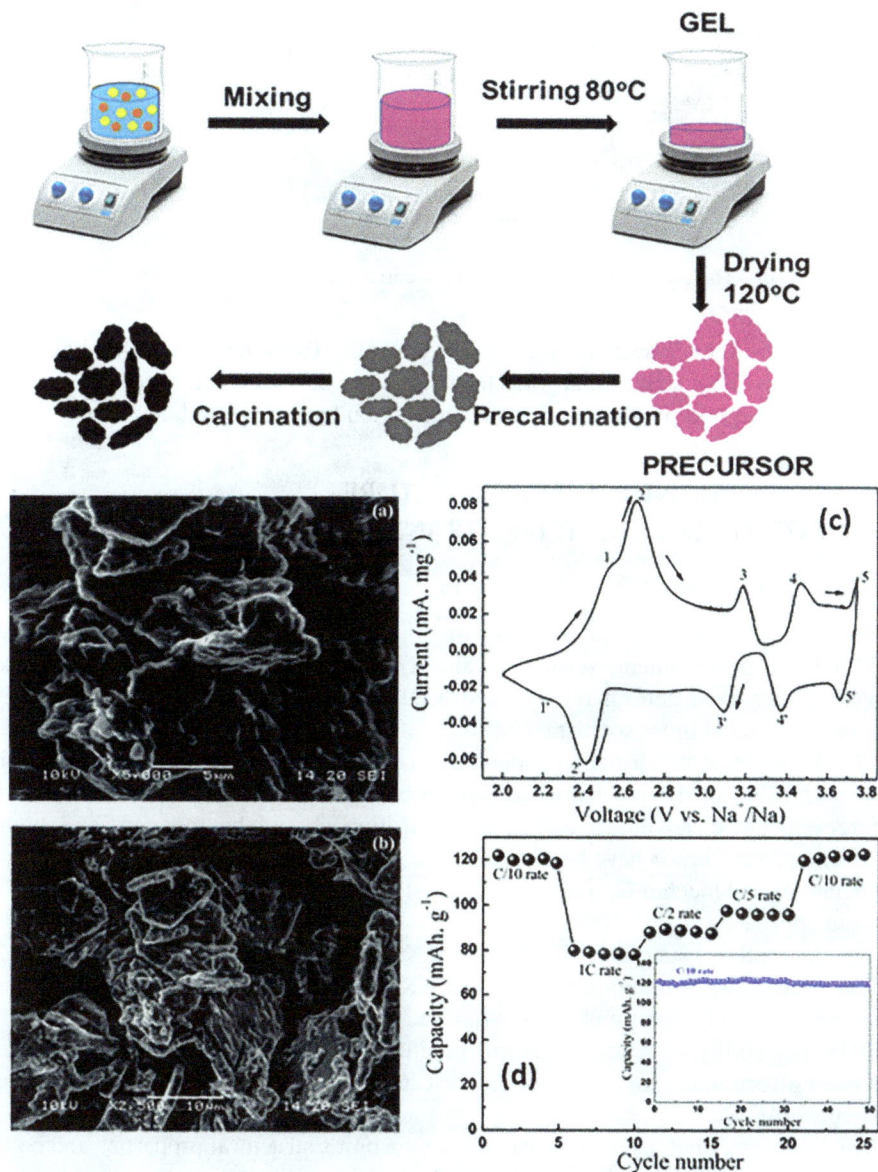

FIGURE 5.9 Procedure for synthesis of NaNMC samples using Sol–gel method. (a, b) SEM images, (c) CV and (d) Cycling performance of $NaMn_{1/3}Ni_{1/3}Co_{1/3}O_2$. (Reproduced with permission [39].)

core to enhance the capacity and high Co, Mn contents in the shell for better Li^+/Na^+ diffusion and stableness. However, the high demand for crystalline compatibility between the core and the shell is required to avoid the volume expansion defects during the charge/discharge processes. Therefore, the gradient concentration has been investigated to solve these drawbacks. In this structure, the concentration of two or

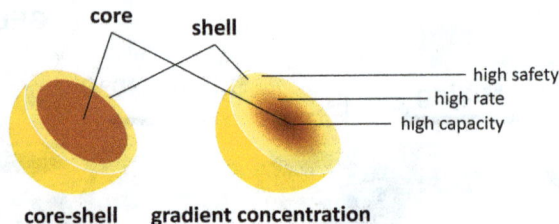

FIGURE 5.10   The core-shell and gradient concentration structures.

more compositions is decreased/increased gradually from the core to the shell. The crystal grows more stable and avoids the crystalline defect caused by volume expansion while the good properties of each composition have remained.

5.4   UNDERSTANDING THE STRUCTURE EVOLUTION AND PHASE TRANSITION

5.4.1   Ex situ and In Situ Techniques

Inspired by the fundamental understanding of electrochemical systems and to guide for material development, several advanced techniques have been developed and upgraded for even getting reliable and accurate information, and therefore, they become more and more sophisticated.

In Na-insertion materials, to understand the effects of phase transition on materials' performance, the mechanisms have to be clearly and completely understood. Moreover, the disadvantages of phase transition should be reduced, and high-performance materials have been developed.

Kinetics and mechanisms have been investigated by using time-resolved in situ or ex-situ approaches.

In-situ methods such as NMR, EPR, and XRD, and ex-situ methods such as SEM/TEM, cryo-TEM, Raman, XPS, and even XRD are powerful and important tools in the intensive studying of materials [44].

The possibility to readily perform in-situ experiments on a time scale makes scattering techniques of great use for investigating the dynamic aspects of many systems [44].

In definition, phases could be separated by a unit's structural properties and composition. Therefore, the phase evolution during the intercalation/deintercalation of sodium ions would be traced using XRD techniques for structural analysis. In addition, sodium cooperation leads to changes in chemical composition, morphology, oxidation state, bond length, and coordination environment which are asserted by EDX, SEM/TEM, XPS, and Raman spectroscopy.

The advantage of the ex-situ methods is simple. However, the disadvantages of this method are requiring a lot of preparation time and many samples, keeping a strictly controlled condition, and low repeatability. The other methods have been improved to perform in-situ investigation. Among them, in-situ XRD is a strong method to study the phase transition of electrode materials since it is fast, accurate, highly reliable and can be achieved during the operator [45]. However, the preparation of instrument and cell measurements is complicated and involves high cost.

In-situ investigations of Li-ion batteries have proven extremely insightful but require the electrochemical cell to be fully compatible with the testing conditions, and therefore, many challenges are often encountered during execution. Advantageously, in the past few years, significant progress has been made with innovative and advanced in situ techniques.

Until now, to improve the accuracy of the ex-situ method, the structure and phase transition of many electrode materials were analyzed by in situ XRD method during the Na^+ insertion.

5.4.2 Ex Situ and In Situ Techniques in the Study of Phase Transition

Phase transition is detrimental to the performance of sodium cathode material when accumulating a big ion size of sodium ions. The evolution of structure is either a monophasic (solid solution) mechanism or a biphasic mechanism. As mentioned, the P2-structure cathodes that undergoes a solid solution insertion in the relevant voltage range.

P2-$NaNi_{1/3}Mn_{2/3}O_2$ is a highly impressive candidate for high-capacity cathode [30]. In situ XRD experience in Figure 5.11 confirmed the structure preservation because a clear absence of additional diffraction peaks beyond the P2 structure is detected in a voltage range below 4 V [3]. The (002) and (004) peaks shift to a lower angle while (100), (102), and (103) peaks shift toward a higher angle, indicating an expansion of c axis and the contraction of ab plane. Due to the structure preservation, the peaks shift back and forth as seen on the XRD diagram, and a revolution of lattice parameters could be obtained [4]. As sodium content lowers or at a higher voltage above 4.1 V, a phase transition P2-O2 occurs. The phase transition was detected by the ex-situ XRD experiment reported by Lee et al. [5] who conducted XRD at different charge/discharge stages, including the high voltage region as shown in Figure 5.11b. The diffraction patterns of the O2 phase are characterized by peaks located at about 21° and 69° (convert to a wavelength of CuKα X-ray source). The accompanied voltage profile reveals that the phase transition P2-O2 begins when sodium content is lower than about 0.3, which results in a significant decrease of c axis or shrinkage of layer slabs and thus destroys the electrochemical activity.

The transportation of sodium ions in the O3 phase is limited kinetically so that phase transition is needed to remain the sodium storage ability. The most common phase transition is from O3 to P3 phase at the early stage of sodium extraction which is discussed later. On the other hand, the O3 tracking preservation has also been seen in various single TM systems Na_xTMO_2 with TM = Fe, Ti, V, Mn [47].

$NaFeO_2$ owns the typical O3-layered structure and has been evaluated as a cathode material for sodium-ion batteries. Lee et al. [46] investigated the stability of $NaFeO_2$ at the charged stage by Mossbauer spectroscopy. The spectrum for the electrode that is charged up to 0.46 SOC indicates the presence of a new emerging Fe^{4+} species as seen by the deconvoluted spectrum into two kinds of doublets related to Fe^{3+} and Fe^{4+}. However, the decrease of a new doublet of Fe^{4+} suggested a parasitic reaction such as SEI formation or a spontaneous reduction of Fe^{4+} back

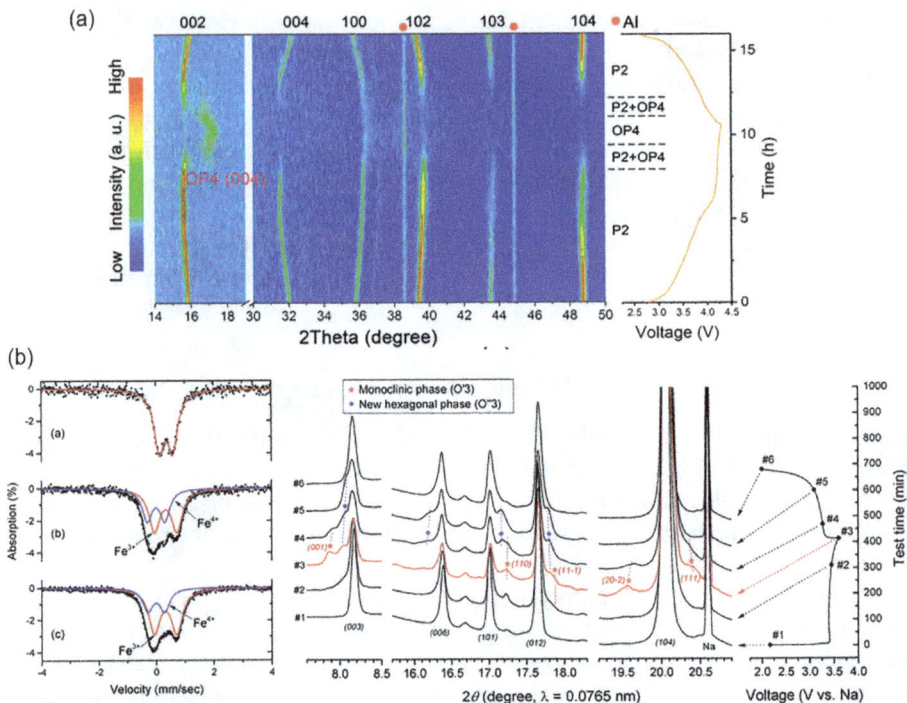

FIGURE 5.11 (a) In situ XRD pattern collected during the first charge (sodiation)-discharge (desodiation) process $P2-Na_{2/3}Ni_{1/3}Mn_{2/3}O_2$ cell between 2.6 and 4.3 V [30]. (b) Ex situ Mössbauer spectra for $NaFeO_2$ electrodes in the pristine state, charged to 0.46 SOC and stored for 2 days under open circuit conditions after being charged to 0.5 SOC and selective in situ SXRD patterns of $NaFeO_2$ during the initial charge–discharge cycle [46]. (Reproduced with permission [30,46].)

to Fe^{3+}. Furthermore, the structure evolution during the charge-discharge cycle was investigated by in-situ synchrotron XRD, and the selective XRD patterns obtained at different states of charge corresponding to the marked points in the voltage curve present in Figure 5.11d. The patterns obtained at the charged state exhibit low-intensity shoulder on the low, and high angle denoted an emerging of a new phase. This phase could be m-$Na_{1e-x}FeO_2$ which has a distortion monoclinic structure due to the Jahn-Teller effect from high-spin Fe^{4+} or a structural compensation for the loss of Na^+ ions. Additionally, the appearance of the third phase other than the O_3-Na_xFeO_2 and m-$N_{a1-x}FeO_2$ was confirmed by the change of peaks in different 2θ regions. The phase was denoted as $O''3$-$Na_{1-x}FeO_2$ and the phase transition is an asymmetry between the charge and discharge stage. The continuous evolution of the three phases during the initial charge-discharge process is represented as both a three-dimensional waterfall view and a two-dimensional bird's eye view in Figure 5.12. It is revealed that the m-$Na_{1-x}FeO_2$ phase is reversible during the charge and discharge process which forms on the charge step and disappears gradually as discharge proceeds whereas otherwise the $O''3$-$Na_{1-x}FeO_2$ displays asymmetric transition and persists in the subsequent cycle.

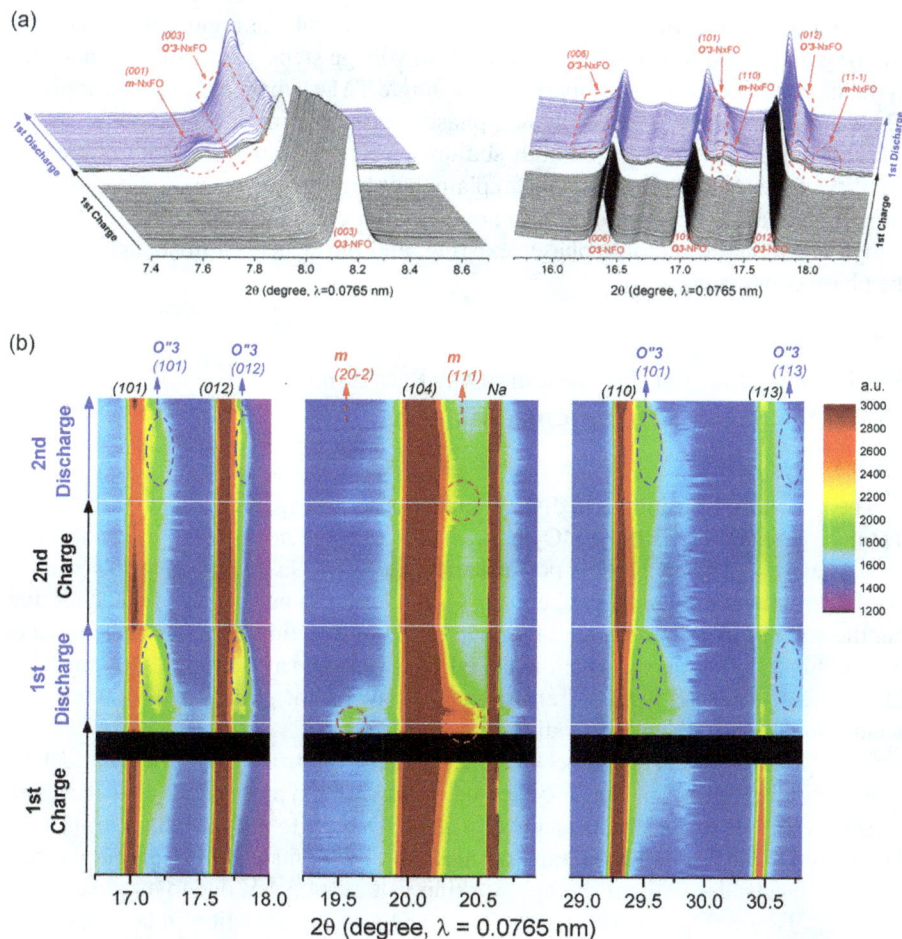

FIGURE 5.12 (a) 3D waterfall view and (b) 2D bird's eye view of *in situ* SXRD patterns at selective 2θ angles during initial cycles of NaFeO₂ cathode. (Reproduced with permission [46].)

The multiple TM systems have the synergic advantage of each element to enhance the electrochemical properties much better. Particularly, the binary system has been synthesized to evaluate the role of each element and the effect of the ratio between them to the electrochemical properties of the cathode materials. The materials display a sodium storage mechanism including an O3-P3 phase transition that enhances the reversibility of the phase transition and improves the rate capability as well because of the higher ionic conductivity of P3 phases compared to that of O3 phases in most cases. For example, $NaFe_{0.5}Co_{0.5}O_2$ is appreciated for its high specific capacity and high working voltage. The material displays phase transition O3-P3-O3 as the voltage increases in which the P3 intermediate phase contributes the majority to the specific capacity. In-situ XRD results of the material exhibited that the P3 phase forms at the early stage of sodium extraction, so the material benefits from the long monophasic intercalation into P3 phase. At the high-voltage region, the structured ordering and phase transition P3-O′3 result in voltage steps which are also detected in the in-situ XRD result. Compared to the single TM-element cathode, the multiple TM material exhibits the well-defined phase transition, simplifies the phase transition mechanism, enabling a smooth sodium migration into the structure, and the stability of the structure which could explain their high performance compared to the individual phases. The evolution was not a smooth or cyclic change but a relatively complex one due to multiple phases existing, and the structural mismatch between the phase component.

5.4.3 Structure and Electrochemistry Properties of Sodium Layered Structure Materials

5.4.3.1 Na$_x$MO$_2$

The simple composition and distinct electrochemical characteristics of the single metal layered materials Na$_x$TMO$_2$ have drawn a lot of attention to study the role of a single element on battery performance. Layered Na$_x$TMO$_2$ materials usually exhibit structure rearrangement between O-type and P-type stacking sequences during the sodium insertion/extraction due to strong interaction between bigger size of the sodium ion and a host, which is called phase transformation, rarely observed in LiTMO$_2$ electrodes. The phase transformations involve the gliding of TMO$_2$ layers to accommodate the presence of sodium ions.

NaFeO$_2$ is a typical O3-type layered structure electrode but owns monoclinic symmetry different from the O-type electrodes and a typical phase transition that preserves O3-type stacking sequence without O-type to P-type phase transition. The material displays a flat voltage profile like olivine material but unique layered structure electrodes with an average working voltage of 3.4 V due to a redox couple Fe^{4+}/Fe^{3+}. However, the experienced O3-O″3-O′3 phase transition induces unfavorable electrode kinetic, cation mixing between sodium ion and iron cations, and the instability of Fe^{4+} ion reduces the voltage range of operation and cyclability of the material. Fe^{4+}/Fe^{3+} redox couple could be used along with another redox couple to increase the average working voltage. Other O3-type electrodes such as NaCrO$_2$, NaTiO$_2$, and NaVO$_2$ also present O3 hexagonal to O′3 monoclinic, and the capacity declines when working in high voltage range [4].

Layered structure NaNiO$_2$ could be obtained in a high oxidative environment of oxygen at relatively low temperature and display disturb O′3 due to Jahn-Teller effect of low spin Ni^{3+} ion [4]. During sodium ion insertion/extraction, the material displays a stair-like voltage profile with many monophasic and binary phasic regions. In detail, an O to P mechanism including a series of disturb phase O3-P′3-P″3-O′3-O‴3 was observed by mean of in-situ XRD in the first charge. In the next discharge, a phase transition occurring in reversed direction revealed a high reversible phase transition in the high voltage. Indeed, a new diffraction peak of a new O‴3 (mark as *) was seen on the XRD pattern when discharge below 2.3 V that indicates that the phase obtained at the end of discharge is O3-type but different from the pristine.

Different from $NaFeO_2$ and $NaNiO_2$ mentioned above, $NaCoO_2$ and $NaMnO_2$ both own P-type and O-type layered structure by varying the sodium content $x \sim 1$ (O3) and $x = 2/3, 0.7, 0.5$ (P2) [19,20,48,49]. Na_xMnO_2 has also displayed a distorted structure, namely O′3 due to the Jahn-Teller effect of Mn^{3+} ion, and two kinds of structures are obtained at low and high temperature namely α-$NaMnO_2$ and β-$NaMnO_2$ respectively [1].

Na_xCoO_2 ($x \sim 1$) is a disturbing O3 layered structure and also displays O to P phase transition. The O3-Na_xCoO_2 delivers high specific capacity but average stability because of the complex phase transition mechanism. Meanwhile, the sodium insertion/disinsertion into the P2-type structure exhibits complex phase transition as well as showed a stair-like voltage profile because of Na^+/vacancy distribution due to a change of sodium composition [1]. The first charge capacity of the P2 phase is lower than that of the O3 phase due to the partial sodium content in the pristine electrode [49,50].

5.4.3.2 $Na_xMM'O_2$

The wide variety of stacking sequences in layered offers high flexibility in choosing suitable transition-metal species. Especially, compared to the single metal layered material, the binary system using two metal corporation in the slabs usually possesses an ideal layered structure without distorting that is contributed to the Jahn-Teller effect because of the synergistic effect of the two elements.

The electrode materials with a composition based on Fe and Mn were desirable to reduce the battery cost but deliver an acceptable energy density and power density. $NaFeO_2$ and $NaMnO_2$ are disadvantageous and cannot satisfy the expectation because the capacity decreases significantly after several cycles.

One of the popular couples is Ni-Mn because of the similar ionic radius of these two elements, so Ni could replace Mn partially in $NaMnO_2$. However, Ni ions seem to capable to replace for manganese ion in the sheets for average concentration that generally below 30%, for example, spinel $LiNi_{0.5}Mn_{1.5}O_4$ [51]. High Ni/Mn ratio in the slab probably prefer uptaking O3 stacking instead of P2 stacking [52–54].

P2-$Na_{2/3}Ni_{1/3}Mn_{2/3}O_2$ is a P2 structure material famous for its high specific capacity and high stability during various Ni/Mn ratios [4]. P2-$Na_{2/3}Ni_{1/3}Mn_{2/3}O_2$ is also an air-stable material due to the ordering distribution of Ni/Mn in the slab which could increase the interaction between adjacent slabs to prevent the intercalation of water molecules. The material delivers the highest capacity of >160 mAh g^{-1} when utilizing high voltage Ni^{4+}/Ni^{3+} redox couples but has poor capacity retention, so the way to improve the stability of high redox couples is to take the benefits from high working voltage and high specific capacity.

The rapid decrease of P2-$Na_{2/3}Ni_{1/3}Mn_{2/3}O_2$ capacity is a result of P2-O2 phase transition, leading to migration of Ni cation into sodium-ion vacancies. Doping could be effective to reduce the disadvantages of the phase transition based on utilization of dopants that have a similar ionic radius and coordination numbers such as Al^{3+} [55], Mg^{2+} [56,57], Zn^{2+} [58], Cu^{2+} [28], and Ti^{4+} [59]. Mg-doping was proven to enhance the capacity stability of the material despite a slightly lower initial capacity [56,57]. Mg ion is electrochemically inactive that uptake sites of transition metal ions in the slabs allowing higher Na^+ ion remains in the prismatic site in the interslab and reduce shrinkage of the lattice so that prevent the P2-O2 transition. In general, the more dopant concentration substitutes, the more the reduction of specific capacity [56,57,59]. In contrast to Mg^{2+}, Cu^{2+} is electrochemically active, so the initial capacity of doped material is less or indifferent compared to the other dopants. P2-$Na_{2/3}Ni_{1/3}Mn_{1/2}Ti_{1/6}O_2$ delivered capacity of 127 mAh g^{-1} at the working voltage of 3.7 V. The voltage profile turns into a sloping feature after Ti substitution [59].

Li$^+$ can also incorporate into transition metal layers, for example, P2-$Na_{0.8}$ [$Li_{0.12}Ni_{0.22}Mn_{0.66}$]$O_2$. Li is an effective dopant to prevent the P2-O2 phase transition and enhance both the working voltage range and specific capacity of the doped material. The utilization of ^{7}Li-NMR spectroscopy allows monitoring of the Li$^+$ migration between the transition metal sites and sodium sites on cycling, especially in the high voltage region of the O2 phase from 4.1 to 4.4 V [60,61].

The material contains Ni and Mn which has an O3-type layered structure that has been investigated such as $NaNi_{1/2}Mn_{1/2}O_2$. The material delivers the highest capacity of 185 mAh g^{-1} in charge up to 4.5 V due to the contribution of Ni^{4+}/Ni^{2+} redox couple. In the low voltage range of 2.2–3.8 V, the only charge of 125 mAh g^{-1} was obtained but with higher reversibility. The capacity reduces seriously due to a complex phase transition of $O'3_{mon}$–$P'3_{hex}$–$P'3_{mon}$–$P3''_{hex}$. The material is also air sensitive. The issue could be solved when introducing a dopant element into the transition metal layers. For example, $NaNi_{0.45}Cu_{0.05}Mn_{0.4}Ti_{0.1}O_2$, which is Cu and Ti co-doped, significantly enhances tremendously the air stability, and the electrochemical activeness is maintained after the material is soaked with water [62].

With only Ti dopant, the slight difference between the ionic radius of Ti^{4+} (60.5 pm) and Mn^{4+} (53 pm) enables the monogenous distribution of charged species in the layers. The best performance obtained for O3-$NaNi_{0.5}Mn_{0.5-x}Ti_xO_2$ with $x = 0.3$ could remain 85% of the initial capacity after 200 cycles at a rate of 1 C and delivered a capacity of 93 mAh g^{-1} at 5 C [24]. Moreover, the complex phase transfer was suppressed so that only O3-P3 phase transfer begins at the first charge and a solid-solution-based intercalation mechanism is found for the rest of the operation voltage, which results in high performance and rate capability.

$NaNiO_2$-$NaFeO_2$ system could form a solid solution material at the ratio upto 1:1, but the best sodium storage performance was obtained by the lower ratio between two compositions [63]. O3-$NaFe_{0.5}Ni_{0.5}O_2$ and O3-$NaFe_{0.3}Ni_{0.7}O_2$ deliver a capacity of 112 and 135 mAh g^{-1},respectively, with the retained capacity of 62 and 100 mAh g^{-1}, and the average working potential of 2.85 and 2.7 V, respectively. Smooth voltage profiles but relatively low performances were obtained as a result of the instability of Fe^{4+} cation and the migration of iron cation into Na$^+$ vacancy.

$NaFeO_2$-$NaCoO_2$ system also forms a solid solution for every Fe/Co ratio but is much more stable and shows high-performance than Fe-Ni system. O3-$NaFe_{0.5}Co_{0.5}O_2$ was first investigated by Komaba group and reported to have the best performance among various Fe/Co ratios with a high specific capacity of about 160 mAh g^{-1} and superior rate capability of up to 30 C compared to $NaCoO_2$ and $NaFeO_2$ [7,64]. The material exhibits phase transformation in series O3-P3-P'3-O3' during sodium extraction, but in most cases, the voltage range limits at 4 V where P'3-O3' intergrowth is present. The phase transition from O to P phase and a long P3 region up to 70% of sodium concentration range explains the high-rate capability of the material

[7,64,65]. Indeed, the presence of stabilizing cobalt ions, as well as dopants such as Mg^{2+} and Cu^{2+}, prevent the migration of iron ions that enhance the sodium storage stability over cycles [65,66].

The incorporation of Mn and Co in the octahedral sites helps to form an ideal layered structure from separate distortion of Na_xMnO_2 or Na_xCoO_2 and reduces the Co content in the electrode composition as well. The investigation of Co/Mn ratio in P2-$Na_{2/3}Mn_yCo_{1-y}O_2$ [67] showed that the initial capacity increases whereas cycling stability declines when y rises. Solid solution P2-$Na_{2/3}Co_{0.5}Mn_{0.5}O_2$ with $y = 0.5$ [37] is attractive for its sloping voltage profile and superior cycling stability. The material delivered a capacity of 147 mAh g^{-1} in a voltage range of 1.5–4.3 V at the rate of 0.1 C and indeed showed excellent capacity retention of roughly 100% at the rate of 1 C and robust structure stability. The phase transfer is almost suppressed in the wide voltage range of sodium intercalation/deintercalation. The disadvantage of the Mn-Co-based system is the air and moisture sensitivity preventing it forming practical applications.

Na_x[Fe, Mn]O_2 material using the combination of Fe and Mn is attractive because of its low cost, abundance, and neutralness of raw sources. A series of Fe/Mn ratios of 2/3, 1/3, and 1/2 of both P2 and O3 phases was investigated. Especially, a certain Fe/Mn ratio could crystallize in both P2 and O3 structures. For example, Gonzalo synthesized P2 and O3 forms of $Na_{2/3}Fe_{2/3}Mn_{1/3}O_2$ material [68,69]. Han also successfully synthesized the two types of $Na_{2/3}Fe_{1/2}Mn_{1/2}O_2$ [6,70] layer structure. The capacity of the material enhances in increasing Mn content, the voltage range of operation which was usually limited below 4.1 V for long cycling and surface morphology stability [69]. Yabuuchi and co-workers synthesized a superior material P2-$Na_{2/3}Fe_{1/2}Mn_{1/2}O_2$ with an excellently high capacity of 190 mAh g^{-1} at the cutoff charge voltage of 4.2 V and revealed that the P2 phase is capable of delivering higher capacity than the O3 phase. On their calculation on half-cell, the energy density could approach the value of 520 Wh kg^{-1} of $LiFePO_4$ [6,69].

5.4.3.3 NaNi$_{1/3}$Mn$_{1/3}$Co$_{1/3}$O$_2$

NaNMC, the counterpart of the commercially $LiNi_{1/3}Mn_{1/3}Co_{1/3}O_2$ (NMC111), appears as an attractive candidate for Na-ion batteries. NaNMC takes advantage of a bigger radius of Na^+ ion than that of 3d metal cations, which alleviates the cation mixing upon cycling. NaNMC material can be prepared by several methods. Co-precipitation is common for the synthesis of multiple metal cathode materials. However, the first NaNMC was synthesized by the sol-gel method by Sathya group [39]. The electrochemical properties of the material were thoroughly investigated.

Although the NaNMC has O3-type layer stacking, most synthesis processes are conducted at a high temperature of 900°C to overcome the high activation energy. However, a common heating procedure including being heated at high temperature in the air seems not to benefit the formation of octahedral coordination of sodium ions, but formed triangular prismatic coordination instead and finally led to multiphase component material. XRD pattern results reveal that the P2 phase dominates whereas the impurity of NiO is also observed [71]. Such a pattern could result from the high synthesis temperature that favors the formation of the P2 phase and gradual sodium element evaporation decreases the sodium content below the needed level to form O3 phase. Additionally, Ni^{2+} is quite difficult to access the octahedral site in the transition metal layers without the support of a high oxidation environment. The electrochemical cycle test showed that the voltage profile looks like the one of $Na_xCo_{0.5}Mn_{0.5}O_2$ cathode material which has a P2 structure [72,73].

Xu et al. [74] performed a comprehensive study on the intercalation mechanism between phase structure, interfacial micro strain, and electrochemical properties of NaNMC. The electrochemical test results showed that the material with an inter-growth P2/O3/O1 structure displays higher reversible capacity, cycle stability, and thermal stability than the one with P2/O3 binary-phase and the only dominant P2 phase, which is related to their distinct interfacial microstructure-strain development during synthesis and charge/discharge stage. During the charge-discharge process, Ni^{2+}-Ni^{3+} contributes mostly to the overall capacity whereas Co and Mn partially contribute. The specific capacity of NaNMC in different phases is in the order of P2/O1/O3 > P2/O3 > P2.

NaNMC material displays good performance with long cycle life regardless of phase composition. Among various phase components, O1 and O3 contribute mainly to the specific capacity of the material, so the content of these phases should be improved to obtain a high specific capacity of the NaNMC material [74]. Control the synthesis procedure is needed to gain their O1 phase dominant or O3 phase dominant. Phung et al. [75] synthesized P2/O1/O3 composite NaNMC that has the same characteristic as the single-phase component electrode.

Through the last work, the appreciated pure O3-NaNMC was obtained by a low temperature of 750°C and a long calcination time of 24 hours [33]. The author synthesized O3-NaNMC with impure NiO by calcination at 900°C in a muffle furnace and cooling naturally in the air [76]. The samples were prepared by the sol-gel method by the following step: the resultant gel was decomposed by heating at 400°C for 24 hours, then the precursor is pelleted. The pellets were annealed at 900°C for 24 hours with a heating rate of 10°C min^{-1} (NMC-9012T1). In addition, the pellets are heated under double-sided pressing to limit the loss of Na ions (NMC-9012T2). Both samples were naturally cooled to room temperature and stored in a desiccator. XRD diagrams heated under different conditions are shown in Figure 5.13. XRD diagram of NMC-9012T1 sample (shown in Figure 5.13a) displays diffraction peaks consistent with the standard diagram of phase P2-$Na_{0.74}CoO_2$ (PDF No: 01-087-0274). The diffraction peaks are clear, narrow, and high intensity, indicating high crystallinity. However, we observed the additional peaks marked with '*' of impure NiO (the peak position at $2\theta = 37°$, $43°$, and $63°$). Moreover, these peaks having high intensity are almost equivalent to that of the main phase, indicating that Ni does not participate in the structure of the P2-phase but exists individually as doping. This reason may be that NiO is quite inert, so it is difficult to engage in P2- or O3-phase structure under normal conditions due to the difference in the lattice type. The formation of Ni-containing materials is often calcined in a strongly oxidized atmosphere of oxygen. Besides, Figure 5.13b illustrated the patterns of NMC-9012T2 sample, exhibiting the P3-phase belonging to the R3m space group (PDF No: 01-071-1281 as standard) while the intensity of the diffraction peaks of NiO has decreased significantly. Therefore, the sample is annealed at 900°C for 12 hours under double side pressing to limit the loss of Na ions, leading to a decrease in the intensity of the peak of NiO.

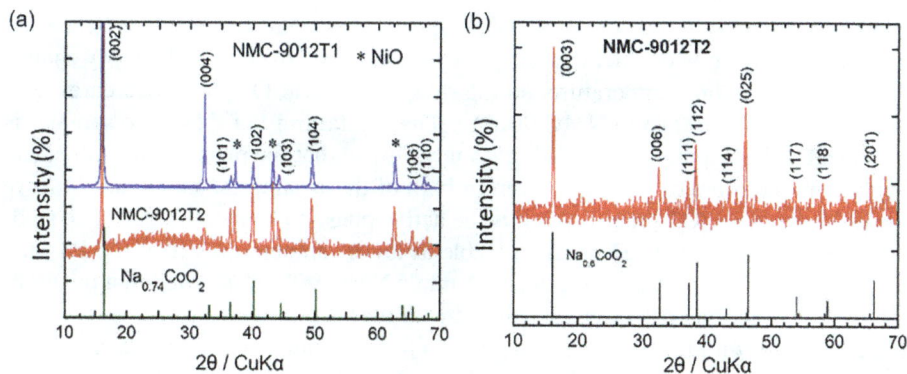

FIGURE 5.13 X-ray diffraction diagram of NMC-9012T1 (a) and NMC-9012T2 (b) samples. (Reproduced with permission.)

The authors continuously investigated the different synthetic conditions of $NaNi_{1/3}Mn_{1/3}Co_{1/3}O_2$ material by the sol-gel method [77]. In this study, the $NaNi_{1/3}Mn_{1/3}Co_{1/3}O_2$ material was decomposed by heating at 300°C for 5 hours before annealing at 900°C for 24 hours. After annealing at 900°C, the pellet sample is annealed without double-side pressed and cooled naturally with Argon gas (NMC-9012T3). Additionally, the sample is annealed under double-sided pressing and rapidly cooled by Argon gas in the pellet sample (NMC-9012T4) and the powder sample (NMC-9012T5). Figure 5.14 showed XRD patterns of synthesized materials at different conditions in order from bottom to top: NMC-9012T3, NMC-9012T4 và NMC-9012T5. NMC-9012T3 and NMC-9012T4 samples displayed diffraction peaks

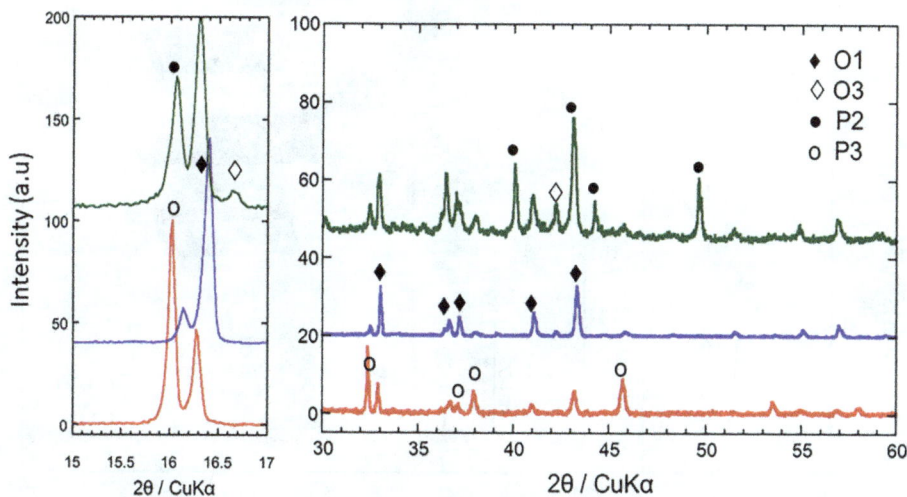

FIGURE 5.14 XRD patterns of synthesized materials at different conditions in order from bottom to top: NMC-9012T3, NMC-9012T4 và NMC-9012T5. (Reproduced with permission [77].)

of phases P3 and O1 marked o and ◊, respectively. It can be seen that when the sample is slowly cooled under the furnace condition, the O1-phase will slowly change to P3-phase at room temperature, so the ratio of P3- and O1-phase decreases as it goes from NMC-9012T3 to NMC-9012T4. The crystallinity of these two samples is improved and compared to the author's study [76] so that the phase composition can be determined more exactly. However, in both of these samples, we cannot observe the presence of the O3-phase, which is the active phase to be synthesized. The studies have shown that the O3-phase is stable at lower temperatures than the P3- and P2-phases [78]. Besides, XRD results of sample NMC-9012T5 showed that the high-intensity diffraction peaks were consistent with the O1 phase structure and other diffraction peaks with lower intensity were also found. Comparing with the previously synthesized samples and the XRD results of the O3 phase [33,74], it is possible to confirm the coexistence of P3- and O3-phase besides O1-phase. Thus, the NMC-9012T6 sample has desired composition of O3-phase although the content is still low. However, the O1-phase is also an active phase, as has been studied previously.

Biphase O3/P2 $NaNi_{1/3}Mn_{1/3}Co_{1/3}O_2$ was synthesized by the sol-gel method by the following two steps: the gel was treated at 400°C for 24 hours and the precursor was heated in a muffle furnace at 900°C for 12 hours [71]. After heating, the material was removed directly from the furnace and transferred into an inert atmosphere glovebox where it was grounded and stored to avoid any reaction with moisture and carbon dioxide. Figure 5.15 presents the XRD results of the composites NaNMC. The patterns of NaNMC powder can be indexed in O3-type (PDF No: 00-054-0887) and P2-type (PDF No: 01-071-1281). Beside the signals of expected phases, we observed the additional peaks marked with '*' of impurity NiO (the peak position at 2θ = 37°, 43°, and 63°). Therefore, the formation of NiO is unavoidable during the synthesis of NaNMC. Based on the peak intensity of Na^+, it can be seen that the O3-phase concentration is

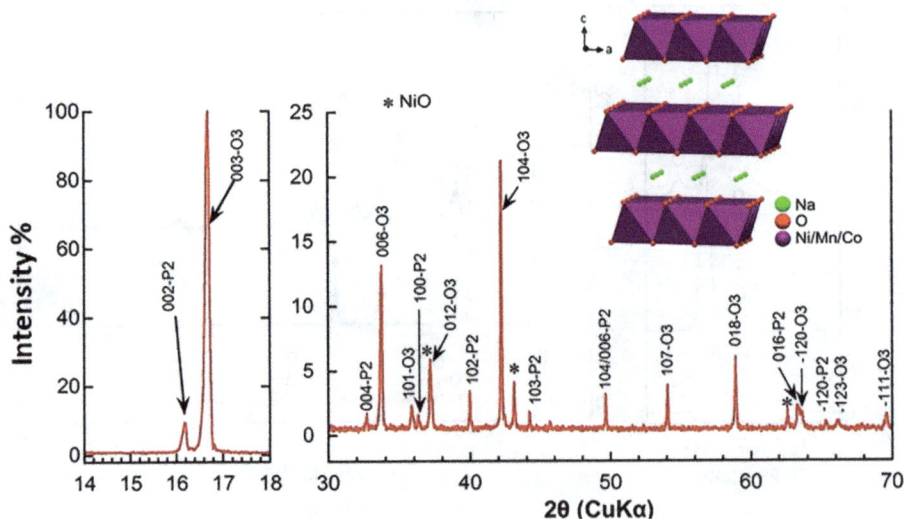

FIGURE 5.15 X-ray diffraction diagram of synthesized biphasic P2/O3 $NaNi_{1/3}Mn_{1/3}Co_{1/3}O_2$. (Reproduced with permission [71].)

higher than the P2-phase. The amount of P2-phase and NiO are relatively low: about 5% based on the relative intensity of the strongest peaks of the main phase. Thus, the synthesized sample has mainly the active component of the O3-phase and the slight P2 amount. Lattice parameters calculated using hexagonal symmetry and R3m space group of O3-phase result in $a = b = 2.9240$ Å; $c = 15.9649$ Å; $V = 118.21$ Å^3.

The electrochemical performance of biphasic P2/O3 NaNi$_{1/3}$Mn$_{1/3}$Co$_{1/3}$O$_2$ is illustrated in Figure 5.16. NaNMC electrode showed similar electrochemical characteristics in the three electrolytes. This resulting shows that NaNMC cathode performed the best in NaClO$_4$/PC + 2% FEC electrolyte in terms of capacity and capacity retention. NaNMC electrode delivers a reversible capacity of 107 mAh g^{-1} when cycling between 4.0 and 2.0 V vs. Na$^+$/Na after the first cycle and the value of 88 mAh g^{-1} after the 100th cycle in NaClO$_4$/PC + 2% FEC electrolyte.

A multiple-phase P2/O1/O3 NaNi$_{1/3}$Mn$_{1/3}$Co$_{1/3}$O$_2$ was synthesized via a sol-gel process [75]. The gel was heated at 300°C for 24 hours to decompose citrate and organic compounds and then calcinated at 900°C for 12 hours. Finally, the sample was taken out of the furnace at 600°C and then immediately transferred into an argon-filled glove box for quenching and storing. XRD patterns in Figure 5.17 reveal the coexistence of different phases in the sol-gel NaNMC sample: at least three intergrowth phases, P2, O1, and O3, were identified under the calcination/quenching condition when using reference data of layered Na$_x$CoO$_2$. An obvious triple-phase coexistence (P2/O3/O1) could be obtained with calcination up to 1,000°C; however, the O3- and O1-phase gradually disappeared when the temperature was reduced. P2 and P3 still dominated in the sample at a lower temperature due to high stability at elevated temperatures and did not induce phase transfer during the cooling step. The lattice parameters and relative percentage of component phases are given in Table 5.1. This result is very helpful for a rational design of the cathode material and unravels the relationship between the phase structures and the electrochemical performance.

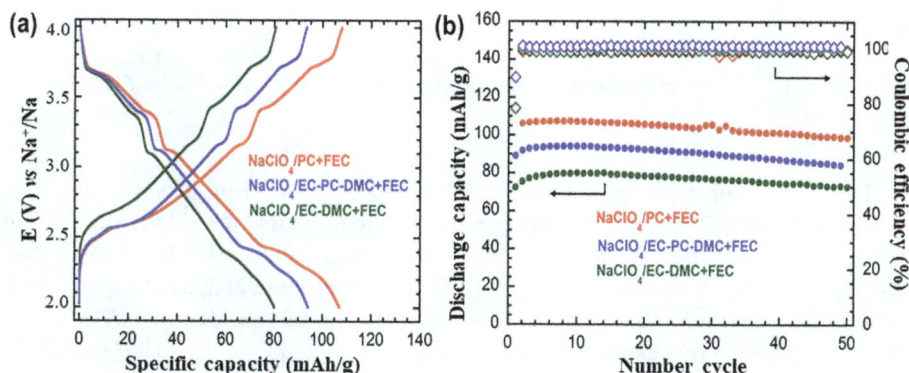

FIGURE 5.16 The initial charge/discharge curves of P2/O3-NaNi$_{1/3}$Mn$_{1/3}$Co$_{1/3}$O$_2$ (a) and Coulombic efficiency with capacity retention for 50 cycles electrodes in different electrolytes (b). (Reproduced with permission [71].)

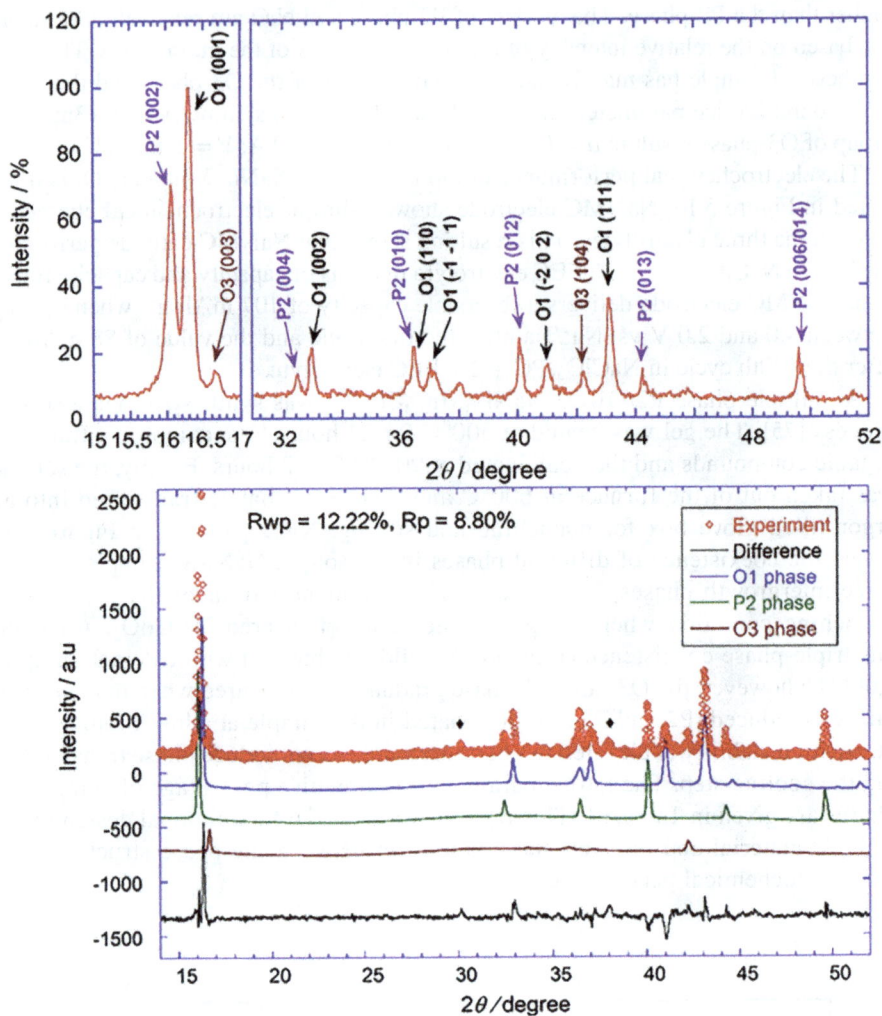

FIGURE 5.17 X-ray diffraction results of synthesized P2/O1/O3- $NaNi_{1/3}Mn_{1/3}Co_{1/3}O_2$. (Reproduced with permission [75].)

The electrochemical properties of P2/O1/O3- $NaNi_{1/3}Mn_{1/3}Co_{1/3}O_2$ are displayed in Figure 5.18. Figure 5.18a consists of many pairs of oxidation-reduction peaks and well-defined phase transitions in the voltage range 2.0–4.0 V vs. Na⁺/Na. Interestingly, the electrochemical activity of our triple-phase NaNMC is quite similar to that of the single-phase NaNMC, as no strange signal was identified. The initial discharge capacity of 135 mAh.g⁻¹ was obtained for triple-phase P2/O1/O3-NaNMC, according to the highest 0.55 ion Na+ incorporated into the structure when cycled between 2–3.85 V. The capacity decrease gradually to 110 mAh.g⁻¹ after 100 cycles which is 80% capacity retention was gained (Figure 5.18b–d). Figure 5.18e exhibits the rate capability of the NaNMC cathode in a voltage range of 2.0–3.85 V at various

TABLE 5.1
Phase Fraction Retrieved from Rietveld Refinement Results of Synthesized P2/O1/O3- NaNi$_{1/3}$Mn$_{1/3}$Co$_{1/3}$O$_2$

Phase ID		Lattice Parameters	Intensity Fraction (%)	Weight Fraction (%)
O1 Phase	C2/m	$a = 4.9768$ Å, $b = 2.9104$ Å, $c = 5.8397$ Å $\alpha = \gamma = 90°$, $\beta = 111.410°$ $V = 78.75$ Å^3 $Z = 2$	67.142	0.04217
P2 Phase	P6$_3$/mmc	$a = b = 2.8457$ Å, $c = 10.8836$ Å $\alpha = \beta = 90°$, $\gamma = 120°$ $V = 76.33$ Å^3 $Z = 2$	27.019	0.04351
O3 Phase	$R\bar{3}m$	$a = b = 2.9254$ Å, $c = 15.9604$ Å $\alpha = \beta = 90°$, $\gamma = 120°$ $V = 118.29$ Å^3 $Z = 3$	5.840	0.04211

Source: Reproduced with permission [75].

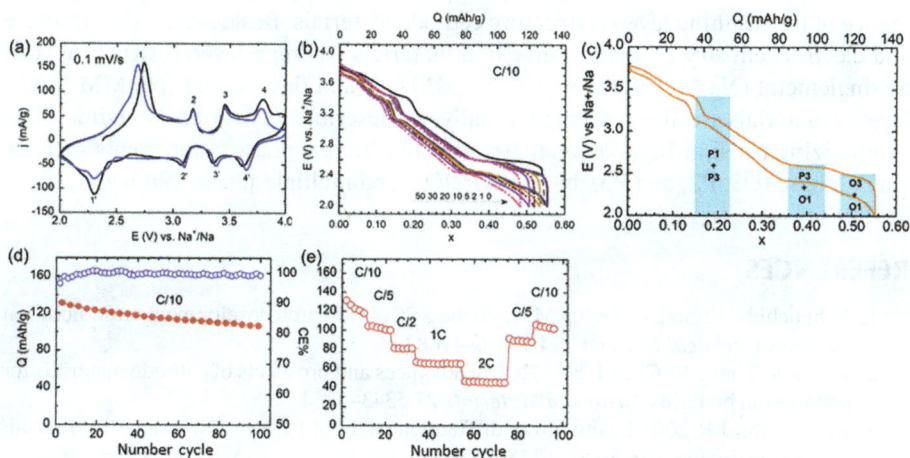

FIGURE 5.18 Cyclic voltammogram of P2/O1/O3-NaNi$_{1/3}$Mn$_{1/3}$Co$_{1/3}$O$_2$ cathode material in NaClO$_4$ 1 M/PC + 2% FEC electrolyte at a scan rate of 100 μV s^{-1} (a); The voltage composition profile of composite material (b); Phase transformation of material occurred during cycling (c); The stability of capacity during cycling at C/10 (d) and the rate capability of a sample at different rates (e). (Reproduce with permission [75].)

current densities ranging from 0.1 to 2 C (10 cycles each). When the rate increased two-fold, the capacity decreased. At the highest rate (2 C), the discharge capacity was only about 40 mAh g^{-1}. In general, the performance of multiple-phase NaNMC was expectedly as good as the single-phase.

5.5 CONCLUDING REMARKS

In this review, the applications, and electrochemical performances of layered cathode materials for SIBs have been systematically summarized. The structures of two common sodium layered metal oxides are O3-type and P2-type. During electrochemical cycling, both P2 and O3 layered phases commonly suffer from a series of phase transitions induced by the extraction of Na ions. Therefore, issues and strategies for phase transition suppression listed and presented in this chapter bring fundamental comprehension of transition mechanisms in each different material. Recently, rapid progress and development in spectroscopic, microscopic, and scattering techniques have provided extensive insight into the nature of the structure evolution and phase transition in battery materials. In this chapter, we also provide a comprehensive overview of both static (ex situ) and real-time (in situ) techniques for in-depth studying of the structure evolution and phase transition of layered structure materials for SIBs.

Since the design and capacity of most SIBs are cathode limited, the key strategy is the controlled design of cathode materials to enhance performance. Hence, we discuss the different synthesis methods utilized for powder cathode materials such as solid-state reaction, sol-gel, and co-precipitation; the latter method has proved to be useful for layered structure material development. Moreover, we emphasize how these methods affect terms of discharge capacity, capacity retention, and rate performance of the resulting layered structure cathode materials. Besides that, the structure and electrochemistry of sodium insertion materials having a layered structure such as singlemetal (Na_xMO_2), two-metal ($Na_xMM'O_2$), and three-metal ($Na_xMM'M''O_2$) layered materials are also presented. Finally, we described in detail the conditions for synthesizing $NaNi_{1/3}Mn_{1/3}Co_{1/3}O_2$ material with different phase components such as: single-phase (O3, P2, and P3), bi-phase P2/O3, and multiple-phase P2/O1/O3.

REFERENCES

1. Yabuuchi N, Kubota K, Dahbi M, Komaba S 2014 Research development on sodium-ion batteries *Chemical Reviews* **114** 11636–11682.
2. Xiang X, Zhang K, Chen J 2015 Recent advances and prospects of cathode materials for sodium-ion batteries *Advanced Materials* **27** 5343–5364.
3. Lu Z, Dahn J R 2001 In situ X-ray diffraction study of P2 $Na_{2/3}[Ni_{1/3}Mn_{2/3}]O_2$ *Journal of the Electrochemical Society* **148** A1225.
4. Wen Y, Wang B, Zeng G, Nogita K, Ye D, Wang L 2015 Electrochemical and structural study of layered P2-type $Na_{2/3}Ni_{1/3}Mn_{2/3}O_2$ as cathode material for sodium-ion battery *Chemistry-An Asian Journal* **10** 661–666.
5. Lee D H, Xu J, Meng Y S 2013 An advanced cathode for Na-ion batteries with high rate and excellent structural stability *Physical Chemistry Chemical Physics* **15** 3304–3312.

6. Yabuuchi N, Kajiyama M, Iwatate J, Nishikawa H, Hitomi S, Okuyama R, Usui R, Yamada Y, Komaba S 2012 P2-type $Na_x[Fe_{1/2}Mn_{1/2}]O_2$ made from earth-abundant elements for rechargeable Na batteries *Nature Materials* **11** 512–517.
7. Yoshida H, Yabuuchi N, Komaba S 2013 $NaFe_{0.5}Co_{0.5}O_2$ as high energy and power positive electrode for Na-ion batteries *Electrochemistry Communications* **34** 60–63.
8. Xiong F, An Q, Xia L, Zhao Y, Mai L, Tao H, Yue Y 2019 Revealing the atomistic origin of the disorder-enhanced Na-storage performance in $NaFePO_4$ battery cathode *Nano Energy* **57** 608–615.
9. Ko J S, Paul P P, Wan G, Seitzman N, DeBlock R H, Dunn B S, Toney M F, Nelson Weker J 2020 NASICON $Na_3V_2(PO_4)_3$ enables quasi-two-stage Na^+ and Zn^{2+} intercalation for multivalent zinc batteries *Chemistry of Materials* **32** 3028–3035.
10. Zuo D, Wang C, Han J, Wu J, Qiu H, Zhang Q, Lu Y, Lin Y, Liu X 2020 Oriented formation of a Prussian blue nanoflower as a high performance cathode for sodium ion batteries *ACS Sustainable Chemistry Engineering* **8** 16229–16240.
11. Kang H, Liu Y, Cao K, Zhao Y, Jiao L, Wang Y, Yuan H 2015 Update on anode materials for Na-ion batteries *Journal of Materials Chemistry A* **3** 17899–17913.
12. Wang P-F, Yao H-R, Zuo T-T, Yin Y-X, Guo Y-G 2017 Novel P2-type $Na_{2/3}Ni_{1/6}Mg_{1/6}Ti_{2/3}O_2$ as an anode material for sodium-ion batteries *Chemical Communications* **53** 1957–1960.
13. Delmas C, Fouassier C, Hagenmuller P 1980 Structural classification and properties of the layered oxides *Physica B+C* **99** 81–85.
14. Wang P F, You Y, Yin Y X, Guo Y G 2018 Layered oxide cathodes for sodium-ion batteries: phase transition, air stability, and performance *Advanced Energy Materials* **8** 1701912.
15. Yabuuchi N, Yoshida H, Komaba S 2012 Crystal structures and electrode performance of alpha-$NaFeO_2$ for rechargeable sodium batteries *Electrochemistry* **80** 716–719.
16. Komaba S, Takei C, Nakayama T, Ogata A, Yabuuchi N 2010 Electrochemical intercalation activity of layered $NaCrO_2$ vs. $LiCrO_2$ *Electrochemistry Communications* **12** 355–358.
17. Hamani D, Ati M, Tarascon J-M, Rozier P 2011 Na_xVO_2 as possible electrode for Na-ion batteries *Electrochemistry Communications* **13** 938–941.
18. Vassilaras P, Ma X, Li X, Ceder G 2012 Electrochemical properties of monoclinic $NaNiO_2$ *Journal of The Electrochemical Society* **160** A207.
19. Ma X, Chen H, Ceder G 2011 Electrochemical properties of monoclinic $NaMnO_2$ *Journal of The Electrochemical Society* **158** A1307.
20. Delmas C, Braconnier J-J, Fouassier C, Hagenmuller P 1981 Electrochemical intercalation of sodium in Na_xCoO_2 bronzes *Solid State Ionics* **3** 165–169.
21. Komaba S, Yabuuchi N, Nakayama T, Ogata A, Ishikawa T, Nakai I 2012 Study on the reversible electrode reaction of $Na_{1-x}Ni_{0.5}Mn_{0.5}O_2$ for a rechargeable sodium-ion battery *Inorganic Chemistry* **51** 6211–6220.
22. Zhao C, Ding F, Lu Y, Chen L, Hu Y S 2020 High-entropy layered oxide cathodes for sodium-ion batteries *Angewandte Chemie International Edition* **59** 264–269.
23. Yao H-R, Lv W-J, Yin Y-X, Ye H, Wu X-W, Wang Y, Gong Y, Li Q, Yu X, Gu L 2019 Suppression of monoclinic phase transitions of O3-type cathodes based on electronic delocalization for Na-ion batteries *ACS Applied Materials Interfaces* **11** 22067–22073.
24. Wang P F, Yao H R, Liu X Y, Zhang J N, Gu L, Yu X Q, Yin Y X, Guo Y G 2017 Ti-substituted $NaNi_{0.5}Mn_{0.5-x}Ti_xO_2$ cathodes with reversible O3– P3 phase transition for high-performance sodium-ion batteries *Advanced Materials* **29** 1700210.
25. Mo Y, Ong S P, Ceder G 2014 Insights into diffusion mechanisms in P2 layered oxide materials by first-principles calculations *Chemistry of Materials* **26** 5208–5214.
26. Liu L, Li X, Bo S H, Wang Y, Chen H, Twu N, Wu D, Ceder G 2015 High-performance P2-type $Na_{2/3}(Mn_{1/2}Fe_{1/4}Co_{1/4})O_2$ cathode material with superior rate capability for Na-ion batteries *Advanced Energy Materials* **5** 1500944.

27. Wang Q-C, Meng J-K, Yue X-Y, Qiu Q-Q, Song Y, Wu X-J, Fu Z-W, Xia Y-Y, Shadike Z, Wu J 2018 Tuning P2-structured cathode material by Na-site Mg substitution for Na-ion batteries *Journal of the American Chemical Society* **141** 840–848.

28. Wang L, Sun Y-G, Hu L-L, Piao J-Y, Guo J, Manthiram A, Ma J, Cao A-M 2017 Copper-substituted $Na_{0.67}Ni_{0.3-x}Cu_xMn_{0.7}O_2$ cathode materials for sodium-ion batteries with suppressed P2-O2 phase transition *Journal of Materials Chemistry A* **5** 8752–8761.

29. Zhou P, Che Z, Ma F, Zhang J, Weng J, Wu X, Miao Z, Lin H, Zhou J, Zhuo S 2021 Designing water/air-stable P2-layered cathodes with delayed P2–O2 phase transition by composition and structure engineering for sodium-ion batteries at high voltage *Chemical Engineering Journal* **420** 127667.

30. Liu H, Gao X, Chen J, Gao J, Yin S, Zhang S, Yang L, Fang S, Mei Y, Xiao X 2021 Reversible OP4 phase in P2–$Na_{2/3}Ni_{1/3}Mn_{2/3}O_2$ sodium ion cathode *Journal of Power Sources* **508** 230324.

31. Hwang J Y, Kim J, Yu T Y, Sun Y K 2019 A new P2-type layered oxide cathode with extremely high energy density for sodium-ion batteries *Advanced Energy Materials* **9** 1803346.

32. West A R. *Solid State Chemistry and Its Applications*: John Wiley & Sons; 2014.

33. Hwang J-Y, Yoon C S, Belharouak I, Sun Y-K 2016 A comprehensive study of the role of transition metals in O3-type layered $Na[Ni_xCo_yMn_z]O_2$ (x= 1/3, 0.5, 0.6, and 0.8) cathodes for sodium-ion batteries *Journal of Materials Chemistry A* **4** 17952–17959.

34. Gupta S, Tripathi M 2012 A review on the synthesis of TiO_2 nanoparticles by solution route *Open Chemistry* **10** 279–294.

35. Tang K, Huang Y, Xie X, Cao S, Liu L, Liu M, Huang Y, Chang B, Luo Z, Wang X 2020 The effects of dual modification on structure and performance of P2-type layered oxide cathode for sodium-ion batteries *Chemical Engineering Journal* **384** 123234.

36. Chen X, Zhou X, Hu M, Liang J, Wu D, Wei J, Zhou Z 2015 Stable layered P3/P2 $Na_{0.66}Co_{0.5}Mn_{0.5}O_2$ cathode materials for sodium-ion batteries *Journal of Materials Chemistry A* **3** 20708–20714.

37. Zhu Y-E, Qi X, Chen X, Zhou X, Zhang X, Wei J, Hu Y, Zhou Z 2016 A P2-$Na_{0.67}Co_{0.5}Mn_{0.5}O_2$ cathode material with excellent rate capability and cycling stability for sodium ion batteries *Journal of Materials Chemistry A* **4** 11103–11109.

38. Yuan D, He W, Pei F, Wu F, Wu Y, Qian J, Cao Y, Ai X, Yang H 2013 Synthesis and electrochemical behaviors of layered $Na_{0.67}[Mn_{0.65}Co_{0.2}Ni_{0.15}]O_2$ microflakes as a stable cathode material for sodium-ion batteries *Journal of Materials Chemistry A* **1** 3895–3899.

39. Sathiya M, Hemalatha K, Ramesha K, Tarascon J-M, Prakash A 2012 Synthesis, structure, and electrochemical properties of the layered sodium insertion cathode material: $NaNi_{1/3}Mn_{1/3}Co_{1/3}O_2$ *Chemistry of Materials* **24** 1846–1853.

40. Kaliyappan K, Liu J, Lushington A, Li R, Sun X 2015 Highly stable $Na_{2/3}(Mn_{0.54}Ni_{0.13}Co_{0.13})O_2$ cathode modified by atomic layer deposition for sodium-ion batteries *ChemSusChem* **8** 2537–2543.

41. Karthikeyan K, Amaresh S, Kim S H, Aravindan V, Lee Y S 2013 Influence of synthesis technique on the structural and electrochemical properties of "cobalt-free", layered type $Li_{1+x}(Mn_{0.4}Ni_{0.4}Fe_{0.2})_{1-x}O_2$ (0<x<0.4) cathode material for lithium secondary battery *Electrochimica Acta* **108** 749–756.

42. Karthikeyan K, Amaresh S, Lee G W, Aravindan V, Kim H, Kang K S, Kim W-S, Lee Y-S 2012 Electrochemical performance of cobalt free, $Li_{1.2}(Mn_{0.32}Ni_{0.32}Fe_{0.16})O_2$ cathodes for lithium batteries *Electrochimica Acta* **68** 246–253.

43. Hou P, Zhang H, Zi Z, Zhang L, Xu X 2017 Core-shell and concentration-gradient cathodes prepared via co-precipitation reaction for advanced lithium-ion batteries *Journal of Materials Chemistry A* **5** 4254–4279.

44. Kjellman T, Alfredsson V 2013 The use of in situ and ex situ techniques for the study of the formation mechanism of mesoporous silica formed with non-ionic triblock copolymers *Chemical Society Reviews* **42** 3777–3791.

45. Zhang J, Wang W, Wang W, Wang S, Li B 2019 Comprehensive review of P2-type $Na_{2/3}Ni_{1/3}Mn_{2/3}O_2$, a potential cathode for practical application of Na-ion batteries *ACS Applied Materials Interfaces* **11** 22051–22066.

46. Lee E, Brown D E, Alp E E, Ren Y, Lu J, Woo J-J, Johnson C S 2015 New insights into the performance degradation of Fe-based layered oxides in sodium-ion batteries: Instability of Fe^{3+}/Fe^{4+} redox in α-$NaFeO_2$ *Chemistry of Materials* **27** 6755–6764.

47. Sun Y, Guo S, Zhou H 2019 Adverse effects of interlayer-gliding in layered transition-metal oxides on electrochemical sodium-ion storage *Energy Environmental Science* **12** 825–840.

48. Mendiboure A, Delmas C, Hagenmuller P 1985 Electrochemical intercalation and deintercalation of Na_xMnO_2 bronzes *Journal of Solid State Chemistry* **57** 323–331.

49. Ding J J, Zhou Y N, Sun Q, Yu X Q, Yang X Q, Fu Z W 2013 Electrochemical properties of P2-phase $Na_{0.74}CoO_2$ compounds as cathode material for rechargeable sodium-ion batteries *Electrochimica Acta* **87** 388–393.

50. Rai A K, Anh L T, Gim J, Mathew V, Kim J 2014 Electrochemical properties of Na_xCoO_2 (x~ 0.71) cathode for rechargeable sodium-ion batteries *Ceramics International* **40** 2411–2417.

51. Manthiram A, Chemelewski K, Lee E-S 2014 A perspective on the high-voltage $LiMn_{1.5}Ni_{0.5}O_4$ spinel cathode for lithium-ion batteries *Energy Environmental Science* **7** 1339–1350.

52. Kalapsazova M, Stoyanova R, Zhecheva E, Tyuliev G, Nihtianova D 2014 Sodium deficient nickel–manganese oxides as intercalation electrodes in lithium ion batteries *Journal of Materials Chemistry A* **2** 19383–19395.

53. Kalapsazova M L, Zhecheva E N, Tyuliev G T, Nihtianova D D, Mihaylov L, Stoyanova R K 2017 Effects of the particle size distribution and of the electrolyte salt on the intercalation properties of P3-$Na_{2/3}Ni_{1/2}Mn_{1/2}O_2$ *The Journal of Physical Chemistry C* **121** 5931–5940.

54. Wang P-F, You Y, Yin Y-X, Guo Y-G 2016 An O3-type $NaNi_{0.5}Mn_{0.5}O_2$ cathode for sodium-ion batteries with improved rate performance and cycling stability *Journal of Materials Chemistry A* **4** 17660–17664.

55. Zhang X-H, Pang W-L, Wan F, Guo J-Z, Lü H-Y, Li J-Y, Xing Y-M, Zhang J-P, Wu X-L 2016 P2–$Na_{2/3}Ni_{1/3}Mn_{5/9}Al_{1/9}O_2$ microparticles as superior cathode material for sodium-ion batteries: Enhanced properties and mechanism via graphene connection *ACS Applied Materials Interfaces* **8** 20650–20659.

56. Singh G, Tapia-Ruiz N, López del Amo J M, Maitra U, Somerville J W, Armstrong A R, Martinez de Ilarduya J, Rojo T, Bruce P G 2016 High voltage Mg-doped $Na_{0.67}Ni_{0.3-x}Mg_xMn_{0.7}O_2$ (x= 0.05, 0.1) Na-ion cathodes with enhanced stability and rate capability *Chemistry of Materials* **28** 5087–5094.

57. Wang P F, You Y, Yin Y X, Wang Y S, Wan L J, Gu L, Guo Y G 2016 Suppressing the P2–O2 phase transition of $Na_{0.67}Mn_{0.67}Ni_{0.33}O_2$ by magnesium substitution for improved sodium-ion batteries *Angewandte Chemie* **128** 7571–7575.

58. Wu X, Guo J, Wang D, Zhong G, McDonald M J, Yang Y 2015 P2-type $Na_{0.66}Ni_{0.33-x}Zn_xMn_{0.67}O_2$ as new high-voltage cathode materials for sodium-ion batteries *Journal of Power Sources* **281** 18–26.

59. Yoshida H, Yabuuchi N, Kubota K, Ikeuchi I, Garsuch A, Schulz-Dobrick M, Komaba S 2014 P2-type $Na_{2/3}Ni_{1/3}Mn_{2/3-x}Ti_xO_2$ as a new positive electrode for higher energy Na-ion batteries *Chemical Communications* **50** 3677–3680.

60. Clément R J, Xu J, Middlemiss D S, Alvarado J, Ma C, Meng Y S, Grey C P 2017 Direct evidence for high Na+ mobility and high voltage structural processes in P2-Na$_x$[Li$_y$Ni$_z$Mn$_{1-y-z}$]O$_2$ (x, y, z ≤ 1) cathodes from solid-state NMR and DFT calculations *Journal of Materials Chemistry A* **5** 4129–4143.
61. Xu J, Lee D H, Clément R J, Yu X, Leskes M, Pell A J, Pintacuda G, Yang X-Q, Grey C P, Meng Y S 2014 Identifying the critical role of Li substitution in P2-Na$_x$[Li$_y$Ni$_z$Mn$_{1-y-z}$]O$_2$ (0 < x, y, z < 1) intercalation cathode materials for high-energy Na-ion batteries *Chemistry of Materials* **26** 1260–1269.
62. Yao H-R, Wang P-F, Gong Y, Zhang J, Yu X, Gu L, OuYang C, Yin Y-X, Hu E, Yang X-Q 2017 Designing air-stable O3-type cathode materials by combined structure modulation for Na-ion batteries *Journal of the American Chemical Society* **139** 8440–8443.
63. Wang X, Liu G, Iwao T, Okubo M, Yamada A 2014 Role of ligand-to-metal charge transfer in O3-type NaFeO$_2$–NaNiO$_2$ solid solution for enhanced electrochemical properties *The Journal of Physical Chemistry C* **118** 2970–2976.
64. Kubota K, Asari T, Yoshida H, Yaabuuchi N, Shiiba H, Nakayama M, Komaba S 2016 Understanding the structural evolution and redox mechanism of a NaFeO$_2$-NaCoO$_2$ solid solution for sodium-ion batteries *Advanced Functional Materials* **26** 6047–6059.
65. Yao H R, Wang P F, Wang Y, Yu X, Yin Y X, Guo Y G 2017 Excellent comprehensive performance of Na-based layered oxide benefiting from the synergetic contributions of multimetal ions *Advanced Energy Materials* **7** 1700189.
66. Le Nguyen M, Tran H P, Tran N T, Le M L P 2021 Cu-doped NaCu$_{0.05}$Fe$_{0.45}$Co$_{0.5}$O$_2$ as promising cathode material for Na-ion batteries: Synthesis and characterization *Journal of Solid State Electrochemistry* **25** 767–775.
67. Wang X, Tamaru M, Okubo M, Yamada A 2013 Electrode properties of P2–Na$_{2/3}$Mn$_y$Co$_{1-y}$O$_2$ as cathode materials for sodium-ion batteries *The Journal of Physical Chemistry C* **117** 15545–15551.
68. Sharma N, Al Bahri O K, Han M H, Gonzalo E, Pramudita J C, Rojo T 2016 Comparison of the structural evolution of the O3 and P2 phases of Na$_{2/3}$Fe$_{2/3}$Mn$_{1/3}$O$_2$ during electrochemical cycling *Electrochimica Acta* **203** 189–197.
69. Gonzalo E, Han M H, Del Amo J M L, Acebedo B, Casas-Cabanas M, Rojo T 2014 Synthesis and characterization of pure P2-and O3-Na$_{2/3}$Fe$_{2/3}$Mn$_{1/3}$O$_2$ as cathode materials for Na ion batteries *Journal of Materials Chemistry A* **2** 18523–18530.
70. Han M H, Acebedo B, Gonzalo E, Fontecoba P S, Clarke S, Saurel D, Rojo T 2015 Synthesis and electrochemistry study of P2-and O3-phase Na$_{2/3}$Fe$_{1/2}$Mn$_{1/2}$O$_2$ *Electrochimica Acta* **182** 1029–1036.
71. Van Nguyen H, Le Nguyen M, Van Tran M, Tran N T, Le P M L 2020 Performance of full-cell Na-ion with NaNi$_{1/3}$Mn$_{1/3}$Co$_{1/3}$O$_2$ cathode material and different carbonate-based electrolytes *Science Technology Development Journal-Natural Sciences* **4** 744–752.
72. Yang P, Zhang C, Li M, Yang X, Wang C, Bie X, Wei Y, Chen G, Du F 2015 P2-NaCo$_{0.5}$Mn$_{0.5}$O$_2$ as a positive electrode material for sodium-ion batteries *ChemPhysChem* **16** 3408–3412.
73. Manikandan P, Heo S, Kim H W, Jeong H Y, Lee E, Kim Y 2017 Structural characterization of layered Na$_{0.5}$Co$_{0.5}$Mn$_{0.5}$O$_2$ material as a promising cathode for sodium-ion batteries *Journal of Power Sources* **363** 442–449.
74. Xu G-L, Amine R, Xu Y-F, Liu J, Gim J, Ma T, Ren Y, Sun C-J, Liu Y, Zhang X 2017 Insights into the structural effects of layered cathode materials for high voltage sodium-ion batteries *Energy Environmental Science* **10** 1677–1693.
75. Van Nguyen H, Nguyen H T N, Huynh N L T, Phan A L B, Van Tran M, Le P M L 2020 A study of the electrochemical kinetics of sodium intercalation in P2/O1/O3-NaNi$_{1/3}$Mn$_{1/3}$Co$_{1/3}$O$_2$ *Journal of Solid State Electrochemistry* **24** 57–67.

76. Nguyễn V H, Nguyễn T T L, Huỳnh L T N, Lê M L P, Trần V M 2017 Tính chất điện hóa của vật liệu NaNi$_{1/3}$Mn$_{1/3}$Co$_{1/3}$O$_2$ tổng hợp bằng phương pháp sol-gel *Vietnam Journal of Chemistry* **55** 105–109.

77. Nguyen V H, Nguyen T N H, Huynh L T N, Tran V M, Le M L P 2018 Sol-gel NaNi$_{1/3}$Mn$_{1/3}$Co$_{1/3}$O$_2$ as potential cathode material for Na-ion batteries: Effect of cooling process on structure and electrochemical properties *Vietnam Journal of Chemistry* **56** 484–490.

78. Kubota K, Kumakura S, Yoda Y, Kuroki K, Komaba S 2018 Electrochemistry and solid-state chemistry of NaMeO$_2$ (Me= 3d transition metals) *Advanced Energy Materials* **8** 1703415.

6 Essential Geometric and Electronic Properties in Stage-n FeCl₃-Graphite Intercalation Compounds

Wei-Bang Li
National Cheng Kung University

Shih-Yang Lin
National Cheng Kung University
National Chung Cheng University

Ming-Fa Lin
National Cheng Kung University

CONTENTS

6.1 INTRODUCTION

Up to now, the 3D graphite [1] systems, regarded as the layered graphene systems [2–6], have drawn researchers' attention in basic science [3,4,6–8], engineering [7], and application [8]. Each graphene layer is formed by a pure carbon honeycomb lattice [2]; for further steps, the graphitic layers are interacted with each other and are attracted together through the weak, but significant van der Waals interactions [9]. The very strong intralayer σ bondings of C-[2s, 2px, 2py] orbitals and the important interlayer C-2pz orbital hybridizations exist in graphene layers: the former determine the most outstanding mechanical material, and the latter dominate the low-lying energy bands and thus the essential physical properties. Very interestingly, the 3D graphite systems might present the AA [10], AB [11], and ABC [12]. All of them

DOI: 10.1201/9781003263807-6

belong to semimetals under the interlayer hopping integrals of C-2pz orbitals. In general, such condensed-matter systems become the n- or p-type metals, depending on the kinds of intercalated atoms or molecules. For example, the intercalation of alkali atoms [13] and $FeCl_3$ [14] into graphite create many free conduction electrons and valence holes, respectively as observed in pure metal, in other words, the former becomes n-type metal and the latter becomes p-type metal.

It is well known that graphite could serve as good electrode materials [15] in the commercialized Li+ based batteries [16–20] due to the lowest cost, the most stable structure for intercalation, and the outstanding ion transport under the charging and discharging processes. The Lithium-ion batteries (LIBs) are important rechargeable batteries and promising for various devices, e.g., cell phones, vehicles, and almost all portable equipment, because of the reversible intercalation/de-intercalation of Li+ into/from the host lattices of either graphite-layered anodes or other layer-based cathodes. When the Li+ ions are released from the cathode, they will transport through the electrolyte and then intercalate into the graphitic system. Most importantly, the flexible interlayer spacings between graphene layers are capable of providing sufficient positions for the various intercalant concentrations [21]. Any intermediate states and meta-stable configurations that are created during the ion/atom intercalation, could survive through the very strong σ bondings in graphitic sheets. As a result, graphite is rather suitable for studying the structural transformation in chemical reactions. For example, the close relations between graphene and intercalant layers in lattice symmetries are expected to present the dramatic transformation before and after the intercalation/de-intercalation processes. Recently, many research papers mainly focused on the distinguished guest atoms/ions like Li, Na, K [22–24], or ion clusters like $AlCl_4$, Al_2Cl_7 [25–27]; also, the battery capacity was gained with ferric chloride–graphite intercalation compounds ($FeCl_3$-GICs) [28,29].

According to the previous work [30], the restacking of graphenes will lower the energy storage in batteries. Therefore, how to enlarge the interlayer distances between graphenes with proper intercalants is a very vital issue. The Fe (iron) plays one potential candidate of anode materials in batteries because of the physical and chemical properties and the great amount of roughly 5% on earth [the top five elements are O, Si, Al, Fe, and Ca with the amount of 48%, 26%, 8%, 5%, and 3%, respectively]; in other words, it is relatively cheap compared to other metals. The $FeCl_3$-GIC is found to possess a high redox potential, reversible capacity, cycling stability, and an outstanding rate, demonstrating its potential as electrode material in LIB. The theoretical value of the capacity of graphite is roughly 372 mAh g^{-1}; however, the $FeCl_3$-GIC exhibits a much higher capacity [28]. Furthermore, based on the in-plane structural point of view, the GICs belong to the class of metal chlorides, and it is characterized by a two-dimensional intercalant with layers. The $FeCl_3$-GICs have attracted many theoretical and experimental works. Most of them are focused on the electrochemical test for stage-1 and stage-2 and how to improve their electrochemical performances.

As an electrode for Lithium-ion batteries, stage-1 $FeCl_3$-GIC possesses reversible capacities of 525 mAh g^{-1} after 100 charge-discharge cycles [29]. This is considerably higher than the capacity for expanded graphite, which is 113 mAh g^{-1}. Moreover, the in-plane conductivity at room temperature for stage-1 $FeCl_3$-GIC is 1.1×10^5 Ω cm^{-1}, which is higher than the 2.5×10^4 Ω cm^{-1} in graphite. Not only that, but the stage-2

$FeCl_3$-GIC also exhibits a much higher reversible capacity with a specific capacity of 719 mAh g^{-1} after 100 charge-discharge cycles, and even 910 mAh g^{-1} after 170 charge-discharge cycles. The special and important performance of $FeCl_3$-GICs undoubtedly results from the unique geometric structure of intercalations and the stage index; in other words, the interlayer spacings of graphite are enlarged because of the intercalations of $FeCl_3$, and the enlarged spacings provide larger channels for the electrochemical reaction while promoting fast Lithium-ion transport.

Up to now, there are many kinds of theoretical and experimental research on the essential properties in graphite-related systems. The former covers the phenomenological models and number simulation methods. For instance, the tight-binding model, the random-phase approximation, and the Kubo formula are, respectively, utilized to investigate their electronic properties & magnetic quantization behaviors [31–34], Coulomb excitations [6,35–38] and impurity screenings [39–41], and optical absorption spectra [42]. Moreover, the first-principles calculations [43] are available in understanding the optimal stacking configurations, intercalant lattices, π, σ and intercalation-induced energy bands, free conduction electron/valence hole densities, and density of states. However, certain critical physical quantities and pictures are absent in the previous studies, such as the atom-dominated band structures, the spatial charge densities between intercalant and graphene layers, the interlayer orbital hybridization of intercalant and carbon atom, the atom- & orbital-projected van Hove singularities, and the spin-dependent state degeneracy/magnetic moment/ density distribution.

This chapter will focus on the rich features of geometric and electronic properties in stage-n $FeCl_3$ graphite intercalation compounds using the first-principles method. Very interestingly, the dependences on the kinds of intercalants and their concentrations are thoroughly explored through delicate calculations and analyses. The stacking configurations of graphene sheets, the lattice symmetries of intercalant layers, the dominances of valence and conduction states by the carbon and/or intercalants $FeCl_3$, the interlayer charge density variations after the $FeCl_3$ intercalations, and atom- and orbital-decomposed density of states. Most important, the orbital-hybridizations in chemical bonds and the atom-induced spin configurations are determined from the abovementioned results. How to simulate VASP band structures by the tight-binding model is worthy of detailed discussion, since the interlayer hopping integrals are not very complicated under the small primitive unit cells of layered graphite compounds. The predicted interlayer distance/intercalant lattice, band overlap, and unusual van Hove singularities could be detected using the high-resolution X-ray diffraction/low-energy electron diffraction [44], angle-resolved photoemission spectroscopy [45], and scanning tunneling spectroscopy [46], respectively.

6.2 THEORETICAL CALCULATIONS

The first-principle density functional theory (DFT) [47–50] calculations are used to investigate the optimal geometric structures, charge distribution, electronic magnetic, and optical characteristics of intercalation compounds. In this chapter, we performed the calculations by Vienna ab initio simulation package (VASP), which evaluates an approximate solution within the DFT by solving the Kohn-Sham equations. The

Perdew-Burke-Ernzerh formula used in VASP, which depends on the local electron density, is utilized to deal with many-particle Coulomb effects. As for the frequent electron-crystal scatterings, they are characterized by the projector-augmented wave pseudopotentials. The electron Bloch wave functions are solved from the linear superposition of plane waves, with the maximum kinetic energy of 500 eV. The current study on stage-n graphite compounds shows that the first Brillouin zone is sampled by $9 \times 9 \times 9$ and $100 \times 100 \times 100$ k-point meshes within the Monkhorst-Pack scheme for the optimal geometry and band structure, respectively. Moreover, the convergence condition of the ground state energy is set to be $\sim 10^{-5}$ eV between two consecutive evaluation steps, where the maximum Hellmann–Feynman force for each ion is below 0.01 eV/Å during the atom relaxations.

The calculated results include the ground state energy, intralayer distance, bond length, planar honeycomb lattice, position and height of intercalants, intercalant-induced energy gap and free carriers, spatial charge distribution, and density of states (DOS). The hybridization of atomic orbitals of carbons and intercalants could be analyzed and clearly identified from the spatial charge distributions, variation of charge distributions, energy bands, and orbital-projected density of states. Such chemical bondings play an important role in the essential properties; in other words, we can explain the rich geometric structures and electronic properties of graphite intercalation compounds.

6.3 THE STAGE-n CRYSTAL STRUCTURES

There exists weak but significant van der Waals interaction between graphitic sheets. The pristine layered distance between neighboring layers is roughly 3.35 and 3.55 Å for AB-stacking and AA-stacking configurations, respectively [51,52]. The flexible spacing can be intercalated by atom, ion, or molecule with rich configurations of stacking and different concentrations [53]. The interlayer distances mightily depend on the species of intercalants, e.g., the distances are lengthened to 3.76 and 5.65 Å for Li and Rb alkali-metal atoms, respectively. However, the bond length between carbons in hexagonal structures is only slightly changed.

The crystal structures of stage-n $FeCl_3$-graphite intercalation compounds possess periodical arrangements of (x, y)-planes along the z-direction [54]. The Moire superlattices of stage-1, stage-2, stage-3, and stage-4 systems have 82, 132, 182, and 232 atoms, respectively, in an enlarged unit cell, shown in Figure 6.1. Specifically, only the stage-1 system displays the AA-stacking configuration, while other systems exhibit AB-stacking configurations. Such a situation has been performed in the previous study, e.g., the stage-n alkali-metal graphite intercalation compounds [53]. The VASP simulation is able to optimize the crystal symmetries. The high-precision results include total ground state energy per unit cell, the bond lengths of C-C/C-Cl/ Fe-Cl/ Cl-Cl, the interlayer distances, and spatial charge distributions.

For the stage-1 $FeCl_3$-graphite intercalation compound, the supercell with lattice constants $a = b = 12.12$ Å, which is constructed by combining 2×2 unit cells of $FeCl_3$ and 5×5 unit cells of graphene as mentioned in previous works [55,56], covers 50 carbon atoms, 8 iron atoms, and 24 chlorine atoms, due to the high symmetry, and is quite big. In addition, the $FeCl_3$ also forms layered structures between the carbon

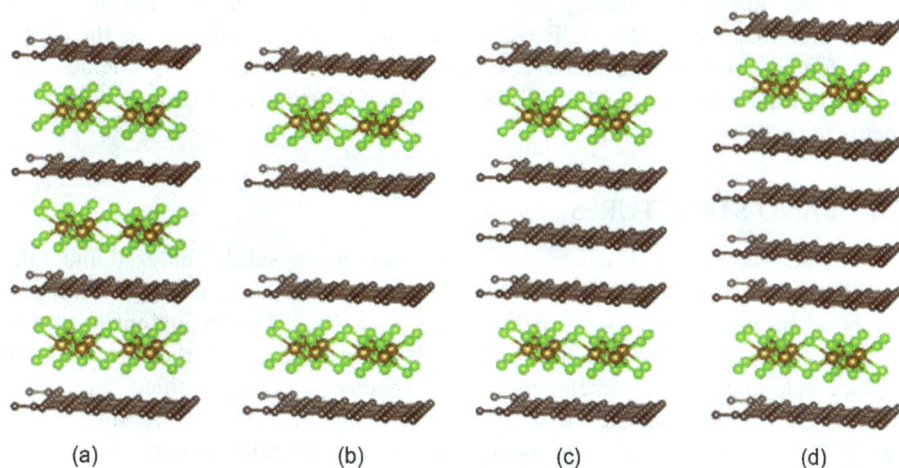

(a)   (b)   (c)   (d)

FIGURE 6.1 Geometric structures of $FeCl_3$-graphite intercalation compounds for (a) stage-1, (b) stage-2, (c) stage-3, and (d) stage-4.

layers. The large spacings of interlayer distances between graphenes allow the guest $FeCl_3$ to be intercalated. Calculation results show that the optimized interlayer distances between the pristine carbon layers are enlarged from 3.55 to 9.200 Å. The calculated results are very similar to the X-ray diffraction experimental observations, which showed an interlayer distance of 9.36 Å [57–59]. Interestingly, the tremendous changes of interlayer distance reveal that the interactions between interlayered 2Pz orbitals become much weaker after intercalation.

For the stage-2, stage-3, and stage-4 cases, the supercells possess atoms as 100/8/24, 150/8/24, 200/8/24 [carbon/iron/chlorine atoms], respectively. Take stage-2 one, for example, the interlayer distances of carbon layers are rather different between carbon layers with and without intercalants. The former is 9.24 Å and the latter is 3.38 Å. However, the C-C bond lengths are similar on each carbon layer. This indicates that the intercalants will easily affect the interaction of 2Pz-2Pz between interlayer carbon atoms. In addition, the interlayer spacings without intercalants are also enlarged to 3.63 and 3.71 Å for stage-3 and stage-4 cases, respectively.

The graphitic sheets consist of honeycomb lattices with very strong σ bondings, which even maintain almost the same after intercalation/de-intercalation, but the interlayer spacing is drastically changed. These features are very important for green energy applications, e.g., the lithium-ion battery and aluminum ion battery, in which the graphite intercalation compounds are widely utilized as electrodes. The theoretical predictions on the crystal structures of stage-n $FeCl_3$-graphite intercalation compounds could be examined by X-ray diffraction patterns. The X-ray diffraction patterns are successful for detecting similar graphite-related systems, e.g., stage-n alkali metal graphite intercalation compounds. That is to say, all the alkali-metal graphite-related compounds present the planar structures of carbon-honeycomb and intercalant lattices. Two periodical intercalant layers cover n-layer graphenes. These high-symmetry layered structures present the planar distributions for the smallest lithium and others, respectively, leading to the well-known chemical structures of $LiC_{\{6n\}}$ and $MC_{\{8n\}}$ [M = Na, K, Rb; Cs]. Obviously, both of them possess the apparent differences in terms of interlayer distances and bond lengths [60]. The accurate X-ray examinations can clarify the quasiparticle properties, the highly non-uniform chemical environment.

6.4 BAND STRUCTURES

The characteristic band structures and wave functions are subtly analyzed under the VASP simulation in which they are clearly exhibited only along the high-symmetry points within the first Brillouin zone, shown in Figure 6.2. The doping-enriched phenomena, which cover the blue shift or redshift of the Fermi level, the asymmetry of valence hole and conduction electron energy spectra [45], the modified Dirac-cone structures [61], the coexistence of the non-crossing/crossing/anti-crossing behaviors [62], the characterizations of the π, σ and intercalant-related energy bands [63], and the specific energy ranges corresponding to the different atom dominances/co-dominances, will be investigated thoroughly.

The pristine graphite possesses an unusual electronic structure with a large or small overlap of conduction bands and valence bands. The band structures are very sensitive to the configurations of stacking, e.g. AA (simple hexagonal), AB (Bernal), and ABC (rhombohedral). The monolayer graphene is well known and possesses zero gap and linear intersection at Dirac point, so the density of states will vanish at the Fermi level. For the AA-stacking graphite (simple hexagonal), the k_z-dependent band structures reveal that there is a band as wide as ~1 eV near the Fermi level. Furthermore, on the (k_x, k_y) plane, the electron (hole) pocket comes to exist near the A (Γ) points. That is to say, the free electrons and holes, which appear in the density of states, are located at the range $0 \leq E \leq 0.5$ eV and -0.5 eV $\leq E \leq 0$, respectively. The low-lying energy bands $\left[\left|E^{c,v}\right| \leq 3 \text{ eV}\right]$ mostly come from π bonding of C-$2p_z$ orbitals.

The AB-stacking graphite (Bernal) possesses two pairs of valence and conduction bands near the K and H valleys. The weak but significant band overlaps are revealed along the K-H line, such as the small co-existence of free electron and hole pockets near the stable K and H valleys. These positive and negative charges have the same contributions to all the essential properties. The electronic states are in charge of most of the important essential physical properties. Specifically, the parabolic energy dispersions belong to the saddle points at the middle points of the first Brillouin zone, e.g., parabolic energy dispersion at the M and L points.

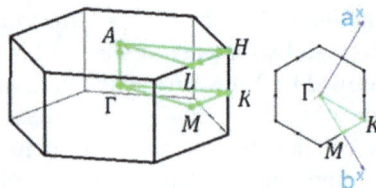

FIGURE 6.2 The first Brillouin zone.

Chemical intercalations can greatly diversify the main features of electronic energy spectra and wave functions. We explore the optimal geometric structures, band structures, density of states, spatial charge distributions, and variation of charge distributions on VASP. The abovementioned properties allow us to analyze the characteristic quasiparticle behaviors, which include various energy dispersions of linear, parabolic wave-vector dependences, band-edged states, the enhanced asymmetric hole and electron band, the redshift of the Fermi level, more valence and conduction bands due to Moire superlattices [64], the normal/undefined π & σ energy bands, and the C-/Fe-/Cl-dominated or co-dominated energy sub-bands.

After intercalation and de-intercalation, the graphite-related systems indicate the unusual energy spectra and wave functions, as shown in Figure 6.7; the calculated band structures of $FeCl_3$-graphite intercalation compounds possess a Dirac cone with a blue shift of Fermi level roughly 0.9, 0.6, 0.3, 0.1 eV, respectively, for stage-1/ stage-2/ stage-3/ stage-4. That is to say, the $FeCl_3$-graphite intercalation compounds turn into p-type systems after intercalations. The complicated horizontal bands mainly arise from the d-orbitals. For the stage-1 case, the bands at 0–1.35 eV present d-orbitals of the iron atoms. Based on the VASP calculations, the band structures of $FeCl_3$-graphite intercalation compounds are sensitive to the concentrations. That is to say, the free valence holes are closely related to the planar structure of $FeCl_3$ distributions, as performed in alkali-metal intercalation compounds [53].

The calculated band structures are also able to provide further information such as atom-decomposed contributions, which are very useful in determining the critical orbital hybridizations of quasiparticle charges. The C, Cl, and Fe atom dominances are shown in Figure 6.7, and the size of colored symbols represent their contributions. The C atoms make the most important contributions with the whole energy range but the other two cases have a partial flat band. For further step, the energy spectra strongly depend on the wave vectors; therefore, the localized states seldom survive even after intercalations. The Cl and Fe co-dominate the occupied and unoccupied near the low-lying energy and perform as linear bands. Also, for deeper-energy states, there exists the Cl- and Fe-related energy sub-bands at $E \leq -2$ eV simultaneously. Take the stage-1 $FeCl_3$-graphite intercalation compounds as an example: the Fe and Cl atoms reveal the partial flat energy bands at 0.7–1.2 eV (1.2–1.7 eV), $-0.5 \sim -1.1$ eV ($0 \sim -0.8$ eV) for spin-up electrons (spin-down electrons) in low-lying energy, and they reveal $-2 \sim 3$ eV, $-4.2 \sim -5.5$ eV for both spin-up and -down cases at deeper energy bands. The C atoms make relatively weak contributions in the low-lying energy relatively to the deeper energy bands. That is to say, the carbons in graphene layers are hard to be affected because of the strong 2s, 2px, and 2py orbitals which form the layered geometric structures. For stage-2 cases, the C atoms exist mainly at the deeper energy bands such as stage-1 case. The Fe atoms reveal the flat energy bands at 0–0.5 eV for spin-up ones and -0.7 ~ 0.2 eV and 1–1.8 eV for spin-down ones. These are in good agreement with Cl cases, which display flat energy bands at 1.1 and 0.4/1.2 eV for spin-up and spin-down cases, respectively. For stage-3 cases, the C atoms behave as the former two cases, and the contributions are much more obvious in deeper energy than the stage-1 and stage-2 cases because of the increment of C atoms. In addition, the Fe and Cl atoms co-dominate at roughly 0—1 eV near low-lying energy range but difficult to be distinguished below -2 eV. The similar localized behaviors are revealed in all the stage-n systems, in which they directly reflect the very strong intra-molecule interactions, but the interactions are relatively weak for the inter-molecules cases.

The high-precision ARPES measurements can be utilized to detect the significant intercalation effects in the current predictions. The semi-metal behavior due to the van der Waals interactions [65] is directly verified from the ARPES examinations. Moreover, the band-overlap-free carriers [electrons and holes] are very sensitive to the thermal excitations so that they exhibit very prolific Coulomb excitations, the single-particle and collective excitations with the unusual temperature- and momentum-dependences under the frequency of ~10–100 meVs [66]. The ARPES observations are rather useful in verifying the blue shift of the Fermi level, the enhanced asymmetry spectra of electron and hole energy spectra, the non-crossing/crossing/mixing behaviors, the well-defined or undefined π & σ bands, intercalant-induced occupied energy bands, as well as their strong dependences on the intercalant concentrations and distributions. Such clarifications could provide partial support on the quasiparticle properties, the important multi-orbital hybridizations [67].

6.5 THE ORBITAL HYBRIDIZATION

The van Hove singularities in density of states (Figures 6.8–6.10), the spatial charge density distributions $\rho(r)$ (Figures 6.3a and b–6.6a and b), and the charge variations after FeCl$_3$-molecule intercalations $\Delta\rho(r)$ (Figures 6.3c and d–6.6c and d) are able to clearly identify the critical single- and multi-orbital hybridizations in active

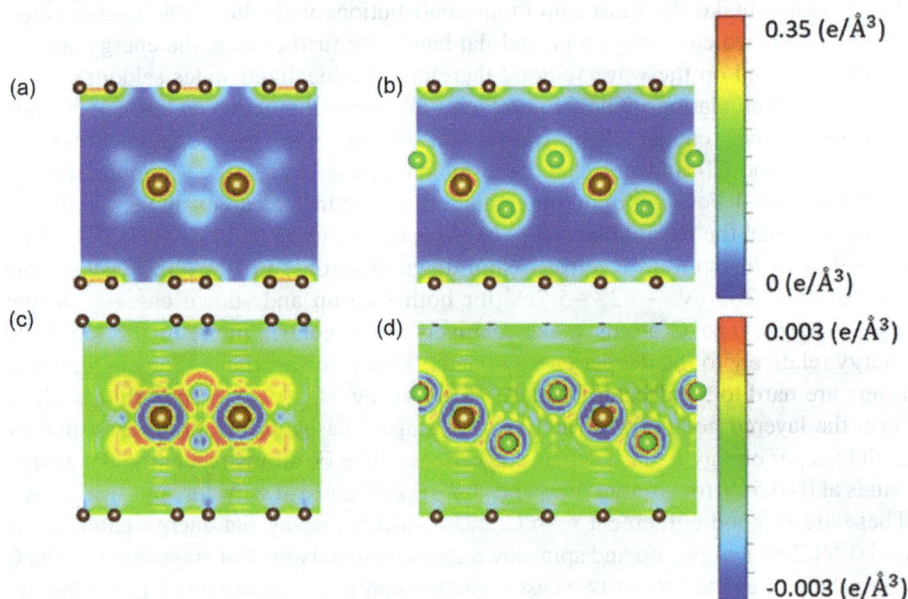

FIGURE 6.3 The spatial charge distributions and variations for stage-1 FeCl$_3$-GIC in x-z plane (a) (b) and y-z plane (c and d).

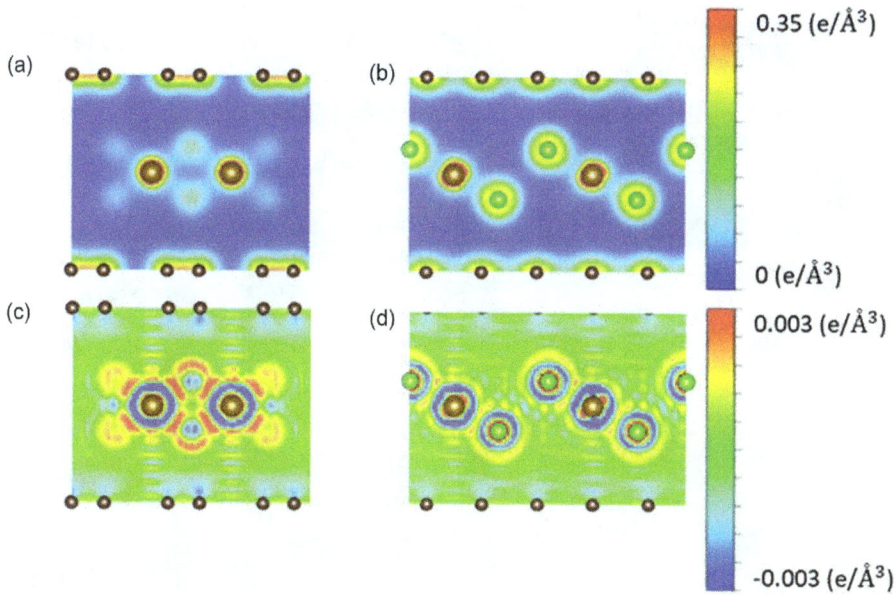

FIGURE 6.4 The spatial charge distributions and variations for stage-2 FeCl$_3$-GIC in x-z plane (a) (b) and y-z plane (c and d).

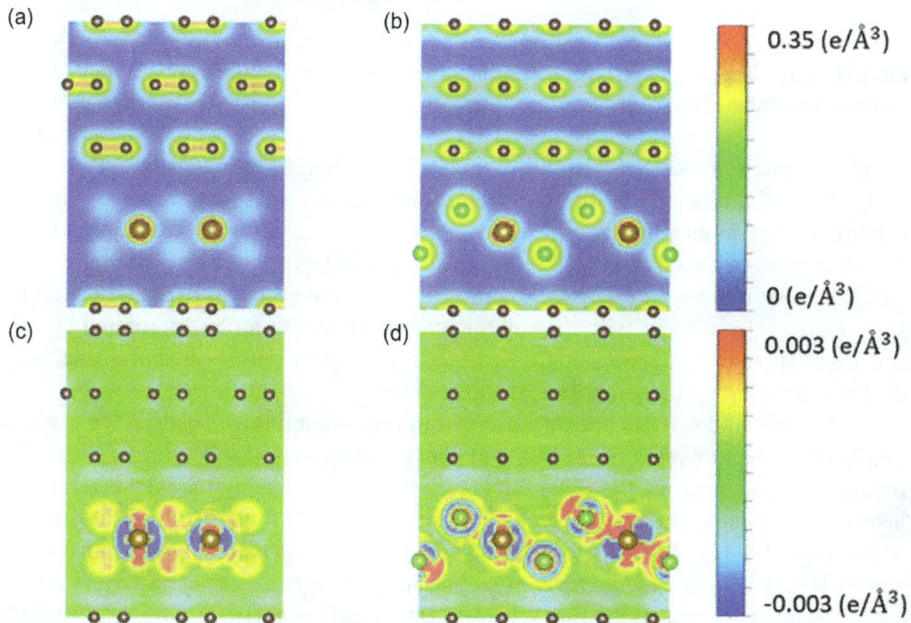

FIGURE 6.5 The spatial charge distributions and variations for stage-3 FeCl$_3$-GIC in x-z plane (a) (b) and y-z plane (c and d).

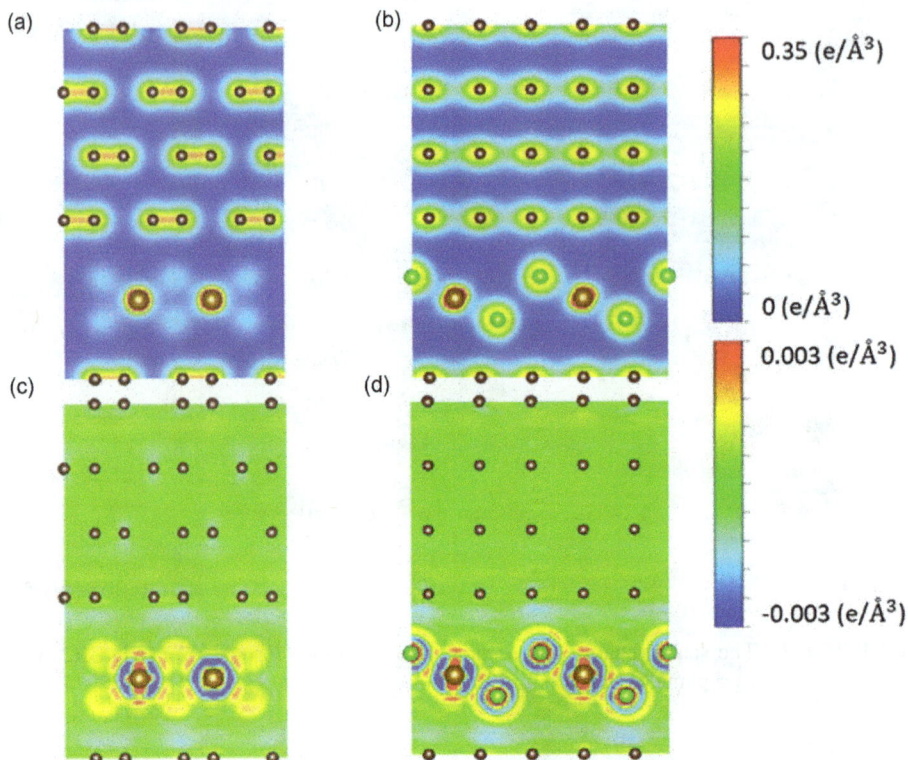

FIGURE 6.6 The spatial charge distributions and variations for stage-4 FeCl$_3$-GIC in x-z plane (a) (b) and y-z plane (c and d).

chemical bonds under the partial support of the atom-dominated energy bands. Only the spatial charge distributions are not sufficient in verifying all the significant orbital mixings associated with the non-negligible chemical bonds. The merged special structures in DOS, which come from the specific atom orbitals simultaneously, indicate their non-negligible hybridizations. We calculate these delicately, step by step. They are consistent with the specific energy ranges of atom-dominated energy spectra (Figure 6.7). The quasiparticle properties will be determined and thus very useful in developing their theoretical framework.

Both $\rho(r)s$ and $\Delta\rho(r)s$ are very useful and important in investigating the concise quasiparticle pictures related to chemical bonds with active orbital hybridizations. By the delicate calculations, the (x, z)- and (y, z)-side views display the intralayer, interlayer, intramolecule and intermolecule orbital mixings. At first, the most outstanding σ bonding of C-[2s, 2px, 2py] orbitals, which exists in a pristine honeycomb lattice, clearly shows very strong covalent bonds between two neighboring carbon atoms [a rather high carrier density by the red color]. Their charge variations are only slightly modified after the FeCl$_3$ being intercalated, as indicated in Figures 6.3c and d, 6.4c and d, 6.5c and d, 6.6a and b, respectively, for the stage-1, stage-2, Stage-3, and stage-4 cases. Furthermore, the π bonding, which is due to the parallel 2pz-orbital

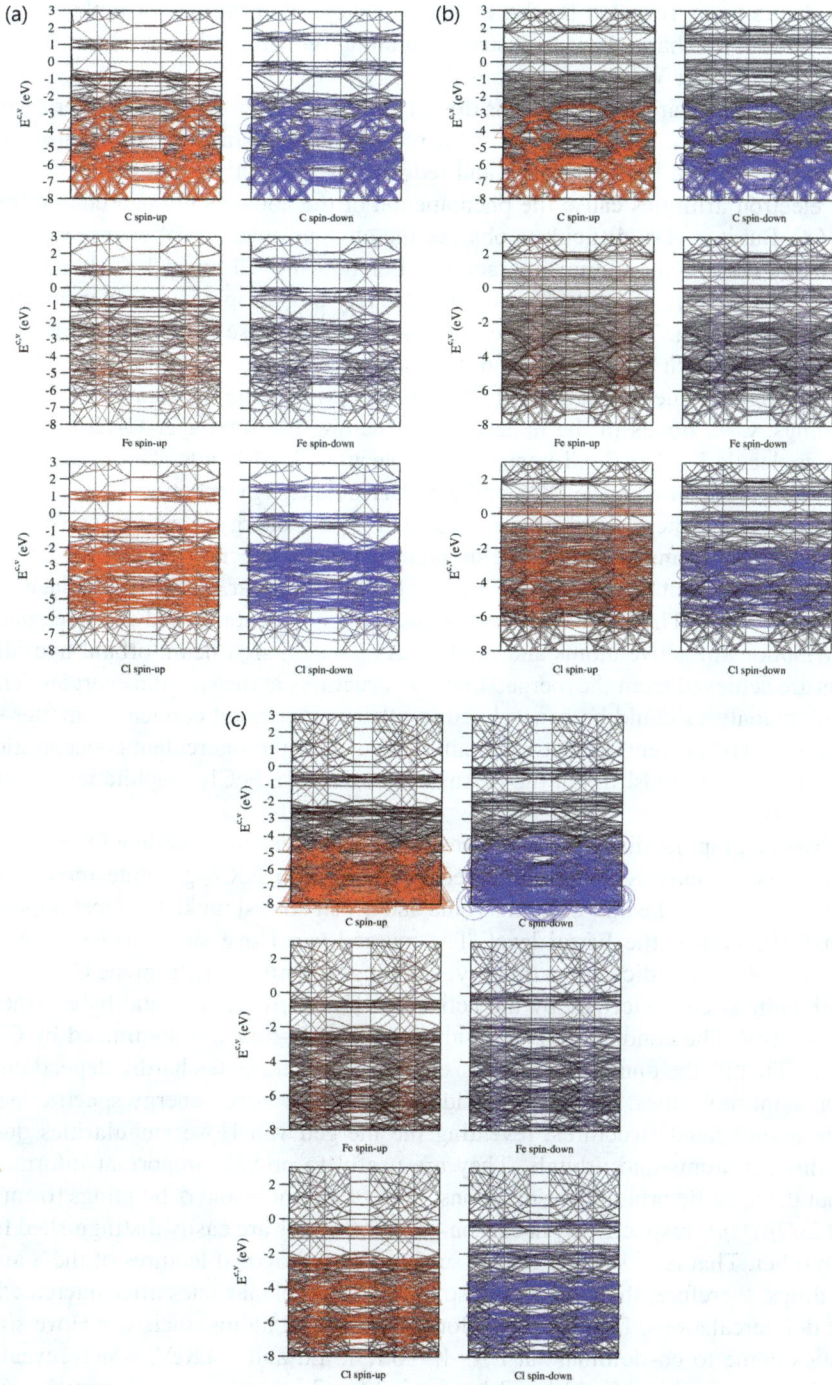

FIGURE 6.7 The band structures for (a) stage-1 (b) stage-2 (c) stage-3.

hybridizations, is revealed by the wave-like charge distributions of both (x, z) and (y, z) planes. Its charge density distribution along the z-direction is extended by the significant van der Waals interactions [9].

The $\Delta\rho(r)s$ near C and Cl atoms directly reflect the significant interlayer carbon-chloride couplings, where there exist the reduced and enhanced charge densities for the former and the latter [the blue and red regions, respectively, in Figure 6.3–6.6]. The electron affinities cause the phenomenon of the abovementioned charge transfer [68]. But it is very difficult to observe the obvious evidence of charge variations about the carbon–aluminum interactions. For both Fe–Cl and Cl–Cl bonds, their features are clearly characterized by the strongly anisotropic charge density distributions near them. The VASP calculations have no precise evidence for Fe-Fe and Fe-C bonds, which hardly appear in the charge density distributions. But the predictions, being identified from both $\rho(r)s$ and $\Delta\rho(r)s$, indicate the significant chemical bondings: C–C bonds in the honeycomb structure, the interlayer C–Cl bonds, the intra-molecule Fe–Cl/Cl–Cl bonds inter-molecule Cl–Cl bonds. These features are in good agreement with the DOS and the electronic energy spectra.

In the DOS, the critical points presented in the energy-wave-vector space are responses for the unusual van Hove singularities. Their special structures are diversified by the different band-edge states of the parabolic, linear, and partially flat energy dispersions [69]. The calculated results are delicately decomposed into the specific contributions of active atoms and orbitals. As a result, significant orbital hybridizations are achieved from the merged unusual structures at the specific energies. These detailed analyses could be generalized to other complicated condensed-matter systems [67]. The current predictions mainly depend on the intercalant concentrations, e.g., the different redshifts of the Fermi level in stage-n $FeCl_3$-graphite intercalation compounds.

Pristine graphite displays the semimetallic properties with the low DOS near the Fermi level [relatively weak compared to the stage-n $FeCl_3$-graphite intercalation compounds]. Both Fe and Cl atoms [blue and green colors] make the most important contributions near the Fermi level. The merged van Hove singularities, shown in Figures 6.8–6.11, indicate that the p-type dopings mainly come from the C–Cl chemical bondings characterized by the active $2pz$-[$3px$, $3py$, $3pz$] orbital hybridizations at $E \sim -2$ eV. The conduction DOS with a range of 2–6 eV are dominated by C-$2pz$ atoms. That is, the unoccupied higher-energy electronic states hardly depend on the intercalant molecules. On the other side, the deeper valence energy spectra display more complicated structures, revealing the merged van Hove singularities due to the distinct atoms and orbitals. They are useful to provide important information about the specific orbital hybridizations. First of all, the π and σ bondings [from $2pz$ and $2s/2px/2py$, respectively] in carbon-related systems are easily distinguished from each other. That is to say, the results suggest the orthogonal features of the π and σ bondings, therefore, the honeycomb structures keep planar ones after intercalations and deintercalations. In addition, as for the Fe and Cl atoms, their van Hove singularities come to co-dominate at $E \sim -1$, -0.1, 1, 1.6 and -4.8 eV, which reveal the Cl-[$3px$, $3py$, $3pz$] and Fe-[$3dxy$, $3dyz$, $3dz^2$, $3dxz$, $3dx^2$] orbital hybridizations. Also, the [$3s$, $3px$, $3py$, $3pz$] orbitals of Cl atoms appear at $-2.2 \sim -3$ eV and $-4.8 \sim -5.2$ eV, revealing the sp^3 bondings in the intra- and inter-molecule Cl-Cl bonds. The van

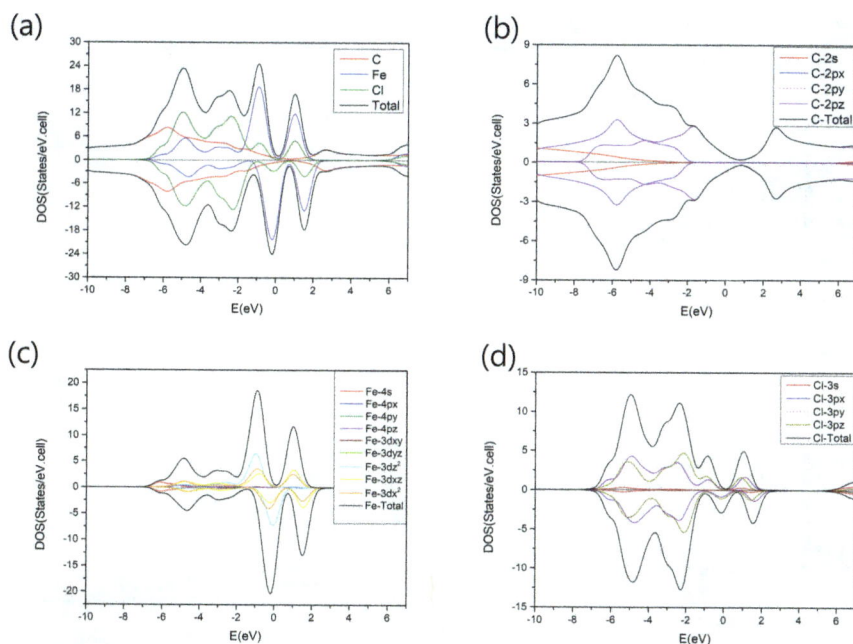

FIGURE 6.8 The atom- and orbital-projected density of states of stage-1 $FeCl_3$-graphite intercalation compounds for (a) total atoms, (b) C atoms, (c) Fe atoms, and (d) Cl atoms.

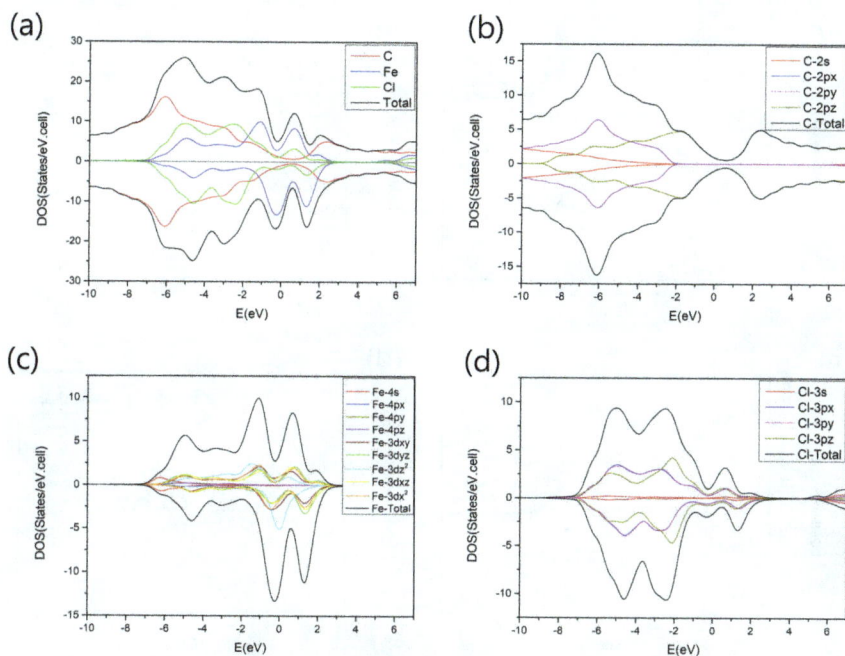

FIGURE 6.9 The atom- and orbital-projected density of states of stage-2 $FeCl_3$-graphite intercalation compounds for (a) total atoms, (b) C atoms, (c) Fe atoms, and (d) Cl atoms.

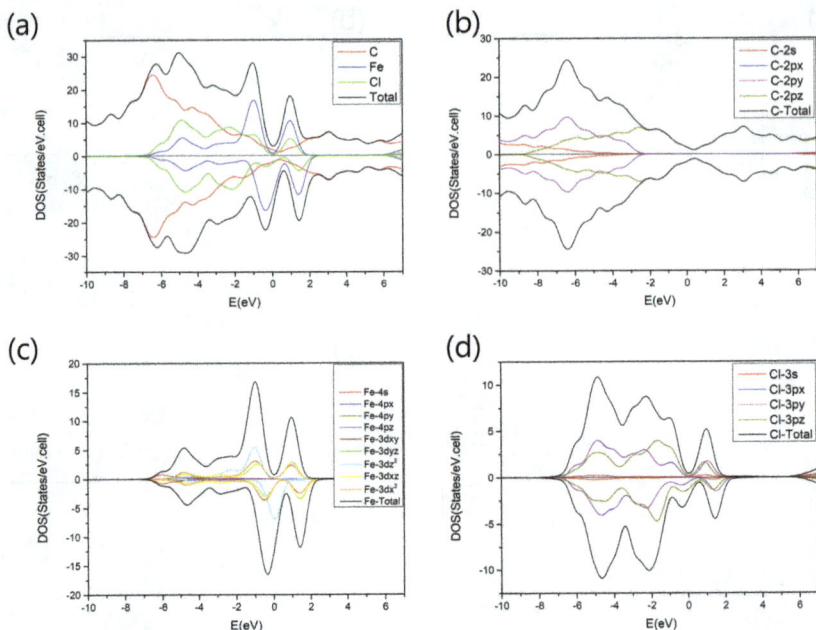

FIGURE 6.10   The atom- and orbital-projected density of states of stage-3 FeCl$_3$-graphite intercalation compounds for (a) total atoms, (b) C atoms, (c) Fe atoms, and (d) Cl atoms.

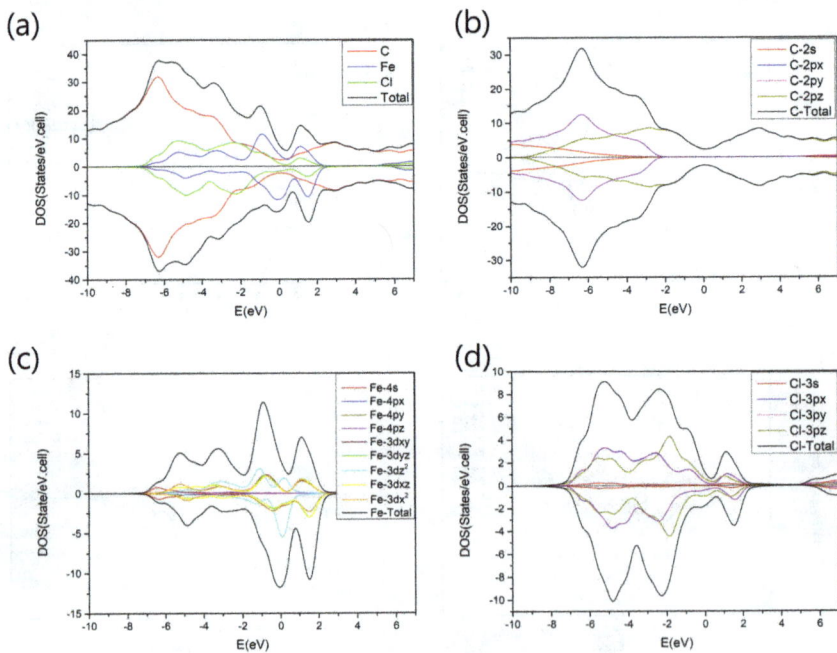

FIGURE 6.11   The atom- and orbital-projected density of states of stage-4 FeCl$_3$-graphite intercalation compounds for (a) total atoms, (b) C atoms, (c) Fe atoms, and (d) Cl atoms.

Hove singularities of Fe atoms mainly arise from the intra-molecule Fe-Cl bonds; that is to say, the Fe-Fe bonds and Fe-C bonds almost vanish.

In brief, the most important quasiparticle property, the charge interactions, could be reached from the delicate analyses on the van Hove singularities (Figures 6.8–6.11), spatial charge density distributions (Figures 6.3–6.6), and atom-dominated band structures [45]. The significant orbital hybridizations in C–C/C–Cl/Fe–Cl/Cl–Cl chemical bonds correspond to [$2s$, $2px$, $2py$]–[$2s$, $2px$, $2py$] & $2pz$-$2pz$/$2pz$-[$3px$, $3py$, $3pz$]/[$3dxy$, $3dyz$, $3dz^2$, $3dxz$, $3dx^2$]-[$3s$, $3px$, $3py$, $3pz$]/[$3s$, $3px$, $3py$, $3pz$]-[$3s$, $3px$, $3py$, $3pz$], respectively. Furthermore, both C-Fe and Fe-Fe bonds are almost absent and do not play a significant role in physical /chemical/material properties. The other intercalants are expected to exhibit the diverse phenomena. The systematic investigations about the various p- and n-type graphite intercalation compounds [70] are required in the further studies.

The energy-dependent van Hove singularities can be directly examined by high-resolution STS measurements only for their concise structures. For instance, such examinations are very successful in exploring the low-lying features of DOS for graphene-related systems, such as [71], few-layer graphenes [72], carbon nanotubes [73], and graphene nanoribbons [74], respectively, with the semi-metallic behavior, metallic or semiconducting properties, and semiconducting one. In addition, the measured special structures strongly depend on dimensionalities and energy dispersions [69]. As for electron doping of stage-n $FeCl_3$ graphite intercalation compounds, the redshift of the Fermi level [53], the Dirac-cone minimum [61], and the strong peak due to the π -electronic saddle-point structure [69] are expected to be observable. However, the deeper- and higher-energy van Hove singularities might be very difficult to clarify their features, mainly owing to the zone-folding effects [75] and the multi-merged structures.

6.6 CONCLUSION

There are certain significant differences among the various graphite intercalation compounds being revealed in the featured crystal symmetries, band structures, orbital hybridizations [chemical bonds], optical absorption spectra, Coulomb excitations, and electrical conductivities. According to the previous experimental measurements and theoretical predictions [76], the diverse quasiparticle phenomena cover the different configuration of stackings/interlayer distances/stage-n dependences, the blue or redshift, due to metal atoms and molecules, and a lot of the dramatic changes in electronic structures. For example, $FeCl_3$ molecules are quite different from alkali atoms in terms of orbital hybridizations after the chemical intercalations, mainly owing to their complicated multi-orbital mixings and the interlayer/intralayer atomic interactions such as charge density distributions and DOS; that is, the essential properties would become more complex under a highly non-uniform chemical environment. Even for alkali-guest graphite intercalation compounds [53], lithium exhibits the highest performance in serving as a transport for ions of batteries because of its many merits [77]. How to achieve the optimal current density and energy capacity remains to be a dominating challenge, i.e., the different critical mechanisms, being unified under a great theoretical framework, need to include in the near-further predictions, such as, the asymmetry-initiated chemical reactions and ion currents. The intercalations of atoms or molecules will be quite different from each other in terms of chemical bondings and low-lying energy essential properties due to the diverse specimen of atoms, configuration of stackings, and concentrations.

Interestingly, the stage-1, stage-2, stage-3, and stage-4 $FeCl_3$-graphite intercalation compounds display the diverse quasiparticle behaviors based on the VASP calculations and analysis [78]. Our work clearly illustrates the theoretical predictions of quasiparticles framework. The orbital hybridizations including the intralayer and interlayer chemical bonds are thoroughly clarified from the highly non-uniform chemical environments of Moiré superlattices, the atom-dominated band structures at specific energy ranges, the spatial charge densities/charge variations after intercalations, and van Hove singularities at the various energies in DOS. The important chemical bondings could contain the intralayer/interlayer C-C bonds, the interlayer C-intercalant bonds, and the intra-molecule/inter-molecule bonds. The observable multi-/single-orbital hybridizations of C-C/C-Cl/Fe-Cl/Cl-Cl chemical bonds cover [2s, 2px, 2py]-[2s, 2px, 2py] &2pz-2pz/2pz-[3px, 3py, 3pz]/[3dxy, 3dyz, 3dz^2, 3dxz, 3dx^2]-[3s, 3px, 3py, 3pz]/[3s, 3px, 3py, 3pz]-[3s, 3px, 3py, 3pz]. However, the evidence almost disappears for the C-Fe and Fe-Fe bonds. Moreover, the strong charge transfers between graphitic layers and molecular intercalants, the zone-folding effects due to the intercalant stackings and arrangements, and the van der Waals interactions are responsible for the featured quasiparticle behaviors. The main features of electronic properties include the semi-metal transitions, the enhanced asymmetry of hole and electron energy spectra, the large modifications about the low-lying energy dispersions/band-edge states, the well-defined or undefined π - and σ -electronic bands, the intercalant-induced/C- and intercalant-co-dominated energy sub-bands, and the greatly diversified van Hove singularities.

The $FeCl_3$-GICs could serve as a promising potential electrode in lithium-ion batteries. The recent experimental results indicate that the remarkable capability and excellent cycling stability of $FeCl_3$-GICs can be attributed to their robust and unique structures. The intercalated $FeCl_3$ increases the interlayer spacing of graphenes and then provides the extra sites for the containing of Li-ions, that is to say, the intercalants causing enlarged spacing, providing more convenient channels for the rapid intercalation/de-intercalation of Li-ion during the electrode reaction. The intercalated $FeCl_3$ also improves the electrical conductivity of pristine graphite to attain a better electrochemical performance [29,57,79–81]. Herein the vital framework of physical and chemical properties is constructed. However, the mentioned electrochemical properties are mainly focused on the low-stage index of few-layer graphene (FLG). It is worthwhile to be examined by delicate experiments for the full-fledged electrochemical performance, e.g., electrical conductivity, specific capacity, cycle stability, etc. on stage-3, stage-4, or even higher stage index.

REFERENCES

1. Lee SM, Kang DS and Roh JS. 2015 Bulk graphite: Materials and manufacturing process *Carbon Lett.* **16**, 135–146.
2. Mecklenburg M and Regan B. 2011 Spin and the honeycomb lattice: Lessons from graphene *Phys. Rev. Lett.* **106**, 116803.

3. Tran NTT, Lin SY, Lin CY and Lin MF 2017 *Geometric and Electronic Properties of Graphene-Related Systems Chemical Bonding Schemes* CRC Press, book, ISBN 9781351368483..

4. Gómez-Santos G 2009 Thermal van der Waals interaction between graphene layers *Phys. Rev. B* **80**, 245424.

5. Chung DDL. 2002 Review graphite *J. Mater. Sci.* **37**, 1475–1489.

6. Ho JH, Chang CP and Lin MF 2006 Electronic excitations of the multilayered graphite *Phys. Lett. A* **352**, 446–450.

7. Zhang Z, Huang H, Yang X and Zang L 2011 Tailoring electronic properties of graphene by π–π stacking with aromatic molecules *J. Phys. Chem. Lett.* **2**, 2897–2905.

8. Cheng Q 2017 Graphene-like-graphite as fast-chargeable and high-capacity anode materials for Lithium ion batteries *Sci. Rep.* 7, 14782.

9. Crowell AD 1965 Van der waals interactions of simple molecules with graphite. *Abstracts of Papers of The American Chemical Society, Apr, C005..*

10. Do TN, Lin CY, Lin YP, Shih PH and Lin MF 2015 Configuration-enriched magneto-electronic spectra of AAB-stacked trilayer graphene. *Carbon*, **94**, 619.

11. Lin CY, Lee MH and Lin MF 2018 Coulomb excitations in trilayer ABC-stacked graphene. *Phys. Rev. B Rapid commun.*, **98**, 041408.

12. Ho JH, Lu CL, Hwang CC, Chang CP and Lin MF 2006 Coulomb excitations in AA- and AB-stacked bilayer graphites. *Phys. Rev. B* **74**(8), 085406.

13. Natori A, Ohno T and Oshiyama, A 1985 Work function of alkali-atom adsorbed graphite *J. Phys. Soc. Jpn.* **54**, 3042–3050.

14. Li Y and Qu Y 2013 First-principles study of electronic and magnetic properties of $FeCl_3$-based graphite intercalation compounds *Phys. Rev. B* **425**, 72–77.

15. Mao C 2018 Selecting the best graphite for long-life, high-energy li-ion batteries *J. Electrochem. Soc.* **165**, A1837–A1845.

16. Seoungwoo B, Joonam P, Williams AA, Ryou MH and Lee YM 2017 The effects of humidity on the self-discharge properties of $Li(Ni_{1/3}Co_{1/3}Mn_{1/3})O_2$/graphite and $LiCoO_2$/graphite lithium-ion batteries during storage *RSC Advances* **7**, 10915.

17. Miriam S, Sebastian R, Norbert Wagner and Friedrich KA 2017 Investigation of the solid electrolyte interphase formation at graphite anodes in Lithium-ion batteries with electrochemical impedance spectroscopy *Electrochem Acta* **228**, 652–658.

18. Frank M, Kindermann, Patrick J, Osswald, Stefan K, Günter E, Jorg S, Andreas N, Simon VE, Wolfgang S and Andreas 2017 Measurements of lithium-ion concentration equilibration processes inside graphite electrodes *J. Power Sources* **342**, 638–642.

19. Marius B, Bernhard R, Stefan S, Peter K, Mario W, Danzer MA and Andreas J 2017 Multi-phase formation induced by kinetic limitations in graphite-based lithium-ion cells: Analyzing the effects on dilation and voltage response *J. Energy Storage* **10**, 1–10.

20. Hewathilake HPTS, Niroshan K, Athula W and Balasooriya NWB 2017 Performance of developed natural vein graphite as the anode material of rechargeable lithium-ion batteries *IONICS* **23**, 1417–1422..

21. Lin SY, Tran NTT, Chang SL, Su WP and Lin MF 2018 *Structure- and Adatom-Enriched Essential Properties of Graphene Nanoribbons* CRC Press, book, ISBN:9780367002299..

22. Ziambaras E, Kleis J, Schröder E and Hyldgaard P 2007 Potassium intercalation in graphite: A van der Waals density-functional study *Phys. Rev. B* **76**, 155425..

23. Persson K, Hinuma Y, Meng YS, Ven AVD and Ceder G 2010 Thermodynamic and kinetic properties of the Li-graphite system from first-principles calculations *Phys. Rev. B* **82**, 125416..

24. Wang ZH, Selbach SM and Grande T 2014 Van der Waals density functional study of the energetics of alkali metal intercalation in graphite *RSC Adv.* **4**, 4069..

25. Bhauriyal P, Mahata A and Pathak B 2017 The staging mechanism of $AlCl_4$ intercalation in a graphite electrode for an aluminium-ion battery *Phys Chem. Chem. Phys.* **19**, 7980..

26. Wu MS, Xu B, Chen LQ and Ouyang CY 2016 Geometry and fast diffusion of AlCl4 cluster intercalated in graphite *Electronchemica Acta* **195**, 158..

27. Wand DY 2017 Advanced rechargeable aluminium ion battery with a high-quality natural graphite cathode *Nature Commun.* **8**, 14283..

28. Qi X 2015 $FeCl_3$ intercalated few-layer graphene for high lithium-ion storage performance *J. Mater. Chem. A* **3**, 15498..

29. Wang L, Zhu YC, Guo C, Zhu X, Liang J and Qian YT 2014 Ferric chloride-graphite intercalation compounds as anode materials for li-ion batteries *ChemSusChem* **7**, 87.

30. Wu Z S, Zhou G, Yin L C, Ren W, Li F and Cheng H M 2012 Graphene/metal oxide composite electrode materials for energy storage *Nano Energy* **1**, 107–131.

31. Ho JH, Lai YH, Tsai SJ, Hwang J, Chang C and Lin MF 2006 Magnetoelectronic properties of a single-layer graphite *J Phys Soc Jap* **75**, 114703.

32. Huang YK, Chen SC, Ho YH, Lin CY and Lin MF 2014 Feature-rich magnetic quantization in sliding bilayer graphenes *Sci. Rep.* **4**, 7509.

33. Koshino M and McCann E 2011 Landau level spectra and the quantum Hall effect of multilayer graphene *Phys Rev. B* **83**, 165443.

34. Wang ZF, Liu F and Chou MY 2012 Fractal landau-level spectra in twisted bilayer graphene *Nano Lett.* **12**, 3833–3838..

35. Lin MF, Chuang YC and Wu JY 2012 Electrically tunable plasma excitations in AA-stacked multilayer graphene *Phys Rev. B* **86**, 125434.

36. Wu JY, Chen SC, Roslyak O, Gumbs G and Lin MF 2011 Plasma excitations in graphene: Their spectral intensity and temperature dependence in magnetic field *ACS Nano* **5**, 1026–1032.

37. Lozovik YE and SokolikAA 2012 Influence of Landau level mixing on the properties of elementary excitations in graphene in strong magnetic field *Nanoscale Lett.* **7**, 134.

38. Ohta T 2007 Interlayer interaction and electronic screening in multilayer graphene investigated with angle-resolved photoemission spectroscopy *Phys Rev. Lett.* **98**, 206802..

39. Laura MR and Pratt Jr GW 1959 A many-body treatment of dielectric screening for impurity states and excitons in semiconductors *J. Phys. Chem. Solids* **8**, 47–49.

40. SchrÄufer K, Metzner C, Hofmann MC and DÄohler GH 1997 Non-linear impurity screening in strongly depleted 2D electron gases *Superlattice Micros.* **21**, 223–230.

41. Mišković ZL, Sharma P and Goodman FO 2012 Ionic screening of charged impurities in electrolytically gated graphene *Phys. Rev. B* **86**, 115437.

42. Ain QT, Al-Modlej A, Alshammari A and Anjum MN 2018 Effect of solvents on optical band gap of silicon-doped graphene oxide *Mater Res. Express* **5**, 035017.

43. Leenaerts O, Partoens B and Peeters FM 2008 Adsorption of H_2O, NH_3, CO, NO_2, and NO on graphene: A first-principles study *Phys. Rev. B* **77**, 125416.

44. Chun JP, Chunze Y, Guanzhou Z, Qian Z, Chen JH, Meng C, Lind, Michael A, Bing JH, Payam K and Hongjie 2018 An operando X-ray diffraction study of chloroaluminate anion-graphite intercalation in aluminum batteries. *PNAS* **115**, 22.

45. Hwang EH and Das Sarma S 2008 Quasiparticle spectral function in doped graphene: Electron-electron interaction effects in ARPES *Phys. Rev. B* **77**, 081412.

46. Matsui T, Kambara H, Niimi Y, Tagami K, Tsukada M and Fukuyama H 2005 STS observations of landau levels at graphite surfaces *Phys. Rev. B* **94**, 26403..

47. Rafiquea M, Unarb MA, Ahmedb I, Chacharc AR and Shuai Y 2018 Ab-initio investigations on physisorption of alkaline earth metal atoms on monolayer hexagonal boron nitride (h-BN) *J Phys. Chem. Solids* **118** ,114–125..

48. Rafique M, Shuaia Y, Tan HP and Muhammad H 2017 Theoretical perspective on structural, electronic and magnetic properties of 3d metal tetraoxide clusters embedded into single and di-vacancy graphene *Appl. Surf. Sci.* **408**, 21–33..

49. Kresse G and Hafner J 1993 Ab initio molecular dynamics for liquid metals *Phys. Rev. B* **47**, 558–561..

50. Kresse G and Hafner J 1994 Ab initio molecular-dynamics simulation of the liquid-metal–amorphous-semiconductor transition in germanium *Phys. Rev. B* **49**, 14251–14269.

51. Lee JK, Lee SC, Ahn JP, Kim SC, Wilson JI and John P 2008 The growth of AA graphite on (111) diamond *J. Chem. Phys.* **129**, 234709.

52. Bhattacharyya S and Singh AK 2016 Lifshitz transition and modulation of electronic and transport properties of bilayer graphene by sliding and applied normal compressive strain *Carbon* **99**, 432438.

53. Li WB, Lin SY, Tran NTT, Lin MF and Lin KI 2020 Essential geometric and electronic properties in stage-n graphite alkali-metal-intercalation compounds *RSC Adv.* **10**, 23573–23581.

54. Wu MS, Xu B, Chen LQ and Ouyang CY 2016 Geometry and fast diffusion of $AlCl_4$ cluster intercalated in graphite *Electrochimica Acta* **195**, 158–165.

55. Zhan D et al 2010 $FeCl_3$-based few-layer graphene intercalation compounds: Single linear dispersion electronic band structure and strong charge transfer doping *Adv. Funct. Mater.* **20**, 3504–3509.

56. Prietsch M, Wortmann G and Kaind G 1986 Mossbauer study of stage-2 $FeCl_3$-graphite *Phys. Rev. B* **33**, 11.

57. Dresselhaus MS and Dresswlhaus G 2002 Intercalation compounds of graphite *Adv. Phys.* **51**, 1–186.

58. Caswell N and Solin SA 1978 Vibrational excitations of pure $FeCl_3$ and graphite intercalation with ferric chloride *Solid State Commun.* **27**, 961.

59. Underhill C, Leung SY, Dresselhaus G and Dresselhaus MS 1979 Infrared and Raman spectroscopy of graphite-ferric chloride *Solid State Commun.* **29**, 769..

60. Kaneko T and Saito R 2017 First-principles study on interlayer state in alkali and alkaline earth metal atoms intercalated bilayer graphene *Surf. Sci.* **665**, 1–9.

61. Malko D, Neiss C, Vines F and Gorling A 2012 Competition for graphene: Graphynes with direction-dependent Dirac cones *Phys. Rev. Lett.* **108**, 086804.

62. Lin CY, Chen RB, Ho YH and Lin MF 2018 *Electronic and Optical Properties of Graphite-Related Systems*. CRC Press, Boca Raton, FL.

63. Tasaki K 2014 Density functional theory study on structural and energetic characteristics of graphite intercalation compounds *J. Phys. Chem. C* **118**, 1443–1450.

64. Patil S, Kolekar S and Deshpande A 2017 Revisiting HOPG superlattices: Structure and conductance properties *Surf. Sci.* **658**, 55–60.

65. Duong DLF, Yun SJ and Lee YH 2017 van der Waals layered materials: Opportunities and challenges *ACS Nano* **11**, 11803–11830..

66. Nozieres P and Pines D 1959 Electron interaction in solids - Characteristic energy loss spectrum *Phys. Rev.* **113**, 1254–1267.

67. Dresselhaus MS and Dresselhaus G 2002 Intercalation compounds of graphite *Adv. Phys.* **51**, 1–186.

68. Lee A C, and Perdew JP 1982 Calculated electron affinities of the elements *Phys. Rev. A* **25**, 1265.

69. Lin SY, Chang SL, Shyu FL, Lu JM and Lin MF 2015 Feature-rich electronic properties in graphene ripples *Carbon* **86**, 207–216.

70. Meng X, Tongay S and Kang J 2013 Stable p- and n-type doping of few-layer graphene/graphite *Carbon* **57**, 507–514.

71. Hu M, Dong X, Wu YJ, Liu LY and Zhao ZS 2018 Low-energy 3D sp(2) carbons with versatile properties beyond graphite and graphene *Dalton Trans-Actions* **47**, 6233–6239.

72. Jung NY, Kim ND, Jockusch S, Turro NJ, Kim P and Brus L 2009 Charge transfer chemical doping of few layer graphenes: Charge distribution and band gap formation *Nano Lett.* **9**(12), 4133–4137.

73. Lien JY and Lin MF 2007 Low-energy electronic properties of a pair of carbon nanotubes *IPSS* **4**, 512–514.

74. Li TS, Huang YC, Chang SC, Chuang YC and Lin MF 2008 Transport properties of AB-stacked bilayer graphene nanoribbons in an electric field *Eur. Phys. J. B* **64**, 73–80.

75. Lin Y, Chen G, Sadowski JT, Li YZ, Tenney SA, Dadap JI, Hybertsen MS and Osgood RM 2019 Observation of intercalation-driven zone folding in quasi-free-standing graphene energy bands. *Phys. Rev. B* **99**, 035428..

76. Marinopoulos AG, Reining L, Rubio A and Olevano V 2004 Ab initio study of the optical absorption and wave-vector-dependent dielectric response of graphite *Phys. Rev. B* **69**, 245419.

77. Gao LJ, Liu SY and Dougal RA 2002 Dynamic lithium-ion battery model for system simulation *IEEE Transac. Comp. Pack. Technol.* **25**, 495–505..

78. Hafner J 2008 Ab-initio simulations of materials using VASP: Density-functional theory and beyond *J. Comput. Chem.* **29**, 2044–2078.

79. Wang G, Shen X and Yao J 2009 Graphene nanosheets for enhanced lithium storage in lithium-ion batteries *Carbon* **47** 2049–2053.

80. Wang F, Yi J, Wang Y, Wang C, Wang J and Xia Y 2014 Graphite intercalation compounds (GICs): A new type of promising anode material for lithium-ion batteries *Adv. Energy Mater.* **4** 1300600.

81. Wu Z S, Ren W, Xu L, Li F and Cheng H M 2011 Doped graphene sheets as anode materials with superhigh rate and large capacity for lithium-ion batteries *ACS Nano* **5** 5463–5471.

7 Studying the Anisotropic Lithiation Mechanisms of Silicon Anode in Li-Ion Batteries Using Molecular Dynamic Simulations

Li-Yi Pan and Chin-Lung Kuo
National Taiwan University

CONTENTS

7.1 INTRODUCTION

Li-ion batteries have become more and more important these days due to their high energy density when compared with other energy storage devices. Increasing the anode weight specific capacity is, therefore, crucial for Li-ion batteries, especially in portable devices. Graphite, having a specific capacity of 372 mAh g^{-1}, has been the most common anode material in the present time. Researchers have found that the silicon anode [1], having a specific capacity of 3,579 mAh g^{-1}, is a promising anode candidate due to its 10-fold specific capacity than graphite. The anode in the Li-ion battery stores electrons from the external circuit with lithium from the electrolyte by forming a chemical bond with lithium. The graphite anode stores lithium by intercalation of lithium in between its interlayer spacing. Unlike the graphite anode, silicon anode stores lithium by forming Li–Si alloy, which is called the *lithiation* of silicon. The silicon anode lithiation from crystalline silicon (c-Si) to amorphous Li–Si alloy (a-Li$_{15}$Si$_4$) undergoes 4-fold volume expansion, which pulverizes the bulk silicon anode and prevents it from further commercial applications.

DOI: 10.1201/9781003263807-7

Therefore, nanostructured silicon proposed by researchers [2] has been used to solve the volume expansion problem. Silicon nanowires, nanoparticles, or thin films are common solutions. Liu et al. and Lee et al. have found that the crystalline silicon nanowire lithiates preferentially along Si⟨110⟩ direction while other facets such as Si⟨100⟩ or Si⟨111⟩ merely move [3,4]. This is called the anisotropic lithiation of silicon. The anisotropic lithiation leads to the stress concentration on the slower moving facet and cracks the silicon anode due to its concaved shape, causing significant capacity loss [3]. Later, Liu et al. have found that, during the lithiation of the silicon nanowire, a 1-nm-thick phase boundary forms between the core Si and the lithiated silicon [5]. Besides, the Si(112) facet is found to undergo a ledge-mechanism by peeling-off Si(111) facet during lithiation and the Si(111) facet is reported to be almost immobile.

Whether the anisotropic lithiation of silicon is thermodynamic or kinetic controlled is still a topic of debate. Jung et al. have calculated the interface energy of Si(100), Si(110), and Si(111) with Li–Si alloy by *ab initio* calculations and stated that the lithiation along Si(110) is the most energetically favorable [6]. Chan et al. have performed another analysis on lithiation voltage for Si(100), Si(110), and Si(111) facets and conclude that the lithiation voltage is higher for Si(110) facets and, therefore, more preferable for lithiation. They have also concluded that the lithium insertion energy barrier is not an issue [7]. These evidences show the thermodynamic preference on the Si(110) lithiation. Meanwhile, Cubuk has calculated the lithium insertion energy barrier into Si(110) and Si(111) facet [8]. They have concluded that the Li insertion energy barrier is higher in the Si(111) facet than Si(110). This shows evidence for the kinetic effect. However, all of these *ab initio* studies do not account for the presence of the phase boundary.

In addition to *ab initio* studies, silicon lithiation is also studied on a larger scale. Kim et al. [9] have performed a classical molecular dynamic study by ReaxFF [10] for the lithiation rate on Si(100), Si(110), Si(111), and *a*-Si and tried to reproduce the anisotropic lithiation. The lithiation rate between Si(110) and Si(100) is clear, but the rate between Si(110) and Si(111) is almost indistinguishable and Si(111) lithiates a bit faster than the Si(110) facet. Later, Choi et al. and Chen et al. have also performed similar slab lithiation [11,12], while Chen et al. have stated that the Si(110) also undergoes Si(111) peeling-off. However, in these studies, the almost immobile Si(111) lithiation as in the experiment cannot be fully reproduced. Besides, Ostadhossein et al. and Ding et al. have also found that the planar compressive stress may slow down the lithiation rate by ReaxFF molecular dynamics simulation [13,14].

In this work, we will use our newly developed ReaxFF model for simulating the lithiation behavior of the silicon anode. We will reproduce the anisotropic lithiation for Si(100), Si(110), Si(111), and Si(112) facet as in the experiment. Later, we will try to explain the mechanism of the anisotropic lithiation, and the stress issue should be the most prominent in the anisotropic lithiation in silicon. Finally, we will conclude whether the anisotropic lithiation should be a thermodynamic or kinetic-dominated process.

7.2 COMPUTATIONAL DETAILS

For studying the dynamic lithiation process of a silicon anode, larger-scale molecular dynamics simulation should be performed. The classical forcefield ReaxFF proposed by van Duin et al. [10] is suitable for large-scale simulations. Besides, it is capable

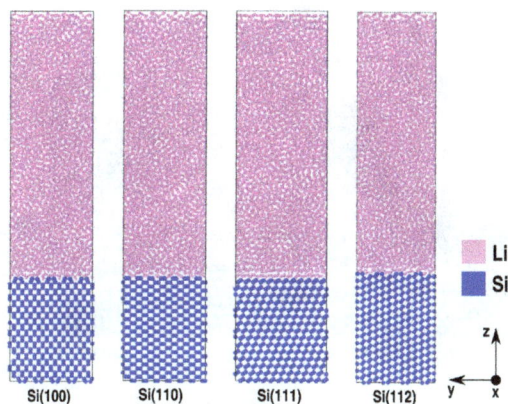

FIGURE 7.1 The initial settings of the slab lithiation in this work. The system contained about 4,000 Si atoms and 9,000 Li atoms, while the cell is about 40 Å × 40 Å × 180 Å. Silicon occupied up to about 50 Å from the bottom and the remaining region is filled with Li with BCC Li density.

of describing the bond-breaking reaction and the charge-variation process, which is important during the lithiation process. We have modified the ReaxFF parameters from Ostadhossein et al. [13] and corrected the lithiation rate problem of Si(111).

In this work, we will use Si(100), Si(110), Si(111), and Si(112) slabs for lithiation as in Figure 7.1. The system contained about 4,000 Si atoms and 9,000 Li atoms. The cell x, y-direction is about 40 Å while z-direction is roughly 50 Å for Si and 130 Å for Li and total length about 180 Å. The molecular dynamic simulation is performed at 800, 900, 1,000, 1,100, and 1,200 K to sample the temperature-dependent anisotropic lithiation effect. The silicon within the bottom 5 Å is fixed to mimic the bulk Si during the molecular dynamic simulation. The simulation is run for 4 ns in total for each slab until the lithiated silicon reaches the top of the simulation cell.

7.3 RESULTS AND DISCUSSIONS

7.3.1 General Lithiation Behavior and Anisotropic Lithiation

The simulation results at 800 K at 4,000 ps are shown in Figure 7.2. The general lithiation behavior for each facet will be clarified: from the bottom up, the *crystalline silicon* lays at the bottom layer. In the second layer, Li insertion into silicon and forms a *phase boundary*. The *lithiated silicon* (Li–Si alloy) is on the third layer and *pure lithium* is on the topmost layer. The lithiation reaction involves three major reactions: (1) Li insertion into phase boundary from the Li–Si alloy layer, (2) Li insertion into silicon from the phase boundary, and (3) Si–Si bond breaking into the Li–Si alloy layer.

In Figure 7.2, a clear anisotropic lithiation is reproduced, and it follows the order of Si(112) > Si(110) > Si(100) > Si(111). First of all, we have observed the Si(112) ledge lithiation mechanism by Si(111) peel-off as observed by Liu et al. in the experiment [5]. A thick phase boundary is observed at this temperature. The Si(110) has the

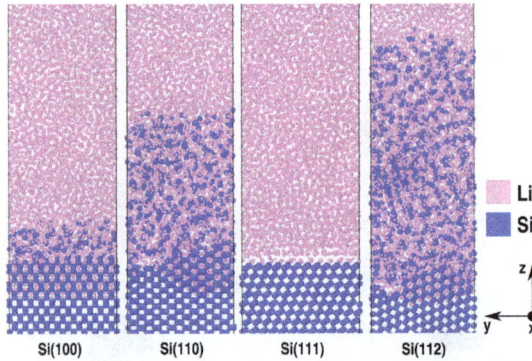

FIGURE 7.2 The Si slab lithiation at 800 K, 4,000 ps in this work. The four major phases from the bottom up are (1) crystalline silicon, (2) phase boundary, (3) lithiated silicon, and (4) pure lithium. A clear anisotropic lithiation is observed at this temperature, and Si(111) is almost immobile as observed in the experiment. Ledge mechanism is observed for Si(112) facet, while Si(110) followed similar peeling-off of Si(111), and a thick phase boundary is observed.

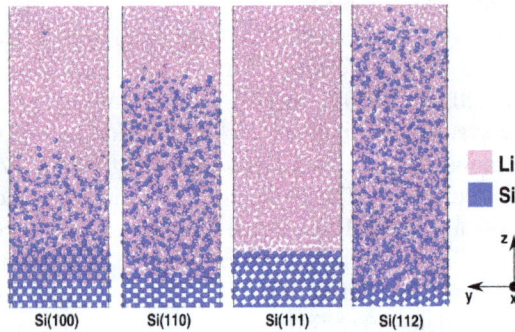

FIGURE 7.3 The Si slab lithiation at 1,000 K, 1,000 ps in this work. Less anisotropy is observed while similar lithiation behavior is found as in the 800 K case. The phase boundary has become thinner at this temperature.

second-fastest lithiation rate and showed a V-shaped phase boundary. It can, therefore, be observed that Si(110) facet followed the same peeling-off of the Si(111) facet. The Si(100) showed a relatively thick and flat phase boundary and the almost immobile Si(111) is reproduced in our study.

When the temperature has increased to 1,000 K, the snapshot at 1,000 ps is shown in Figure 7.3. The lithiation follows the same order, but the anisotropic lithiation is not as clear as in the 800 K case. The Si(112) facet with Si(111) ledge peeling-off behavior is again clear in this figure. The Si(110) comes with a closer lithiation rate as the Si(112) facet than the 800 K case. Besides, the phase boundary in Si(110) becomes thinner than the one at 800 K.

At our highest temperature as 1,200 K, the snapshot at 300 ps is shown in Figure 7.4. The lithiation becomes more isotropic at this temperature and the Si(112),

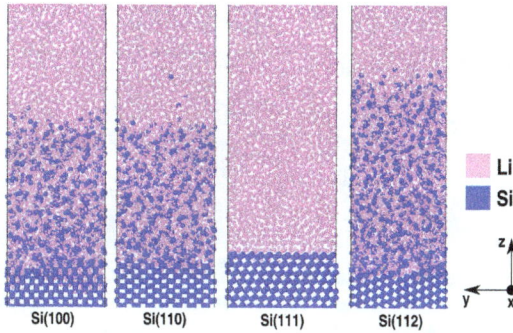

FIGURE 7.4 The Si slab lithiation at 1,200 K, 300 ps in this work. More isotropic lithiation is observed and Si(111) starts to lithiate very slowly at this temperature. The Si(100) turned to a V-shaped phase boundary rather than a flat one at the lower temperature.

Si(110), and Si(100) have nearly the same lithiation rate. The Si(111) started to lithiate very slowly. The phase boundary thickness also becomes thinner for the Si(100) and shows a V-shaped lithiation.

The lithiation rate is quantitatively compared by counting the silicon atoms flux lithiated into Li–Si alloy and its Arrhenius plot is also compared in Figure 7.5. The anisotropic lithiation is clear at the lower temperature while it becomes more isotropic at the higher temperature. This implies that the energy barrier may dominate the anisotropic lithiation process. The activation energy is calculated as 0.85, 0.59, 1.19, and 0.52 eV for Si(100), Si(110), Si(111), and Si(112), respectively. The phase boundary thickness is calculated in Figure 7.6. In general, we have found that the facet with a higher lithiation rate will have a thinner phase boundary.

7.3.2 Stress Influence on the Anisotropic Lithiation

In order to find out the mechanism of the anisotropic lithiation, we have performed stress analysis during the lithiation at 800, 1,000, and 1,200 K as in Figures 7.7, 7.8, and 7.9, respectively. We have found that at the lower temperature of 800 and 1,000 K, the Si(100) facet undergoes high compressive stress in the phase boundary. This phenomenon disappeared at the higher temperature of 1,200 K. This is in the same trend as the relative lithiation rate increase in Si(100) and implies that it is the high compressive stress that impedes the lithiation. More specifically, the high compressive stress developed on the *surface* of the Si(100) facet impedes Li insertion, therefore causing a slower lithiation rate for Si(110) than Si(100).

As for the almost immobile Si(111) facet, we have performed another analysis to emphasize the effect of the Si(111) facet. We have used the same amount of Li and approached the clean Si(100), Si(110), Si(111), and Si(112) surface. Their induced stress on the surface is averaged in Figure 7.10. We have shown that the Si(111) has the highest compressive stress on the surface and, therefore, it will have the highest impedance on the Li insertion. This is not very surprising since Si(111) is the densest facet in the FCC structure.

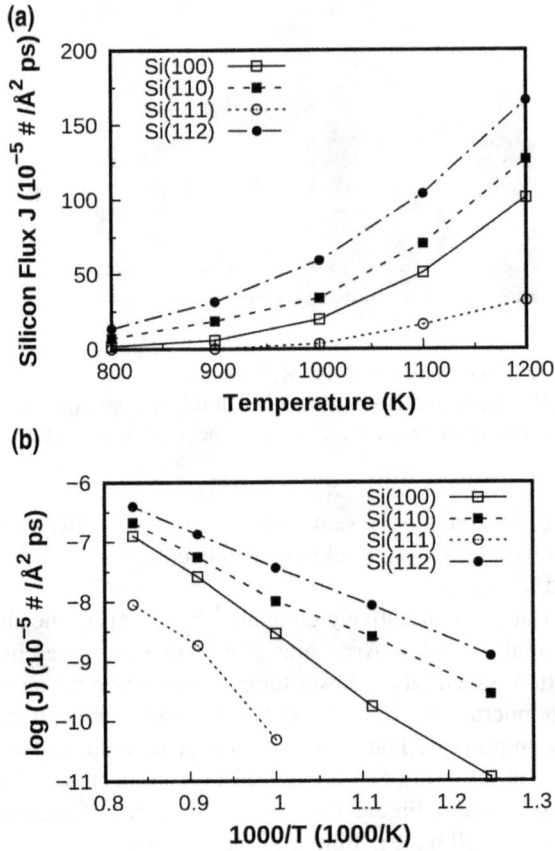

FIGURE 7.5 The (a) lithiation rate and (b) its Arrhenius plot. The anisotropic lithiation is again found to be prominent at the lower temperature.

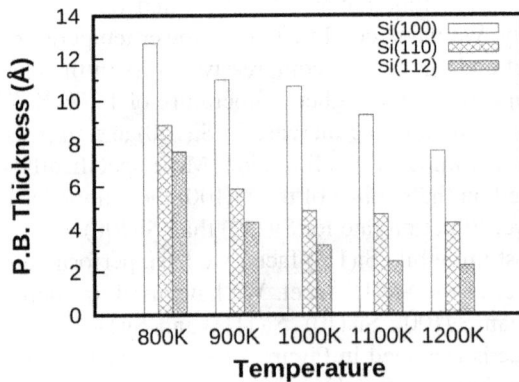

FIGURE 7.6 The phase boundary thickness (unit: Å) of each facet and temperatures. The phase boundary thickness is found to be in reverse order with the lithiation rate.

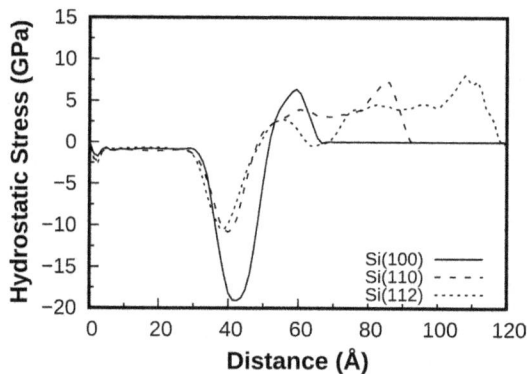

FIGURE 7.7 The Si only stress of the phase boundary region at 800 K, 3,000 ps. High compressive stress is observed in the Si(100) facet than other facets.

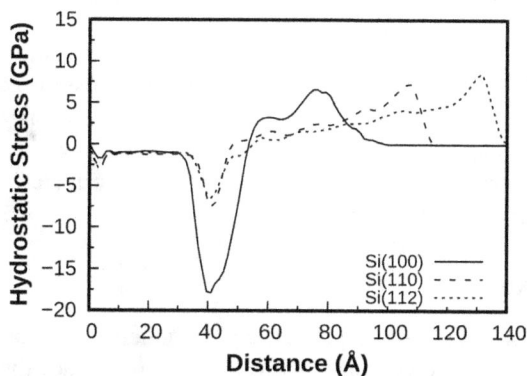

FIGURE 7.8 The Si only stress of the phase boundary region at 1,000 K, 800 ps. A similar high compressive stress is observed in the Si(100) facet.

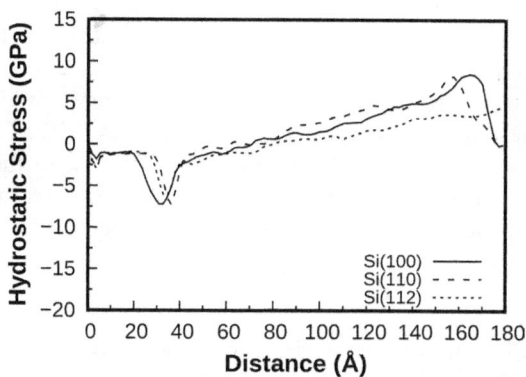

FIGURE 7.9 The Si only stress of the phase boundary region at 1,200 K, 800 ps. The stress of Si(100) is roughly the same as Si(110) and Si(112), which followed the same trend as in Figure 7.5.

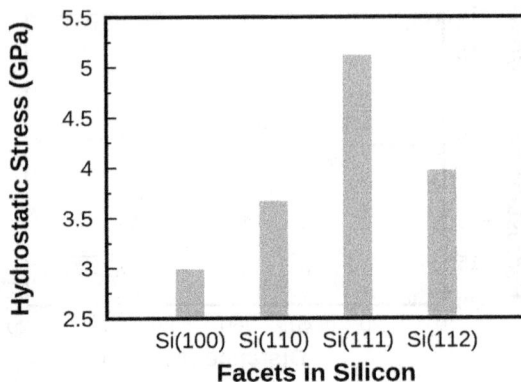

FIGURE 7.10 The surface Si stress by the same amount of Li approaching Si(100), Si(110), Si(111), and Si(112) surface. A much higher compressive stress is observed in Si(100) and therefore impedes Li insertion the most.

In summary, we have concluded that the anisotropic lithiation is a compressive stress induced effect. Different degrees of compressive stress developed on each surface induced different kind of impedance on the Li insertion energy barrier, therefore causing the anisotropic lithiation. These Li insertion energy barrier related effects show that the anisotropic lithiation should be kinetic-dominated rather than a thermodynamically controlled process.

7.4   CONCLUSIONS

In this work, we have newly developed a ReaxFF model from Ostadhossein et al. and corrected some problems, especially the disability to reproduce the almost immobile Si(111) lithiation. We have, therefore, performed the lithiation on Si(100), Si(110), Si(111), and Si(112) slabs and successfully reproduced the experimentally observed anisotropic lithiation. The lithiation rate order is shown as Si(112) > Si(110) > Si(100) > Si(111), and the anisotropy decreased at the higher temperature. The phase boundary is shown in reverse order with the lithiation rate. The stress analysis is performed, and we have found that the compressive stress for the surface Si on Si(100) is higher than Si(110) and Si(112), while Si(111) is shown to have an even higher compressive stress. This compressive stress corresponds to the increase of the Li insertion energy barrier into silicon for Si(100) and Si(111) and the decrease of their lithiation rate. Finally, these Li insertion energy barrier related processes show that the anisotropic lithiation is a kinetic dominated rather than a thermodynamic dominated process.

REFERENCES

1. S. Bourderau; T. Brousse and D. Schleich Amorphous silicon as a possible anode material for li-ion batteries. 1999 *J. Power Sources* **81–82** 233–236. doi:10.1016/s0378-7753(99)00194-9.
2. C. K. Chan; H. Peng; G. Liu; K. McIlwrath and X. F. Zhang *et al.* 2007 High-performance lithium battery anodes using silicon nanowires. *Nat. Nanotechnol.* **3** (1) 31–35. doi:10.1038/nnano.2007.411.

3. X. H. Liu; H. Zheng; L. Zhong; S. Huang and K. Karki *et al.* Anisotropic swelling and fracture of silicon nanowires during lithiation. 2011 *Nano Lett.* **11** (8) 3312–3318. doi:10.1021/nl201684d.

4. S. W. Lee; M. T. McDowell; J. W. Choi and Y. Cui Anomalous shape changes of silicon nanopillars by electrochemical lithiation. 2011 *Nano Lett.* **11** (7) 3034–3039. doi:10.1021/nl201787r.

5. X. H. Liu; J. W. Wang; S. Huang; F. Fan; X. Huang et al. 2012 In situ atomic-scale imaging of electrochemical lithiations in silicon. *Nat. Nanotechnol.* **7** (11) 749–756. doi:10.1038/nnano.2012.170.

6. S. C. Jung; J. W. Choi and Y.-K. Han 2012 Anisotropic volume expansion of crystalline silicon during electrochemical lithium insertion: An atomic level rationale. *Nano Lett.* **12** (10) 5342–5347. doi:10.1021/nl3027197.

7. M. K. Y. Chan; C. Wolverton and J. P. Greeley 2012 First principles simulations of the electrochemical lithiation and delithiation of faceted crystalline silicon. *J. Am. Chem. Soc.* **134** (35) 14362–14374. doi:10.1021/ja301766z.

8. E. D. Cubuk; W. L. Wang; K. Zhao; J. J. Vlassak and Z. Suo *et al.* 2013 Morphological evolution of si nanowires upon lithiation: A first-principles multiscale model. *Nano Lett.* **13** (5), 2011–2015. doi:10.1021/nl400132q.

9. S. P. Kim; D. Datta and V. B. Shenoy 2014 Atomistic mechanisms of phase boundary evolution during initial lithiation of crystalline silicon. *J. Phys. Chem. C* **118** (31) 17247–17253. doi:10.1021/jp502523t.

10. A. C. T. van Duin; S. Dasgupta; F. Lorant and W. A. Goddard ReaxFF: A reactive force field for hydrocarbons. 2001 *J. Phys. Chem. A* **105** (41) 9396–9409. doi:10.1021/jp004368u.

11. D. Choi; J. Kang; J. Park and B. Han First-principles study on thermodynamic stability of the hybrid interfacial structure of LiMn2o4 cathode and carbonate electrolyte in li-ion batteries. 2018 *Phys. Chem. Chem. Phys.* **20** (17) 11592–11597. doi:10.1039/c7cp08037a.

12. S. Chen; A. Du and C. Yan 2020 Molecular dynamic investigation of the structure and stress in crystalline and amorphous silicon during lithiation. *Computational Materials Science* **183** 109811. doi:10.1016/j.commatsci.2020.109811.

13. A. Ostadhossein; E. D. Cubuk; G. A. Tritsaris; E. Kaxiras and S. Zhang *et al.* 2015 Stress effects on the initial lithiation of crystalline silicon nanowires: Reactive molecular dynamics simulations using ReaxFF. *Phys. Chem. Chem. Phys.* **17** (5) 3832–3840. doi:10.1039/c4cp05198j.

14. B. Ding; H. Wu; Z. Xu; X. Li and H. Gao Stress effects on lithiation in silicon. 2017 *Nano Energy* **38** 486–493. doi:10.1016/j.nanoen.2017.06.021.

8 Optical Properties of Monolayer and Lithium-Intercalated HfX$_2$ (X = S, Se, or Te) for Lithium-Ion Batteries

Thi My Duyen Huynh, Hai Duong Pham, and Ming-Fa Lin
National Cheng Kung University

CONTENTS

8.1 INVESTIGATING ORIENTATION IN BATTERIES OF GROUP IV TMDs

To express the challenges arising from global climate change, electrochemical energy storage (EES) systems including capacitors and batteries have been extensively applied in a plenty of fields, especially for electric vehicles, smart electrical grids, and electronic devices. Among EES systems, batteries which are commonly composed of flexible electrode materials with higher theoretical capacity [1] with the high specific energy density and low self-discharge memory effect [2,3] become a numerous power supplies for electronics. In a wide range of batteries, lithium-ion batteries (LIBs) and sodium-ion batteries (SIBs) emerge as an outstanding electrochemical performance [4]. Lithium compound in batteries is selected as the electrode material, and their working principle is based on the chemical reaction of lithiation and delithiation process [5]. These batteries have the high energy and power density, high operating voltage, low self-discharge rate, and long-term cycling stability [6,7].

DOI: 10.1201/9781003263807-8

To satisfy particular requirements of flexible electronics in tough environment situations, flexible LIBs are referred to conventional LIBs [6]. Electrochemical active materials, electrolytes, and collectors are three common parts which are widely concerned in flexible LIBs [8]. Due to the wider availability, lower cost, and similar insertion chemistry properties of SIBs compared to LIBs, they become promising candidate for large-scale energy storage [9], which has attracted the attention. Sodium element becomes the advantageous alternative for batteries besides lithium which exhibits a rather negative redox potential [9]. However, radius of sodium is 55% larger than that of lithium [10], which induces volume change causing the challenge to form staged intercalation compounds with graphite. Furthermore, in enhancement of the flexible LIBs and SIBs for energy storage, a numerous of developments on batteries have been considered, such as flexible magnesium ion batteries [11], potassium ion batteries [12], lithium-O2 batteries [13], and lithium sulfur batteries [14]. Recently, most of the concerns in batteries are related to electrode materials which dominate in enhancing the electrochemical properties of batteries [6,7,15]. 2D materials including graphene and graphene-like such as borophene, silicene, germanene, phosphorene, hexagonal boron nitride, carbon nitride, transition metal oxide (TMOs), transition metal dichalcogenides (TMDs), and so on [16–21] offer an appealing promise for the development of high-performance electrode materials for flexible energy storage [22,23]. These materials display an enhanced specific capacity, superior cycle stability, rate capacity, and flexible.

With regard to applying in LIBs, the enhancement of lithium ions shown in Figure 8.1 illustrates the pathways for diffusion and transfer in electric vehicles which gives a perspective of LIBs applications. Three types of 2D materials including graphene, TMOs, and TMDs have been explored promising for electrodes [24–27]. Due to the super-high theoretical surface area and exceptional room-temperature electron mobility, graphene reveals large interfacial lithium storage and fast charge/carrier transport. However, pristine graphene is only chemically active at the edges that cannot provide a stable potential output [28]. In contrast, TMOs and TMDs are

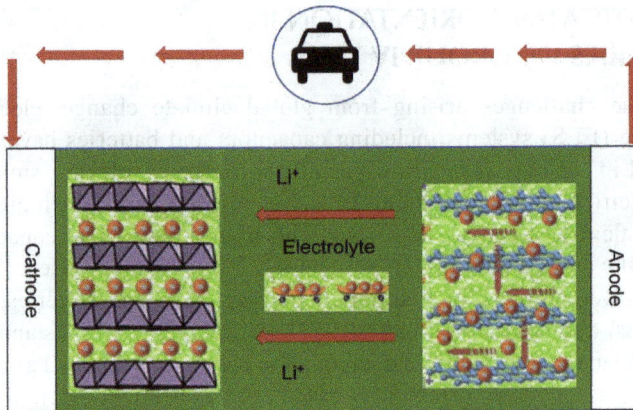

FIGURE 8.1 Schematic demonstration of pathways for lithium-ion diffusion and transfer [22].

fundamentally inferior in terms of electronic conductivity. Most TMOs have superior storage performance because of their outstanding ion transport kinetics and chemically active interfaces [29–31]. To form graphene-based composites, various TMOs such as TiO$_2$ [32,33], Cr$_2$O$_3$ [34], Fe$_2$O$_3$ [35], ZnMn$_2$O$_4$ [36] or TMDs like MoS$_2$ [37–39], WS$_2$ [40,41] nanosheets have been implemented to integrate with graphene. Besides, TMDs, especially MoS$_2$ [42], were also reported as high-performance electrode materials. With the similar structure as graphite, MoS$_2$ allows reversible Li-ion intercalation or extraction without an ultrahigh specific capacity [43,44]. Based on MoS$_2$, WS$_2$ is an attractive material for anode due to its theoretical specific capacity which is higher than that of graphene [41]. On the subject of applying in SIBs, various electrode materials had been explored [45–48]. The transparent freestanding MoS$_2$ films had prepared due to its superior flexibility [49]. In addition to MoS$_2$, WS$_2$ acts as anode material for SIBs [50] which demonstrates the superior property. WSe$_2$ had also reported as anode for sodium storage [51] which simultaneously displays an excellent electrochemical performance and reversible specific capacity. Furthermore, based on the weak van der Waals interaction and the larger interlayer distances [52], ReS$_2$ is a potential candidate to intercalate and deintercalate for sodium ions. Hence, 2D materials, particularly TMDs, promote the lithium or sodium storage capability, rate performance, and cycling stability.

Moreover, these materials were found in other applications of supercapacitors, and solar cell leading to apply in optoelectronic devices or photodetectors. The theoretical and experimental study on MoS$_2$/graphene membrane showed the high electrochemical performance of specific capacitance [53] to get the composite membrane. For ultrasensitive monolayer MoS$_2$ photodetector [54], a photoresponsivity was reported about six orders of magnitude which is higher than that of graphene. By using the high-resolution transmission electron microscopy and selected area electron diffraction, VS$_2$ were analyzed indicating the great crystallinity and the microcosmic orientation [55]. VS$_2$ thin film also showed high conductivity and high specific area for forming an in-plane supercapacitors [55]. Additionally, WS$_2$ [56] and WSe$_2$ [57] also exhibited greatly enhanced capacitance for construction of supercapacitors. For solar cell, monolayer TMDs-based heterojunctions could be the promising candidates [58,59].

Via performances in LIBs, SIBs, and others, some kinds of TMDs promise the practical applications in EES. Figure 8.2 shows the probability to apply TMDs in batteries. Thus, other TMDs based on these materials might have broad influences on batteries by further improving optimization. However, almost concerning has been placed on Group VI, particularly MoS$_2$. Expanding to other groups based on MoS$_2$ recently becomes trend in investigating new TMD materials to open up opportunities for device applications. Following up this orientation, Group IV TMDs have been explored revealing that there exists the decomposition and dissolution because metal dichalcogenides were mostly unstable upon electromechanical treatment [60]. It indicates that the purity of these materials is obviously crucial for high catalytic performance. In this group, HfX$_2$ (X = S, Se, or Te) which has strong covalent bonds in layers is a new type of 2D TMDs. The theoretical investigation on 1T-HfX$_2$ [61] suggests that the trends in the band structures can be attributed to both the type of atoms and the geometric properties. The perspective of features should be examined

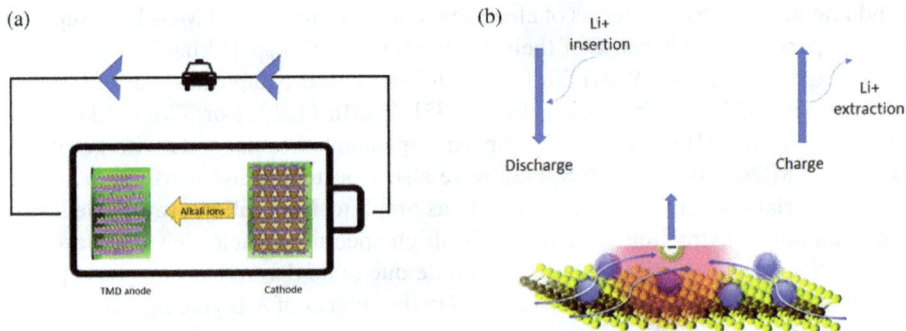

FIGURE 8.2 (a) TMD electrodes in alkali-ion batteries [68]. (b) Enhanced ion transport mechanism of the $MoS_2/MoSe_2$ heterojunction structure in LIBs [68,69].

and defined to provide detailed information about their properties. However, there has been a limitation of theoretically and systematically clarified understanding of electronic, optical, and electrochemical behavior of these materials. Especially, lithium-intercalated/adsorbed/decorated [62,63] HfX_2 has a lack of detailed information that should be clarified to figure out their advantages in optoelectronic devices, that suggest further and expanded applications for batteries.

In general, EES, particularly LIBs and SIBs batteries, has attracted concerning in research community that has been extensively applied in a wide range of smart electrical grids and electronic devices. Although significant progress has been implemented on 2D materials, such as graphene, graphene-like, TMOs, or TMDs, to improve and enhance applications for batteries, some major challenges still exist for construction of these materials. Thus, expanding investigations and applications, the newborn 2D materials become the potential trend in this field. As previously mentioned, Group IV TMD, especially HfX_2, is one of the newborn 2D materials promising for development of high performance in batteries. However, to evaluate the satisfaction in application of this material group, the background of electronic and optical properties needs to analyze to provide the perspective of their properties. Obviously, optical properties connected to structural and electronic properties of these materials have a lack of detailed information. Furthermore, lithium-intercalated HfX_2 might cause a significant effect on electronic structure, leading to high probability to apply in LIBs. In this work, we therefore focused on summary of the geometric, electronic, and optical properties to figure out useful information about these materials. Thus, the lithium intercalation was investigated promise for high conductance of electrode in LIBs.

8.2 STRUCTURAL AND ELECTRONIC PROPERTIES OF MONOLAYER HfX₂

8.2.1 STRUCTURAL PROPERTIES

Monolayer TMDs (MX_2) form a sandwiched structure of X-M-X which has only two polymorphs, namely the T and H phases depending on their hexagonal or trigonal

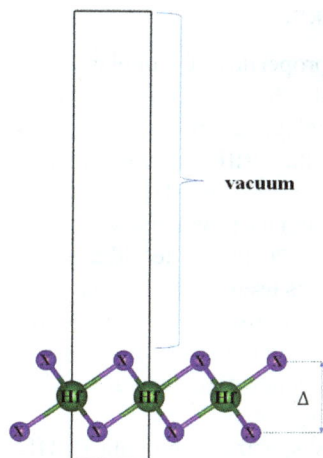

FIGURE 8.3   Structural optimization of monolayer HfX$_2$.

prismatic coordination. According to the stability which is confirmed in our previous report, the T phase is the stable one that is also favorable of monolayer HfX$_2$. With layered structures in this work, monolayer HfX$_2$ is only conducted in the T phase as shown in Figure 8.3. To avoid the interaction between layers, a vacuum of 15 Å is added to this structure. Moreover, monolayer HfX$_2$ exists in a bucked form (delta) (Figure 8.3) as shown in Table 8.1 which attributes to the sandwiched structure. The chemical bonding of Hf-X contributes to electronic properties that depend on electron configuration and species of chalcogen atoms. Moreover, the atomic configuration changes based on the transition metal in the compound when chalcogen atoms involve in chemical bonding. The chalcogen atoms with sp3 hybridized and atomic orbitals of hafnium characterize the geometry and electronic properties of these materials. Besides, the Hf-X interaction suppresses the height fluctuation of the hafnium atom in these monolayers. These height fluctuations are presented at equilibrium that provides the broad distribution of height displacement. This phenomenon predicts the phenomenological theories of thermal fluctuations in flexible membranes.

TABLE 8.1

Lattice Constant (a, (Å)), Chemical Bonding (d_{M-X} (Å)), Buckling (Δ, (Å)), and Bandgap (E_g(eV)) of Monolayer HfX$_2$

HfX$_2$	a (Å)	d_{M-x}(Å)	Δ(Å)	E_g(eV) PBE	HSE	SCAN
HfS$_2$	3.643	2.552	2.892	1.2	–	1.16
HfSe$_2$	3.781	2.684	3.122	0.86	–	0.47
HfTe$_2$	3.976	2.896	3.531	−0.01	−0.004	−0.48

8.2.2 ELECTRONIC PROPERTIES

To understand the optical properties of monolayer HfX_2, the electronic properties should be analyzed to clarify the characteristic in energy structure. The electronic band structures have been calculated along with the high symmetry point Γ-M-K-Γ of the hexagonal lattice in the Brillouin zone. The bandgaps of monolayer HfX_2 are shown in Table 8.1. Both HfS_2 and $HfSe_2$ are semiconducting, which respectively belong to middle-gap and narrow-gap one, while $HfTe_2$ is a gapless semimetal. To further verify the semimetal properties, the hybrid functional Heyd-Scuseria-Ernzerhof (HSE) calculation is used for $HfTe_2$ to make the comparison between two theoretical methods. This value listed in Table 8.1 shows that the HSE calculation causes the opened-gap compared to Perdew-Burke-Ernzerhof (PBE) one indicated in the smaller overlap between conduction and valence bands. However, it still remains the property of semimetal.

The system band energy spectra of monolayer HfX_2 are shown in Figure 8.4, which displays the atom dominances of hafnium and chalcogens. As shown in this figure, the valence band maximum (VBM) is located at Γ point ($k=0, 0, 0$), and the conduction band minimum (CBM) is at M point ($k=\frac{1}{2}, 0, 0$) of the high symmetry k-points. They share a common feature that the unoccupied conduction bands are asymmetric to the occupied valence bands about the Fermi level. A strong dispersion relation with the parabolic dispersion is being exhibited in both the conduction and valence bands of most energy bands. An indirect bandgap with the direction of the gap shifted from Γ point to M point is exhibited in semiconductor subgroup, and an overlap between conduction and valence bands exists in semimetallic one. The occupied valence band is strongly asymmetric to the unoccupied conduction band. The low-lying energy valence bands also change into parabolic bands. The parabolic dispersions have band-edge states at M point belonging to saddle points, which features in van Hove singularities (vHs).

However, each material is predicted to have an unusual energy band. There are many distinguishing features among HfX_2 family in the electronic band structures. In Figure 8.4, the maroon circles describe hafnium atoms, and the cyan and green circles, respectively, represent sulfur and selenium atoms. The gap shifted from Γ point to M point demonstrates an indirect bandgap semiconductor. The occupied whole band is strongly asymmetric to the unoccupied electron band. The σ and σ^* bands of HfS_2 initiate from Γ point at -2.5, 0, and 2 eV, while the σ and σ^* of

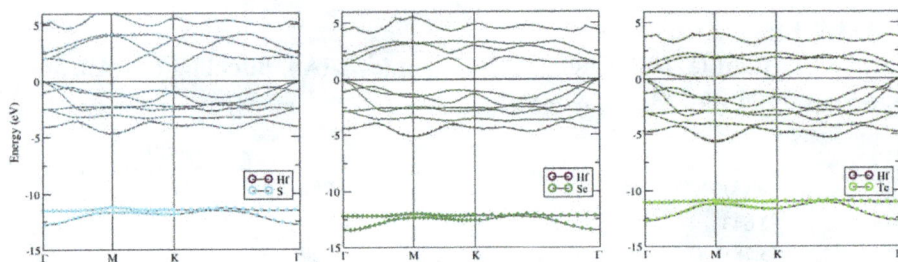

FIGURE 8.4 Electronic band energy with atom dominance of monolayer HfX_2.

HfSe$_2$ are at -3, 0, and 1.5 eV. The valence band shows not only the special valleys located at band edges but also flat bands in the low-lying energy. Simultaneously, energy dispersion in this band changes into parabolic bands. Furthermore, the crossing bands are found in low-lying valence band between Γ and M points as well as Γ and K points; particularly, the crossing point is found at K point. With regard to atom dominance in bands, the sulfur and selenium atoms dominate in valence band ranging from 0 to -15 eV, especially in low-lying energy, while the contribution of hafnium atoms is smaller. In contrast, hafnium contributes to the conduction band more than the valence band. Hafnium, sulfur, and selenium contribute to the whole band with ranges from -15 to 6 eV, but two chalcogen atoms show a sharper dominance in valence band while a sharper dominance of hafnium is found in conduction band.

On the other hand, a semimetallic property of HfTe$_2$ has been shown, which is in good agreement with previous studies [64–66]. As shown in Figure 8.4, the maroon circles describe hafnium atoms, and the slight green circles, respectively, represent tellurium atoms. Apart from the special valleys located in the low-lying valence band that is similar to HfS$_2$ and HfSe$_2$, there exist flat bands in the low-lying valence bands. There are σ and σ^* that initiate from Γ point at -3, 0, and 0.5 eV. The difference between HfTe$_2$ and both HfS$_2$ and HfSe$_2$ is that all of the VBMs and CBMs are nearly at the Fermi level, which attributes to its semimetallic feature. Besides, tellurium atoms dominate in valence band ranging from 0 to -15 eV, while the contribution of hafnium atoms is smaller than in this band. In contrast, hafnium contributes to the conduction band more than the valence band. In general, these band energy spectra exhibit a lot of asymmetric peaks due to the parabolic form of energy bands. A pair of asymmetric peaks appears near the Fermi level that defines the indirect energy gaps. Energy dispersion in their low-lying energy of each material is different although they are in the same group. Additionally, most vHs come from the parabolic energy dispersions. The peaks near the Fermi level are found to arise from the band-edge states mainly which could be a signature characteristic unique to vHs.

To further verify the configuration effect in electronic properties, the orbital-projected density of states (DOSs) are used as described in Figure 8.5 indicating the local DOSs with multi-orbital hybridization monolayer HfX$_2$. As shown in this figure, the main contributions are due to s, d-orbitals of hafnium atoms and s, p-orbitals of chalcogens atoms. With regard to multi-hybridization of the orbital-projected DOS, monolayer HfX$_2$ shows that $5d_{xy}$, $5d_{z^2}$, and $5d_{x^2-y^2}$ orbitals of hafnium and ($3p_x$, $3p_y$, $3p_z$), ($4p_x$, $4p_y$, $4p_z$), and ($5p_x$, $5p_y$, $5p_z$) of sulfur, selenium, tellurium, respectively, dominate in conduction band and valence band. In addition, there is the dominance of 3s, 4s, and 5s of sulfur, selenium, and tellurium, respectively, located at low-lying energy. Generally speaking, their contributions are comparable to each other in that the feature-rich energy bands are mainly determined by the d and p orbitals. The strong hybridizations of ($4d_{xy}$, $4d_{z^2}$, $4d_{x^2-y^2}$), ($5d_{xy}$, $5d_{z^2}$, $5d_{x^2-y^2}$) Hf orbitals, and ($3p_x$, $3p_y$, $3p_z$), ($4p_x$, $4p_y$, $4p_z$) and ($5p_x$, $5p_y$, $5p_z$) chalcogen orbitals exist in DOS demonstrating the multi-hybridization of their orbitals. The p-orbitals of the chalcogens make significant contributions to the bonding between hafnium and chalcogen atoms in sandwiched structure. In addition, the DOS ranged from -4 to 4 eV and shows a

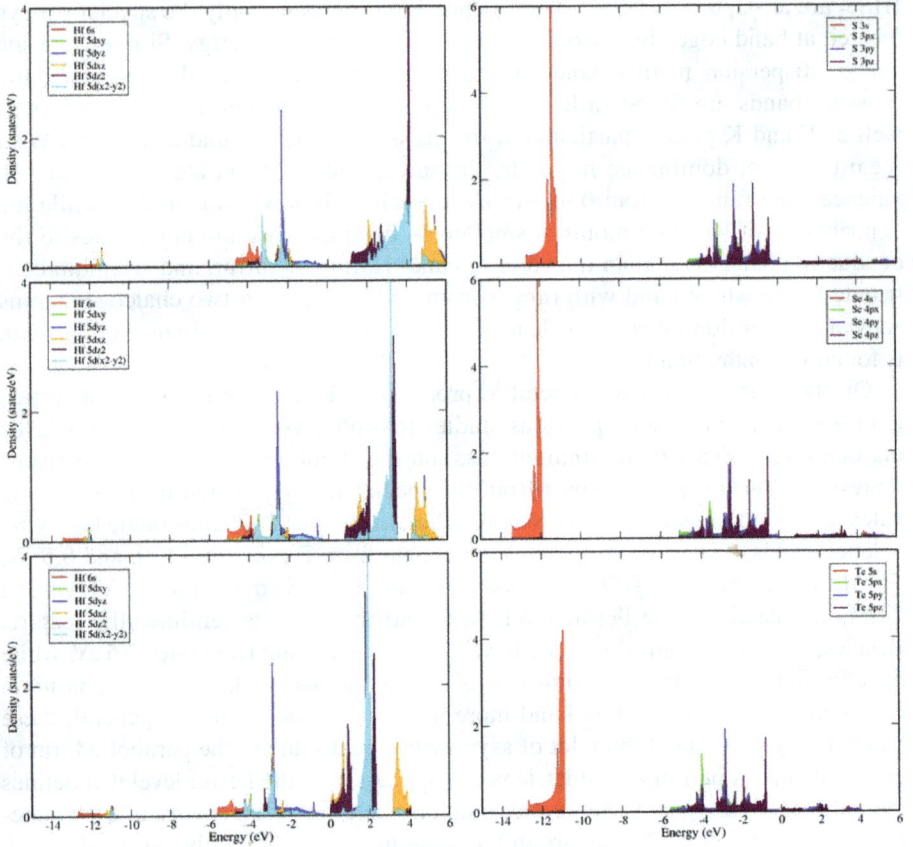

FIGURE 8.5 Orbital-projected DOS of monolayer HfX$_2$.

pair of asymmetric peaks centered at the Fermi level, which characterizes the energy gap and the direction of bandgap. The large energy difference between the valence and conduction peaks obviously demonstrates the asymmetry of the two bands. These structures originate from linear bands, parabolic bands near saddle points, and initial band-edge states of the parabolic bands. Also, DOS exhibits a logarithmic divergence at middle energy and shoulder structures at deeper and higher energy. Partially, flat energy is exhibited in the valence bands of all structures and in the conduction bands of HfSe$_2$ and HfTe$_2$, while shoulder structures evidently appear in the conduction band of HfS$_2$. Moreover, the linear straights in the DOS near the Fermi level which are parallel lines indicate the feature of vHs accompanied with the parabolic energy dispersions and saddle points. This vHs is peculiar in the corresponding region of Brillouin zone as neither conduction band nor valence band have singularities in the DOSs.

To improve upon the accuracy of a computationally expensive hybrid functional, at almost GGA, strongly constrained and appropriately normed (SCAN) is used which is expected to have broad impact on these materials. Bandgap using SCAN

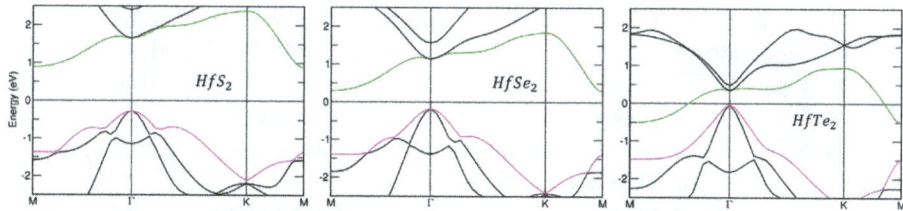

FIGURE 8.6   Band energy spectra with scan test.

test of monolayers HfX_2 is also listed in Table 8.1. Compared to PBE, these values are significantly smaller ones that might give more realistic bandgaps. However, the features of semiconductors and semimetal are still remained. Furthermore, the band energy spectra of these materials with SCAN are shown in Figure 8.6 to further provide information about the impact of SCAN. To clearly describe the features of band energy, band energy spectra are plotted along M-Γ-K-M in the range of (−2.5, 2.5) eV. As shown in this figure, the direction of bandgap is still shifted from M to Γ point for semiconductors while there exists the strong overlap between conduction and valence bands of semimetal that are consistent with PBE bands (Figure 8.4). Apart from the characteristics which are discussed in band PBE, these bands show the saddle points at Γ points belong to vHs feature. This point indicates the significant properties of vHs characterized by the parabolic bands in valence bands near the Fermi level. Hence, monolayer HfX_2 demonstrates for the signature of vHs characteristic. With regard to semimetal $HfTe_2$, the overlap between conduction and valence bands is obviously displayed in this figure in which the minimum conduction band is at the below Fermi level. In short, SCAN gives a theoretical method to further check the feature of material sciences. These results accompany with PBE or HSE calculations that are again affirmed the properties of these materials.

8.3   OPTICAL PROPERTIES

To determine the optical properties, two parts of relative permittivity (dielectric constant) are first considered and described in Figure 8.7. The magenta and green curves respectively represent the imaginary (Im ε) and real parts (Re ε). As shown in this

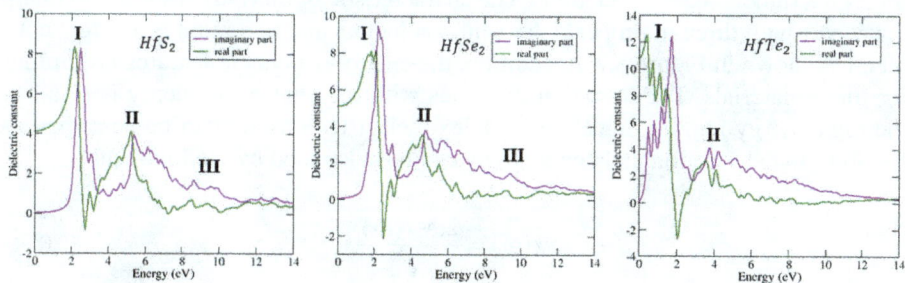

FIGURE 8.7   Dielectric constant with imaginary and real parts of monolayer HfX_2.

figure, there are three main parts and two main parts of both imaginary and real ones for two semiconductors and semimetal, respectively, labeled I, II, and III. Here, the imaginary part is first considered. Similar to band energy, because of standing in the same group, three materials also share the common features for optical properties. The shift of all structures is toward lower energies when going from sulfur to selenium to tellurium. This feature can be attributed to the change of energy gap as replacing sulfur by selenium or tellurium. II and III are remarkably weaker than I in HfS_2 and $HfSe_2$ while III part in $HfTe_2$ is almost absent from both Im ε and Re ε. This weakness could be caused by the scale of imaginary part as $1/\omega^2$ from the following equations of relative permittivity.

$$\varepsilon = \varepsilon_1 + i\varepsilon_2 \tag{8.1}$$

$$\varepsilon_2(\omega) = \frac{4\pi e^2}{m^2\omega^2} \sum_v \sum_c \left| \vec{\varepsilon}.\overline{M_{cv}}\left(\vec{k}\right) \right|^2 G_{cv}(\hbar\omega) \tag{8.2}$$

where ε_1, ε_2 are, respectively, real and imaginary parts. Furthermore, these peaks are related to the vHs of high symmetry k-points that is consistent with the feature of band energy spectra related to vHs.

For each material, the distinct features are remarkable and unique properties demonstrating a wide range of their properties. Im ε of monolayers HfS_2 and $HfSe_2$ is similar with strong peak at I and two noticeably weaker peaks at II and III. This characteristic is in good agreement with the experimental data [67]. However, the calculated height of peaks is much larger than the measured height. Moreover, the calculated dielectric constant shows more peaks compared to the experimental study. With regard to monolayer $HfTe_2$, the height peak at I is much larger than that of HfS_2 and $HfSe_2$ while the peak at II is weaker and peak at III is almost absent from the fluctuation of imaginary part. Since it is semimetal, the Drude term has induced as

$$\varepsilon_2 = \frac{\tau\omega_p^2}{\omega\left(1+\omega^2\tau^2\right)} \tag{8.3}$$

On the topic of real part, the result is plotted by green curves that could be obtained through Kramers–Kronig relations. The characteristic of this part is similar to imaginary part with three main peaks for semiconductors and two main peaks for semimetal as shown in Figure 8.7. In addition, the energy-dependent features are similar for three materials. The vibration of values with the change of energy is small in the high-energy region. Since the complex dielectric constant can be described as Eq. (8.1), the absorption and conductivity can be calculated by the following:

$$\alpha(\omega) = \frac{2\omega\varepsilon_2(\omega)}{c} \tag{8.4}$$

$$\sigma_r(\omega) = \varepsilon_0\omega\varepsilon_2 \tag{8.5}$$

Additionally, the peak positions corresponding to energy of imaginary and real parts for these materials are also listed in Table 8.2 to further define the location and the height of peaks. According to this table, there are 11, 8, 8 peaks of imaginary part

TABLE 8.2

The Peak Position (eV) of Imaginary and Real Parts for Monolayer HfX$_2$ in Range of (0,8) eV of Energy

HfX$_2$	Peak Order	Imaginary Part	Real Part
HfS$_2$	1	2.48	2.15
	2	3.07	3.0
	3	3.72	3.66
	4	4.31	4.24
	5	4.7	4.57
	6	5.42	5.22
	7	6.14	5.81
	8	6.33	6.53
	9	6.85	6.79
	10	7.18	7.11
	11	7.7	7.64
HfSe$_2$	1	2.19	1.8
	2	2.96	2.06
	3	3.6	2.89
	4	4.18	4.12
	5	4.76	4.63
	6	5.72	5.27
	7	6.56	5.53
	8	7.72	6.43
	9		7.14
	10		7.65
HfTe$_2$	1	1.72	0.37
	2	3.74	0.74
	3	4.23	1.1
	4	5.15	1.41
	5	6.08	2.52
	6	6.69	3.56
	7	7.18	4.11
	8	7.67	4.66
	9		4.91
	10		5.46
	11		5.83
	12		6.32
	13		6.63
	14		7.06
	15		7.61

and 11, 10, 15 peaks of real part for HfS_2, $HfSe_2$, and $HfTe_2$, respectively. The strong peak labeled I in Figure 8.7 is located at 2.48, 2.19, 1.72 eV for imaginary part and 2.15, 1.8, 0.37 eV for real part of HfS_2, $HfSe_2$, and $HfTe_2$, respectively.

In order to provide more information about optical properties, the absorption, reflectivity, and refractive properties have been calculated based on the dielectric constants. The absorption coefficient of three compounds is shown in Figure 8.8. As shown in this figure, all monolayer structures start to absorb photons in the ultraviolet regions of the spectrum. By replacing the chalcogen atoms, a change of absorption property is accordingly appeared that is similar to the behaviors of other properties. A downward is shifted in absorption energy when moving from sulfur to selenium to tellurium. The monolayer starts to absorb photons at an energy of approximately 1.25, 1.2, and 0.4 eV for HfS_2, $HfSe_2$, and $HfTe_2$, respectively. Moreover, the first two peaks of three compounds shift in absorption spectra with decreasing energy as going from sulfur to selenium to tellurium attributing to the decrease in bandgaps of these structures. It should be note that a critical trend is observed that the position of the first peak is shifted toward lower energy when going down the chalcogen atoms in the group of the periodic table. In addition, these structures have strong optical absorption in the visible region of the spectra.

On the subject of reflectivity and refractive depending on ω, Figure 8.9 displays these factors of three materials. As shown in Figure 8.9a, a strong reflectivity minimum between 2.5 and 4 eV for all compounds indicates a collective plasma resonance [61] that is determined by the imaginary part of dielectric function. The imaginary part also defines the depth of this plasma minimum. Furthermore, when changing from sulfur to selenium to tellurium, the plasma minimum point shifts toward lower energies with increasing the values of the plasma minimum. On the other words, the plasma minimum of $HfTe_2$ is larger than that of $HfSe_2$ and HfS_2. Simultaneously, each material is typified by the broad maxima, especially in low energies 1–3 eV, corresponding to the strong peaks in dielectric constant. In addition,

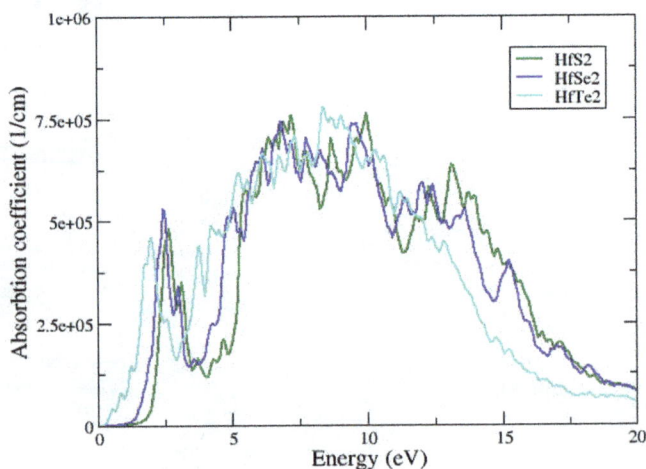

FIGURE 8.8 Calculated absorption coefficient of monolayer HfX_2.

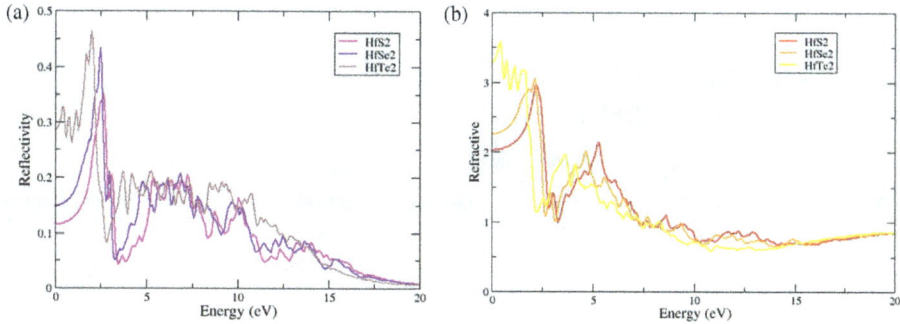

FIGURE 8.9   Calculated (a) reflectivity and (b) refractive of monolayer HfX$_2$.

the refractive of these compounds also described in Figure 8.9b shows the consistence with other optical properties and dielectric constant. The maximum refractive index is found in lower energy of approximately 2.5, 2.4, and 0.4 eV for HfS$_2$, HfSe$_2$, and HfTe$_2$, respectively. The change of refractive is also shifted toward lower energy when replacing the chalcogen atoms as previously mentioned trend. When increasing the energy with the increase of wavelength, the refractive index accordingly decreases indicating the consistence of this factor and two parts of dielectric constant. The trend of increasing refractive index when changing chalcogen atoms can be observed in the low energies. All optical factors such as absorption, reflectivity and refractive are completely consistent with the imaginary and real parts of dielectric function demonstrating the essential influence of the dielectric function in optical properties.

In short, monolayer HfX$_2$ (X = S, Se, or Te) reveals the sensitive optical properties with various factors such as absorption spectra, reflectivity and refractive index are determined and examined. Based on the imaginary part of dielectric function, the features of optical properties are verified indicating the consistence of structural and optical characteristics. Similar to the relation of geometry and electronic properties, optical properties of these materials are consistent with geometric properties, which are indicated in the trend of absorption, reflectivity, and refractive factors when going from sulfur to selenium to tellurium.

8.4   LITHIUM-INTERCALATED 1T-HfX$_2$

Based on HfX$_2$ and LiHfX$_2$ crystallization in the 1T phase, the electronic and optical properties of HfX$_2$ are the sake of consistency and become fundamental to create lithium-intercalated HfX$_2$ for LIBs. The uniform structure of monolayer HfX$_2$ is remained due to the same bond length of Hf-X and Li-X as shown in Table 8.3. After lithium intercalation, the bond lengths do not change despite Li-X interactions. This indicates that the high geometric symmetry of HfX$_2$ system be not affected by lithium intercalation. However, lithium intercalation affects the magnetism of systems, illustrating in the translation from nonmagnetic to ferromagnetic of all system. Table 8.3 lists magnetic moment of three systems, exhibiting the contribution and

TABLE 8.3

Ground State Energy (eV), Chemical Bonds (Å), Charge Transfer, and Magnetic Moment of Lithium-Intercalated HfX$_2$ Systems

HfX$_2$	Ground State Energy (eV)	Bond Length (Å)		Bader Charge Transfer (e)	Magnetic Moment
		Hf-S	Li-S		
HfS$_2$	−27.83	2.56	2.56	0.79	0.35
HfSe$_2$	−25.74	2.69	2.69	0.792	0.31
HfTe$_2$	−23.37	2.89	2.89	0.81	0.18

effect of lithium atom on system characteristics. The magnetic feature might depend on radius of chalcogen atoms; for example, LiHfTe$_2$ reveals the smallest magnetic moment while LiHfS$_2$ shows the largest one. Furthermore, one of the most important characteristics to evaluate the transfer of electrons in system and examine interacted properties is the charge transfer (see Table 8.3). It could show how much contribution of transferred electrons between lithium and HfX$_2$. As listed in Table 8.3, these values are, respectively, 0.79, 0.792, and 0.81 e, indicating the strong contribution of lithium electrons and strong interaction between lithium and HfX$_2$. In addition, this feature bases on the size of chalcogen atoms, demonstrating that in bigger atom, system has larger value and the electrical conductance is therefore expected to behave. The lithium-intercalated HfX$_2$ might be promising materials in LIBs.

The main features of electronic properties after lithium intercalation depend on interaction of Li-X bonds. Band energy spectra of three systems are shown in Figure 8.10. They all reveal metallic characteristic with lithium intercalation, demonstrated by the crossing of occupied to unoccupied band. This crossing occurs between M and K points, as well as K and Γ points. Moreover, the strong dispersions are exhibited with various subbands and a pair of asymmetric occupied and unoccupied band. These features indicate the effects of lithium intercalation in band structure, especially energy band near the Fermi level. σ and σ^* in valence and

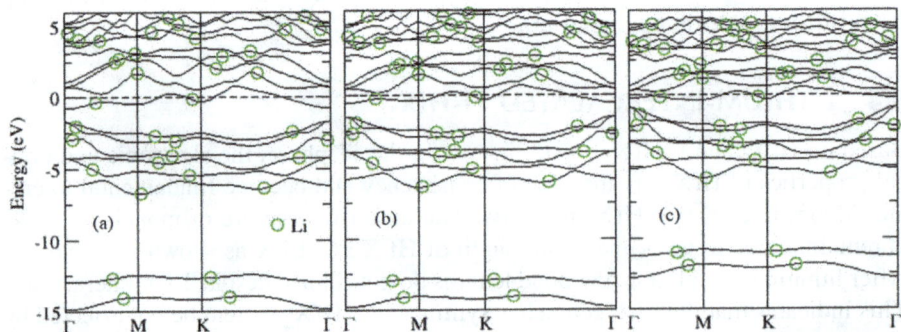

FIGURE 8.10 Band energy spectra of lithium-intercalated (a) 1T-HfS$_2$, (b) 1T-HfSe$_2$, and (c) 1T-HfTe$_2$.

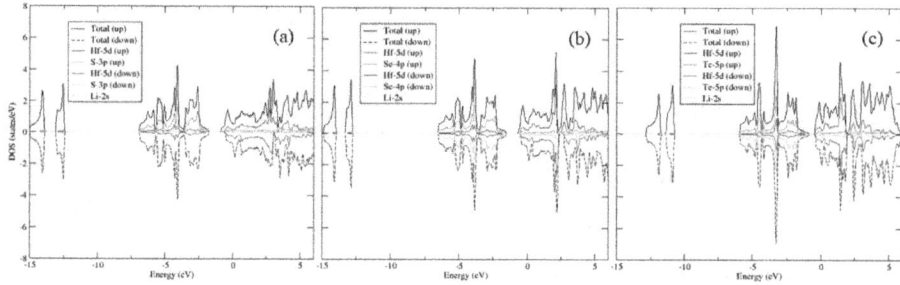

FIGURE 8.11 Orbital-projected DOS of lithium-intercalated (a) 1T-HfS$_2$, (b) 1T-HfSe$_2$, and (c) 1T-HfTe$_2$.

conduction bands could be clearly observed, for example, two σ bands and σ^* of LiHfS$_2$ initiate from Γ point at -2, -4.3, and 0.7 eV, respectively. These σ bands initiate from Γ point at -4, -1.5, 0.5 eV and -3.5, -1, 0.2 eV in LiHfSe$_2$ and LiHfTe$_2$, respectively. In addition, large spacing locates in low-lying energy being similar with pristine HfX$_2$, indicating that lithium atoms do not affect in this region. Lithium interaction reveals strong effect in energy bands near the Fermi level and high energy in conduction bands as shown in the figure with lithium atom dominance (green circles).

The orbital-projected DOSs further provide useful information about electronic properties after intercalations. As shown in Figure 8.11, orbital-projected DOSs with spinup and spindown configurations demonstrate the ferromagnetic feature in lithium-intercalated HfX$_2$. The main contributions are also due to s, d-orbitals of hafnium atoms and s, p-orbitals of chalcogen atoms. This indicates orbital hybridization in HfX$_2$ is independent of lithium intercalation. The system reveals strong dominance of X-p- and Hf-d-orbitals (mentioned in DOS of monolayer HfX$_2$) is remained, corresponding to the weak contribution of Li-s orbital (cyan curves). However, this hybridization is significantly manipulated by Li-X interactions, illustrated in stronger dominance of X-p (green and orange curves) and Hf-d orbitals (red and blue curves) compared to pristine cases. In addition, the DOS shows many asymmetric peaks, which characterizes the metallic features and strong dispersion in band energy. The large energy difference between the valence and conduction peaks further demonstrates this asymmetry and strong dispersion. These structures demonstrate the effect of lithium intercalation and strong interaction of Li-X. Generally, lithium intercalation changes in the band structure resulting in ferromagnetic configuration and band crossing. The strong orbital hybridization of X-p and Hf-d orbitals is enhanced after intercalation, indicating strong effect of lithium atoms in the compounds. Importantly, lithium-intercalated HfX$_2$ exhibits sensitive charge transfer, supporting for conductance in lithium batteries. These materials promise for electrode in LIBs.

8.5 CONCLUSION

In conclusion, monolayer HfX$_2$ ($X =$ S, Se, or Te) shows the essential geometric, electronic, and optical properties. Changing of chalcogen atoms from top to bottom in periodic table causes a reduction of bandgap in HfS_2, $HfSe_2$, and an overlap between conduction and valence bands of $HfTe_2$. This shift also attributes to multi-orbital hybridization as illustrated in orbital-projected DOS. Detailly, the degree of this hybridization between chalcogen-p and hafnium-d strongly depends on the chalcogen atoms. Instantaneously, the trends in band structures of these compounds can be attributed to both atomic species and structure parameters that indicate the electronic property dependences on configuration of the structure. The optical-properties-dependent dielectric function shows not only good agreement with experimental data but also the effect of imaginary and real parts on optical characteristics. Commonly, the vibration of optical and electronic properties can be explained by the features of geometry such as the species of chalcogen atoms, the size of these atoms, the electron configurations corresponding to chemical bonding, buckling, and orbital hybridizations. The results could promote the understanding of monolayer HfX_2 and development of these materials in optoelectronics. Beyond the understanding of monolayer HfX_2 system, lithium intercalation is examined indicating the possible applied progress of these materials in LIBs due to sensitive change of band, DOS, and charge transfer. Their features might make some of new-phase candidates for fast ionic conductors, promising electrode materials in LIBs.

REFERENCES

1. Wang X L and Shi G Q 2015 Flexible graphene devices related to energy conversion and storage *Energy & Environmental Science* **8** 790–823.
2. Lu X M and Xia Y N 2006 Electronic materials - Buckling down for flexible electronics *Nature Nanotechnology* **1** 163–4.
3. Gelinck G H, Huitema H E A, Van Veenendaal E, Cantatore E, Schrijnemakers L, Van der Putten J, Geuns T C T, Beenhakkers M, Giesbers J B, Huisman B H, Meijer E J, Benito E M, Touwslager F J, Marsman A W, Van Rens B J E and De Leeuw D M 2004 Flexible active-matrix displays and shift registers based on solution-processed organic transistors *Nature Materials* **3** 106–10.
4. Deng J W, Chen L F, Sun Y Y, Ma M H and Fu L 2015 Interconnected MnO2 nanoflakes assembled on graphene foam as a binder-free and long-cycle life lithium battery anode *Carbon* **92** 177–84.
5. Li L, Wu Z, Yuan S and Zhang X B 2014 Advances and challenges for flexible energy storage and conversion devices and systems *Energy & Environmental Science* **7** 2101–22.
6. Miao Y E, Huang Y P, Zhang L S, Fan W, Lai F L and Liu T X 2015 Electrospun porous carbon nanofiber@MoS2 core/sheath fiber membranes as highly flexible and binder-free anodes for lithium-ion batteries *Nanoscale* **7** 11093–101.
7. Hu X B, Ma M H, Mendes R G, Zeng M Q, Zhang Q, Xue Y H, Zhang T, Rummeli M H and Fu L 2015 Li-storage performance of binder-free and flexible iron fluoride@graphene cathodes *Journal of Materials Chemistry A* **3** 23930–5.
8. Lee S Y, Choi K H, Choi W S, Kwon Y H, Jung H R, Shin H C and Kim J Y 2013 Progress in flexible energy storage and conversion systems, with a focus on cable-type lithium-ion batteries *Energy & Environmental Science* **6** 2414–23.
9. Slater M D, Kim D, Lee E and Johnson C S 2013 Sodium-ion batteries *Advanced Functional Materials* **23** 947–58.

10. Xiao L F, Cao Y L, Xiao J, Wang W, Kovarik L, Nie Z M and Liu J 2012 High capacity, reversible alloying reactions in SnSb/C nanocomposites for Na-ion battery applications *Chemical Communications* **48** 3321–3.

11. Liu B, Luo T, Mu G Y, Wang X F, Chen D and Shen G Z 2013 Rechargeable Mg-ion batteries based on WSe2 nanowire cathodes *ACS Nano* **7** 8051–8.

12. Luo W, Wan J Y, Ozdemir B, Bao W Z, Chen Y N, Dai J Q, Lin H, Xu Y, Gu F, Barone V and Hu L B 2015 Potassium ion batteries with graphitic materials *Nano Letters* **15** 7671–7.

13. Freunberger S A, Chen Y H, Peng Z Q, Griffin J M, Hardwick L J, Barde F, Novak P and Bruce P G 2011 Reactions in the rechargeable lithium-O-2 battery with alkyl carbonate electrolytes *Journal of the American Chemical Society* **133** 8040–7.

14. Zhang Q, Tan S J, Kong X, Xiao Y and Fu L 2015 Synthesis of sulfur encapsulated 3D graphene sponge driven by micro-pump and its application in Li-S battery *Journal of Materiomics* **1** 333–9.

15. Wang X F, Lu X H, Liu B, Chen D, Tong Y X and Shen G Z 2014 Flexible energy-storage devices: Design consideration and recent progress *Advanced Materials* **26** 4763–82.

16. Mannix A J, Zhou X F, Kiraly B, Wood J D, Alducin D, Myers B D, Liu X L, Fisher B L, Santiago U, Guest J R, Yacaman M J, Ponce A, Oganov A R, Hersam M C and Guisinger N P 2015 Synthesis of borophenes: Anisotropic, two-dimensional boron polymorphs *Science* **350** 1513–6.

17. Zhao J J, Liu H S, Yu Z M, Quhe R G, Zhou S, Wang Y Y, Liu C C, Zhong H X, Han N N, Lu J, Yao Y G and Wu K H 2016 Rise of silicene: A competitive 2D material *Progress in Materials Science* **83** 24–151.

18. Li L F, Lu S Z, Pan J B, Qin Z H, Wang Y Q, Wang Y L, Cao G Y, Du S X and Gao H J 2014 Buckled germanene formation on Pt(111) *Advanced Materials* **26** 4820–24.

19. Carvalho A, Wang M, Zhu X, Rodin A S, Su H B and Neto A H C 2016 Phosphorene: From theory to applications *Nature Reviews Materials* **1** 1–16.

20. Sun Z Q, Liao T, Dou Y H, Hwang S M, Park M S, Jiang L, Kim J H and Dou S X 2014 Generalized self-assembly of scalable two-dimensional transition metal oxide nanosheets *Nature Communications* **5** 1–9.

21. Wang Q H, Kalantar-Zadeh K, Kis A, Coleman J N and Strano M S 2012 Electronics and optoelectronics of two-dimensional transition metal dichalcogenides *Nature Nanotechnology* **7** 699–712.

22. Mei J, Zhang Y W, Liao T, Sun Z Q and Dou S X 2018 Strategies for improving the lithium-storage performance of 2D nanomaterials *National Science Review* **5** 389–416.

23. Liu J X, Cao H, Jiang B, Xue Y H and Fu L 2016 Newborn 2D materials for flexible energy conversion and storage *Science China-Materials* **59** 459–74.

24. Su D W, Dou S X and Wang G X 2015 Ultrathin MoS2 nanosheets as anode materials for sodium-ion batteries with superior performance *Advanced Energy Materials* **5** 1401205.

25. ten Elshof J E, Yuan H Y and Rodriguez P G 2016 Two-dimensional metal oxide and metal hydroxide nanosheets: Synthesis, controlled assembly and applications in energy conversion and storage *Advanced Energy Materials* **6** 1600355.

26. Choi W, Choudhary N, Han G H, Park J, Akinwande D and Lee Y H 2017 Recent development of two-dimensional transition metal dichalcogenides and their applications *Materials Today* **20** 116–30.

27. Huang X, Zeng Z Y, Fan Z X, Liu J Q and Zhang H 2012 Graphene-based electrodes *Advanced Materials* **24** 5979–6004.

28. Liao T, Sun C H, Du A J, Sun Z Q, Hulicova-Jurcakova D and Smith S 2012 Charge carrier exchange at chemically modified graphene edges: A density functional theory study *Journal of Materials Chemistry* **22** 8321–6.

29. Dou Y H, Wang Y X, Tian D L, Xu J T, Zhang Z J, Liu Q N, Ruan B Y, Ma J M, Sun Z Q and Dou S X 2017 Atomically thin Co3O4 nanosheet-coated stainless steel mesh with enhanced capacitive Na+ storage for high-performance sodium-ion batteries *2d Materials* **4** 015022.

30. Ni J F, Zhao Y, Li L and Mai L Q 2015 Ultrathin MoO2 nanosheets for superior lithium storage *Nano Energy* **11** 129–35.

31. Liu Y, Elzatahry A A, Luo W, Lan K, Zhang P F, Fan J W, Wei Y, Wang C, Deng Y H, Zheng G F, Zhang F, Tang Y, Mai L Q and Zhao D Y 2016 Surfactant-templating strategy for ultrathin mesoporous TiO_2 coating on flexible graphitized carbon supports for high-performance lithium-ion battery *Nano Energy* **25** 80–90.

32. Yang S B, Feng X L and Mullen K 2011 Sandwich-like, graphene-based titania nanosheets with high surface area for fast lithium storage *Advanced Materials* **23** 3575–9.

33. Li W, Wang F, Liu Y P, Wang J X, Yang J P, Zhang L J, Elzatahry A A, Al-Dahyan D, Xia Y Y and Zhao D Y 2015 General strategy to synthesize uniform mesoporous TiO_2/graphene/mesoporous TiO_2 sandwich-like nanosheets for highly reversible lithium storage *Nano Letters* **15** 2186–93.

34. Zhao G X, Wen T, Zhang J, Li J X, Dong H L, Wang X K, Guo Y G and Hu W P 2014 Two-dimensional Cr_2O_3 and interconnected graphene-Cr_2O_3 nanosheets: Synthesis and their application in lithium storage *Journal of Materials Chemistry A* **2** 944–8.

35. Wang X, Tian W, Liu D Q, Zhi C Y, Bando Y and Golberg D 2013 Unusual formation of alpha-Fe_2O_3 hexagonal nanoplatelets in N-doped sandwiched graphene chamber for high-performance lithium-ions batteries *Nano Energy* **2** 257–67.

36. Xiong P, Liu B R, Teran V, Zhao Y, Peng L L, Wang X and Yu G H 2014 Chemically integrated two-dimensional hybrid zinc manganate/graphene nanosheets with enhanced lithium storage capability *ACS Nano* **8** 8610–6.

37. Jing Y, Ortiz-Quiles E O, Cabrera C R, Chen Z F and Zhou Z 2014 Layer-by-layer hybrids of MoS_2 and reduced graphene oxide for lithium ion batteries *Electrochimica Acta* **147** 392–400.

38. Teng Y Q, Zhao H L, Zhang Z J, Li Z L, Xia Q, Zhang Y, Zhao L N, Du X F, Du Z H, Lv P P and Swierczek K 2016 MoS_2 nanosheets vertically grown on graphene sheets for lithium-ion battery anodes *ACS Nano* **10** 8526–35.

39. Jiang L F, Lin B H, Li X M, Song X F, Xia H, Li L and Zeng H B 2016 Monolayer MoS_2-graphene hybrid aerogels with controllable porosity for lithium-ion batteries with high reversible capacity *ACS Applied Materials & Interfaces* **8** 2680–7.

40. Chen D Y, Ji G, Ding B, Ma Y, Qu B H, Chen W X and Lee J Y 2013 In situ nitrogenated graphene-few-layer WS2 composites for fast and reversible Li+ storage *Nanoscale* **5** 7890–6.

41. Liu Y, Wang W, Wang Y W and Peng X S 2014 Homogeneously assembling like-charged WS2 and GO nanosheets lamellar composite films by filtration for highly efficient lithium ion batteries *Nano Energy* **7** 25–32.

42. Yang L C, Wang S N, Mao J J, Deng J W, Gao Q S, Tang Y and Schmidt O G 2013 Hierarchical MoS_2/polyaniline nanowires with excellent electrochemical performance for lithium-ion batteries *Advanced Materials* **25** 1180–4.

43. Chen Y, Song B H, Tang X S, Lu L and Xue J M 2014 Ultrasmall Fe_3O_4 nanoparticle/MoS_2 nanosheet composites with superior performances for lithium ion batteries *Small* **10** 1536–43.

44. Hu S, Chen W, Zhou J, Yin F, Uchaker E, Zhang Q F and Cao G Z 2014 Preparation of carbon coated MoS_2 flower-like nanostructure with self-assembled nanosheets as high-performance lithium-ion battery anodes *Journal of Materials Chemistry A* **2** 7862–72.

45. Barik G and Pal S 2019 Defect induced performance enhancement of monolayer MoS_2 for Li- and Na-ion batteries *Journal of Physical Chemistry C* **123** 21852–65.

46. Tang C, Min Y X, Chen C Y, Xu W W and Xu L 2019 Potential applications of hetero-structures of TMDs with MXenes in sodium-ion and Na-O-2 batteries *Nano Letters* **19** 5577–86.

47. Han D L, Zhang J, Weng Z, Kong D B, Tao Y, Ding F, Ruan D B and Yang Q H 2019 Two-dimensional materials for lithium/sodium-ion capacitors *Materials Today Energy* **11** 30–45.

48. Yang E, Ji H and Jung Y 2015 Two-dimensional transition metal dichalcogenide mono layers as promising sodium ion battery anodes *Journal of Physical Chemistry C* **119** 26374–80.

49. Li Y Y, Zhu H L, Shen F, Wan J Y, Lacey S, Fang Z Q, Dai H Q and Hu L B 2015 Nanocellulose as green dispersant for two-dimensional energy materials *Nano Energy* **13** 346–54.

50. Choi S H and Kang Y C 2015 Sodium ion storage properties of WS2-decorated three-dimensional reduced graphene oxide microspheres *Nanoscale* **7** 3965–70.

51. Share K, Lewis J, Oakes L, Carter R E, Cohn A P and Pint C L 2015 Tungsten disel-enide (WSe$_2$) as a high capacity, low overpotential conversion electrode for sodium ion batteries *RSC Advances* **5** 101262–7.

52. Zhang Q, Tan S J, Mendes R G, Sun Z T, Chen Y T, Kong X, Xue Y H, Rummeli M H, Wu X J, Chen S L and Fu L 2016 Extremely weak van der Waals coupling in vertical ReS2 nanowalls for high-current-density lithium-ion batteries *Advanced Materials* **28** 2616–23.

53. Bissett M A, Kinloch I A and Dryfe R A W 2015 Characterization of MoS$_2$-graphene composites for high-performance coin cell supercapacitors *ACS Applied Materials & Interfaces* **7** 17388–98.

54. Lopez-Sanchez O, Lembke D, Kayci M, Radenovic A and Kis A 2013 Ultrasensitive photodetectors based on monolayer MoS$_2$ *Nature Nanotechnology* **8** 497–501.

55. Feng J, Sun X, Wu C Z, Peng L L, Lin C W, Hu S L, Yang J L and Xie Y 2011 Metallic few-layered VS2 ultrathin nanosheets: High two-dimensional conductivity for in-plane supercapacitors *Journal of the American Chemical Society* **133** 17832–8.

56. Ratha S and Rout C S 2013 Supercapacitor electrodes based on layered tungsten disulfide-reduced graphene oxide hybrids synthesized by a facile hydrothermal method *ACS Applied Materials & Interfaces* **5** 11427–33.

57. Chakravarty D and Late D J 2015 Microwave and hydrothermal syntheses of WSe2 micro/nanorods and their application in supercapacitors *RSC Advances* **5** 21700–9.

58. Pan C F, Niu S M, Ding Y, Dong L, Yu R M, Liu Y, Zhu G and Wang Z L 2012 Enhanced Cu2S/CdS coaxial nanowire solar cells by piezo-phototronic effect *Nano Letters* **12** 3302–7.

59. Shao Z B, Jie J S, Sun Z, Xia F F, Wang Y M, Zhang X H, Ding K and Lee S T 2015 MoO3 nanodots decorated CdS nanoribbons for high-performance, homojunction pho-tovoltaic devices on flexible substrates *Nano Letters* **15** 3590–6.

60. Toh R J, Sofer Z and Pumera M 2016 Catalytic properties of group 4 transition metal dichalcogenides (MX2; M = Ti, Zr, Hf; X = S, Se, Te) *Journal of Materials Chemistry A* **4** 18322–34.

61. Reshak A H and Auluck S 2005 Ab initio calculations of the electronic and optical properties of 1T-HfX2 compounds *Physica B-Condensed Matter* **363** 25–31.

62. Reshak A H, Kityk I V and Auluck S 2008 Electronic structure and optical properties of 1T-TiS$_2$ and lithium intercalated 1T-TiS$_2$ for lithium batteries *Journal of Chemical Physics* **129** 074706.

63. Habenicht C, Simon J, Richter M, Schuster R, Knupfer M and Buchner B 2020 Potassium-intercalated bulk HfS$_2$ and HfSe$_2$: Phase stability, structure, and electronic structure *Physical Review Materials* **4** 064002.

64. El Youbi Z, Jung S W, Mukherjee S, Fanciulli M, Schusser J, Heckmann O, Richter C, Minar J, Hricovini K, Watson M D and Cacho C 2020 Bulk and surface electronic states in the dosed semimetallic HfTe$_2$ *Physical Review B* **101** 235431.

65. Mangelsen S, Naumov P G, Barkalov O I, Medvedev S A, Schnelle W, Bobnar M, Mankovsky S, Polesya S, Nather C, Ebert H and Bensch W 2017 Large nonsaturating magnetoresistance and pressure-induced phase transition in the layered semimetal HfTe$_2$ *Physical Review B* **96** 205148.

66. Aminalragia-Giamini S, Marquez-Velasco J, Tsipas P, Tsoutsou D, Renaud G and Dimoulas A 2017 Molecular beam epitaxy of thin HfTe$_2$ semimetal films *2d Materials* **4** 015001.

67. Bayliss S C and Liang W Y 1985 Reflectivity, joint density of states and band-structure of group IVB transition-metal dichalcogenides *Journal of Physics C-Solid State Physics* **18** 3327–35.

68. Zhang Y. Z L, Tuan L. V., Chu P. K. and Huo K. 2020 Two-dimensional transition metal chalcogenides for Alkali metal ions storage *ChemSusChem* **13** 1114–54.

69. Zhao X, Sui J H, Li F, Fang H T, Wang H E, Li J Y, Cai W and Cao G Z 2016 Lamellar MoSe$_2$ nanosheets embedded with MoO2 nanoparticles: Novel hybrid nanostructures promoted excellent performances for lithium ion batteries *Nanoscale* **8** 17902–10.

9 Mn-Based Oxide Nanocomposite with Reduced Graphene Oxide as Anode Material in Li-Ion Battery

Sanjaya Brahma and Shao-Chieh Weng
National Cheng Kung University

Chia-Chin Chang
National University of Tainan

Jow-Lay Huang
National Cheng Kung University

CONTENTS

9.1 INTRODUCTION

As we know that our main energy policy is still based on burning fossil fuels. And burning fossil fuels not only causes CO_2 emissions, resulting in the greenhouse effect, air pollution, and water pollution, but also affects human's health. So, we need renewable, sustainable, and clean energies technology, such as solar energy, hydroelectricity, and wind energy. Due to exploitation of renewable, alternative, green, energy sources

DOI: 10.1201/9781003263807-9

(solar, wind, geothermal), it is now generally accepted that among the various possible choices, the most suitable are electrochemical batteries. Lithium-ion battery (LIB) is one of the electrochemical systems that can efficiently store and deliver energy, and it also plays a crucial role in this field [1].

Nanomaterials of metal oxide (MO), such as ZnO [2], RuO_2 [3], NiOx [4], SnO_2 [5], and MnO_2 [6], have been studied as anode materials for LIBs making an effort to achieve higher specific capacity and better cyclic charge/discharge stability. Compared to those MO materials, Mn_3O_4 [7–10] is a potential anode material for LIBs because of the excellent physical and chemical properties, environmentally compatible, low cost, and abundant in nature. The theoretical capacity of Mn_3O_4 is ~936 mAh g^{-1}, which is two times higher than graphite. However, the extremely low electrical conductivity (~10^{-7}–10^{-8} S cm^{-1}) [11], volume change due to lithiation/ delithiation process, and fast capacity fading are major pros of Mn_3O_4. Recently, the combination of Mn_3O_4 and carbon-based nanomaterials like carbon nanotubes (CNTs) and graphene (oxide) has proved to be a boosting route to enhance the electrical conductivity and the strong adhesion between carbon-based materials at nanoscale [12,13]. Graphene is a single layer of carbon atoms bonded into two-dimensional (2D) hexagonal networks. It has high surface area (theoretically: 2,620 m^{-2} g^{-1}, experimentally: 1,500 m^{-2} g^{-1}), high electron mobility (15,000 cm^{-2} V^{-1} s^{-1} at room temperature), and large Young's modulus (1 TPa and intrinsic strength of 130 GPa), and it is a semiconductor with zero bandgap [14]. A single-layer graphene sheet has theoretical capacity of 744 mAh g^{-1}, if lithium ions are attached to both sides of the graphene sheets. Graphene-based materials already attracted tremendous attention in replacing commercial graphite and applying for the active materials of anode for LIBs [15–17]. As we know, graphene derivate with different oxygen-containing groups such as graphene oxide (GO) and reduced graphene oxide (rGO) can serve as electrode materials with higher capacity. Graphene-based materials can be used as 2D buffer layer for the anisotropic growth of various metal (M)/MOs nanoparticles (NPs) and effectively suppressed the aggregation of these NPs [18–20].

Herein, the purpose of synthesizing Mn_3O_4/rGO nanocomposite is to obtain the synergistic effects of their respective advantages, which is under consideration as a possible strategy for obtaining improved anode materials for high-power LIBs applications.

9.2 EXPERIMENTAL

9.2.1 Fabrication of GO, Manganese Dioxide (MnO_2), MnO_2/rGO, and Mn_3O_4/rGO Nanocomposites

GO, MnO_2, and MnO_2/rGO nanocomposites were prepared by modified Hummers' method and chemical reaction, respectively [19]. For synthesizing Mn_3O_4/rGO nanocomposite, first of all, MnO_2/rGO were added in 100-mL deionized (DI) water, and magnetic stirring was for 30 minutes until uniformly suspending in the DI water. Subsequently, sodium borohydride ($NaBH_4$) (1 g) was dispersed in previous solution and magnetic stirring for 10 minutes. And next, the mixture was heated to 100°C and maintained at this temperature for 10 hours. In that reaction, the color changed from dark brown to light brown. The obtained Mn_3O_4/rGO nanocomposite powder was washed repeatedly by ethanol and DI water and finally collected by centrifugation. Finally, Mn_3O_4/rGO nanocomposite powder was dried at 55°C in an oven.

9.2.2 CHARACTERIZATION OF GO, MnO₂, MnO₂/ rGO, AND Mn₃O₄/rGO NANOCOMPOSITES

X-ray diffraction (XRD, Model No: GADDS/D8 DISCOVER diffractometer) was used to determine the crystallinity and crystallographic structure of GO and its nanocomposite by using Cu Kα ($\lambda = 1.54$ Å) radiation at an angular speed of 3° (2θ) min^{-1} with 2θ from 10° to 80°. The morphology of the GO and its nanocomposite were observed by ultrahigh-resolution field emission scanning electron microscopy (UHFE-SEM, Model: AURIGA-39–50, EHT = 5 kV) and transmission electron microscopy (TEM, JEOL JEM-2100F CS STEM). The powder samples were dispersed in alcohol and dropped on TEM grids for further characterization. The energy level of the carbon and Mn of the GO and nanocomposite were confirmed by electron spectroscopy for chemical analysis (ESCA, XPS). Thermogravimetric analysis (TGA, Perkin Elmer, Pyris 1 TGA) was carried out within 20°C–800°C (heating rate 15°C per min) in N_2 atmosphere for analyzing the ratio of carbon and MnOx.

9.2.3 ELECTROCHEMICAL ANALYSIS

Electrochemical measurements were carried out by using CR2032-type coin cells at room temperature [19]. The working electrode was prepared by mixing four kinds of powders: (1) active material 80 wt.%, (2) 10 wt.% of Super P as a conductive additive, (3) 5 wt.% of LiOH, and (d) 5 wt.% of polyacrylic acid (PAA) as a binder of the total electrode mass [21–23]. PAA was used as a water-soluble binder for anodes in LIBs. The PAA-based binders were lithiated by titration using aqueous LiOH (Li-PAA) and controlled by adjusting the pH value of the solution [21]. The four components were mixed with a suitable amount of DI water to produce a slurry. This was then uniformly loaded on a copper foil with a doctor blade as a current collector. The sample was cut into circular electrodes and dried for 30 minutes at 70°C in an electric oven. The cells were assembled in an Ar-filled glove box with lithium foil as the counter electrode, and a solution of 1.0 M $LiPF_6$ dissolved in 1:1 (v/v) EC/DEC as the electrolyte. Galvanostatic Li$^+$ charge/discharge analysis was carried out using a WonATech WBCS3000 automatic battery cycler. All electrochemical measurements were conducted in the potential range from 0.002 to 3 V (vs. Li$^+$/Li).

9.3 RESULTS AND DISCUSSION

Figure 9.1 shows the XRD pattern of GO, manganese dioxide (MnO_2), MnO_2/rGO nanocomposites, and trimanganese tetraoxide/reduced graphene oxide (Mn_3O_4/rGO) self-assembled hexagonal sheet nanocomposites. Figure 9.1a shows the standard XRD card number of graphite (JCPDS #41-1488); the standard card number of graphite is compared with the self-synthesized GO XRD pattern (Figure 9.1b); and we can observe the GO prepared by the modified Hummers' method in this laboratory has a diffraction peak at

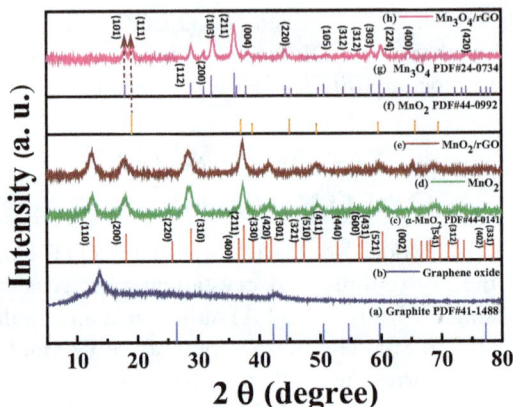

FIGURE 9.1 XRD patterns of (a) graphite PDF#41-1488, (b) GO, (c) α-MnO$_2$ PDF #44-0141, (d) MnO$_2$ nanoneedle, (e) MnO$_2$/rGO nanocomposite, (f) MnO$_2$ PDF#44-0992, (g) Mn$_3$O$_4$ PDF#24-0734, and (h) Mn$_3$O$_4$/rGO nanocomposite.

a low angle of $2\theta = 10.13°$, which is different from that of natural graphite, which has a characteristic hexagonal hexagonal stack (2-hexagonal: 2H) at 26°. However, this broad diffraction peak at a low angle of $2\theta = 10.13°$ can be attributed to the presence of oxygen-containing functional groups; embedding the water molecules separates the graphite layer and has a larger layer spacing, which also corresponds to the diffraction peak of GO (001) crystal plane; and the diffraction peak at 42.5° corresponds to the GO (100) crystal plane. GO does not have a standard card number (JCPDS cards). The degree of oxidation of GO can only be defined by collecting information from references; therefore, when the diffraction peak is shifted from the natural graphite characteristic peak by 26° to a lower angle, at that time, it means that the carbon layer has been sufficiently oxidized. It can be observed from the XRD results that the GO prepared via modified Hummers' method has been fully oxidized, and its d-spacing is about 0.8623 nm, which is sufficient space provided by the interlayer spacing of the oxygen-containing functional groups. The activation position enables the manganese ions to be attracted by the nutrient functional groups of the carbon layer in the subsequent nanocomposite manufacturing process, thereby facilitating the formation of the composite material.

Figure 9.1c shows the XRD standard card number of α-MnO$_2$ (JCPDS#44-0141). Figure 9.1d and e shows the XRD diffraction pattern of MnO$_2$ nanoneedle and MnO$_2$/rGO nanocomposite, respectively. Comparing Figure 9.1c–e, it can be observed that the diffraction peaks are located at $2\theta = 12.78°$, 18.11°, 25.71°, 28.84°, 36.70°, 37.52°, 39.01°, 41.23°, 41.97°, 46.04°, 47.37°, 49.86°, 52.86°, 56.37°, 60.27°, 65.11°, 69.71°, 72.71°, 77.16°, and 78.59°, fully conform to the crystal face on the standard card number of manganese dioxide: (110), (200), (220), (310), (400), (211), (330), (420), (301), (321), (510), (411), (440), (600), (431), (521), (002), (541), (312), (402), and (332). The result of XRD analysis showed that the preparation process of MnO$_2$ nanoneedle and MnO$_2$/rGO nanocomposite was carried out only at 83°C without high-temperature heat treatment process, and we can synthesize manganese oxide with excellent crystallinity. The product has a crystal structure conforming to the tetragonal-structure α-type MnO$_2$ (I4/m, JCPDS card NO.: 44-0141).

Figure 9.1f–h shows XRD standard card numbers for MnO_2 (JCPDS#44-0992), XRD standard card numbers for Mn_3O_4 (JCPDS#24-0734), and XRD diffraction patterns for Mn_3O_4/rGO nanocomposites. Comparing Figure 9.1f–h, it can be observed that in the diffraction pattern of Mn_3O_4/rGO nanocomposites, except for the crystal plane (111) located at $2\theta = 19.11°$ which belongs to the cubic-phase MnO_2 structure (Fd-3m, cubic phase, JCPDS card NO.: 44-0992), the remaining diffraction peaks are at $2\theta = 18°$, 28.88°, 31.02°, 32.32°, 36.09°, 37.98°, 44.44°, 50.71°, 53.86°, 56.01°, 58.51°, 59.84°, 64.65°, and 73.43°, all belong to tetrahedral (I41/amd, tetragonal structure) Mn_3O_4 manganese hausmannite structure (lattice constants: $a = 5.76$ Å, $c = 9.47$ Å; JCPDS card NO.: 24-0734).

The chemical composition of manganese oxide will be converted from α-type MnO_2 to Mn_3O_4/MnO_2. The main reason is the synthesized MnO_2/rGO nanocomposite and the reducing agent borohydride (NaBH4) in DI water. After mixing, the temperature is increased to 100°C, and the reaction is carried out for 10 hours. The chemical reaction of the process is shown in Eqs. (9.1)–(9.3):

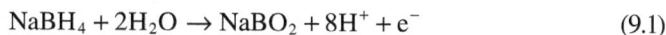

$$NaBH_4 + 2H_2O \rightarrow NaBO_2 + 8H^+ + e^- \qquad (9.1)$$

$$8M^+ + 8e^- \rightarrow 8M^0 \qquad (9.2)$$

$$3MnO_2 \rightarrow Mn_3O_4 + O_2 \qquad (9.3)$$

In addition, some literatures have pointed out that sodium borohydride can reduce not only metal particles but also GO. Therefore, in the reduction reaction, MnO_2 would be reduced to a hybrid structure of Mn_3O_4/MnO_2, and GO would be reduced and removed from the remaining oxygen-containing functional groups, too. Next, the ESCA is used to determine the surface elements and analyze the differences and changes of binding energy for the four materials.

Figure 9.2 shows the result of the ESCA of GO and MnO_2 nanoneedle, MnO_2/rGO nanocomposite, and Mn_3O_4/rGO nanocomposite. Comparing the four maps of Figure 9.2a–d, it can be found that the biggest difference between GO and manganese oxide series materials is the peaks at binding energy = 642 eV and binding energy = 654 eV. The meaning of the peaks is $Mn_{2p_{3/2}}$ and $Mn_{2p_{1/2}}$ binging energy. The difference between the two peaks is due to the difference in element valence or reduction/oxidation state. The observed position is at the C 1s position of the binding energy = 292–295 eV. If a peak appears, it is the carbon element correction signal from $(CF_2)_n$.

Figure 9.3 shows the results of C 1s element ESCA of GO, MnO_2 nanoneedle, MnO_2/rGO nanocomposite, and Mn_3O_4/rGO nanocomposite. In Figure 9.3a–d, only the C 1s peak in Figure 9.3b is due to the carbon element correction signal. Comparing the three spectrums of Figure 9.3a–d, all of the ESCA spectrums have three main peaks at 285, 287, and 289 eV, respectively, which represent the meaning of carbon–carbon single and double bonds (C=C/CC(sp2)), carboxyl single bond (C(O)OH), and carbonyl group (OC=O), respectively. Comparing the intensity of C(O)OH and OC=O binding energy of the three materials, it can be found that when we combine GO with manganese ions forming MnO_2/rGO nanocomposites, the

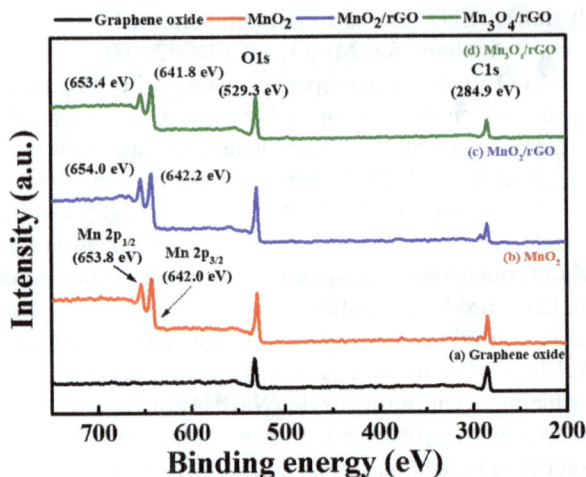

FIGURE 9.2 ESCA spectrum of surface state of (a) GO, (b) MnO$_2$ nanoneedle, (c) MnO$_2$/rGO nanocomposite, and (d) Mn$_3$O$_4$/rGO nanocomposite.

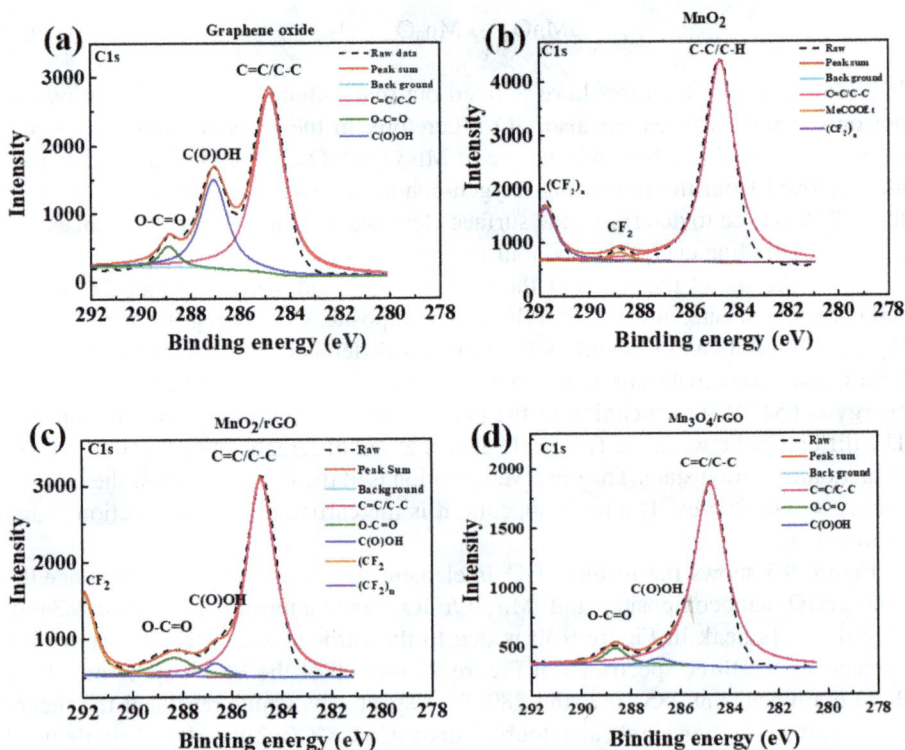

FIGURE 9.3 ESCA spectrum for C1s of (a) GO, (b) MnO$_2$ nanoneedle, (c) MnO$_2$/rGO nanocomposite, and (d) Mn$_3$O$_4$/rGO nanocomposite.

bond strength of C(O)OH and OC=O has a tendency to decrease significantly. It was demonstrated that as a result of the synthesis process, the oxygen-containing functional groups of the GO react with manganese ions, thereby indirectly reducing the GO. Comparing the MnO_2/rGO and Mn_3O_4/rGO nanocomposites, it is obviously to observe that the intensity of OC=O and C(O)OH decreased significantly; it is because the boron hydride anion of the sodium borohydride mainly reacts with carboxyl group and hydroxyl group to an epoxy group (C–O–C, 286.5 eV); and the result also proved that the addition of the reducing agent sodium borohydride does contribute to the removing oxygen-containing functional groups.

Figure 9.4 shows the surface ESCA Mn2p spectrum of GO and MnO_2 nanoneedle, MnO_2/rGO nanocomposite, and Mn_3O_4/rGO nanocomposite. The GO doesn't have the signal of manganese. It is known from the literature that the binding energy of Mn2p (Mn2p3/2 and Mn2p1/2) of manganese dioxide is located at 642.2 and 653.9 eV, and from Figure 9.4b–d, it can be observed that the binding energies of Mn2p3/2 and Mn2p1/2 of MnO_2 nanoneedle, MnO_2/rGO nanocomposite, and Mn_3O_4/rGO nanocomposite are located at 642/653.8, 642.2/654, and 651.8/653.4 eV, respectively. Among them, the peak of Mn2p3/2 of MnO_2 nanoneedle and MnO_2/rGO

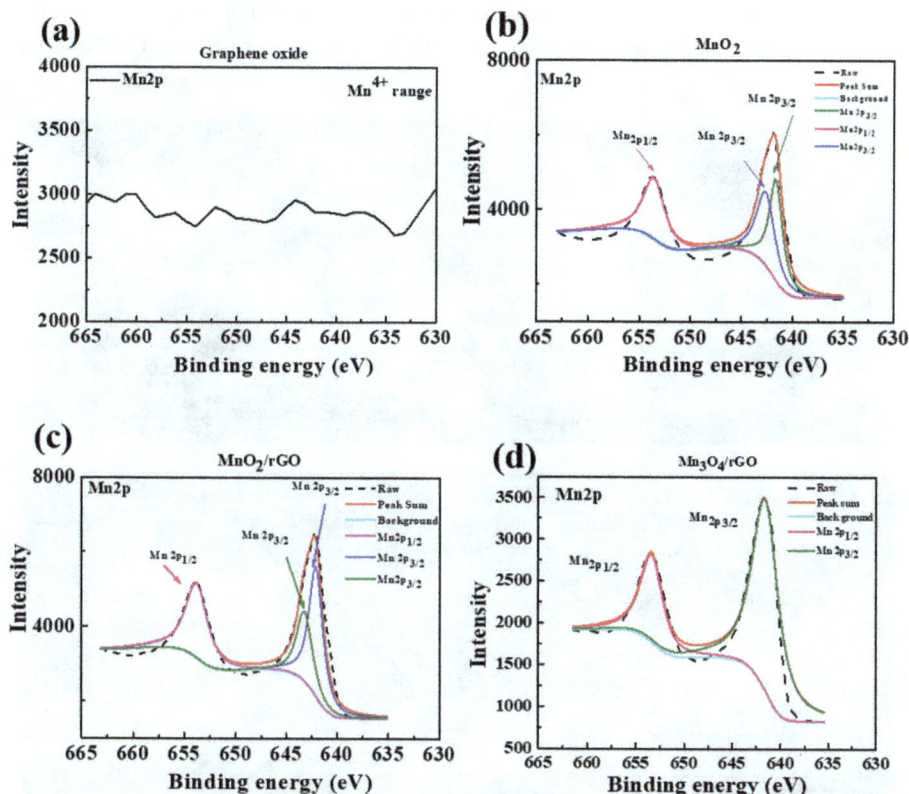

FIGURE 9.4 ESCA spectrum of Mn2p of (a) GO, (b) MnO_2 nanoneedle, (c) MnO_2/rGO nanocomposite, and (d) Mn_3O_4/rGO nanocomposite.

nanocomposite are formed if two separated peaks, it is due to the manganese position is located at the octahedral position and the two Mn^{3+} and Mn^{4+} bonding type existing in the structure, simultaneously. Continuing to observe the spectrum of Mn_3O_4/rGO nanocomposite, we can observe that the binding energy of Mn2p3/2 and Mn2p1/2 falls at 641.8 and 653.4 eV, respectively. The binding energy is obviously red-shifted compared with the other materials. The binding energy of the metal (Mn2p3/2 = 638.8 eV, Mn2p1/2 = 650.06 eV) is lower than that of the MnO_2, and it confirms that reducing MnO_2/rGO nanocomposite not only removes the oxygen-containing functional group but also converts it to the chemical composition of Mn_3O_4.

Figure 9.5 shows a surface morphology and microstructure of GO, MnO_2 nanoneedle, MnO_2/rGO nanocomposite, and Mn_3O_4/rGO nanocomposite. Figure 9.5a shows the SEM image of GO, it is mainly composed of a large layered structure, and its stable sp2-layered structure is destroyed by oxygen-containing functional groups, showing the surface morphology with wrinkle and disordered. In Figure 9.5b, it can be observed that the MnO_2 nanoneedle has a one-dimensional needle-like structure, and the length and diameter of the needle-like structure are about 450 ± 40 nm and 20 ± 10 nm, respectively.

Figure 9.5c shows a FESEM surface analysis of the MnO_2/rGO nanocomposite. It can be observed that the one-dimensional structure of MnO_2 nanoneedle is uniformly and densely adsorbed on the wrinkled GO sheet structure. Although

FIGURE 9.5 SEM surface images of (a) GO, (b) MnO_2 nanoneedle, (c) MnO_2/rGO nanocomposite, and (d) Mn_3O_4/rGO hexagonal-flat structure nanocomposite.

manganese dioxide is coated on the surface of GO, its structure still maintains a one-dimensional needle-like structure, and the diameter and length knots are consistent with Figure 9.5b.

In this study, the MnO_2/rGO nanocomposite was reduced by the reducing agent sodium borohydride (NaBH4), and its morphology changed greatly. It changed from the original one-dimensional nanoneedle to a two-dimensional hexagonal sheet structure. We can observe the image of each hexagonal sheet structure, and each hexagonal sheet consists of amount of zero-dimensional NPs. The width of the hexagonal sheet-like structure is about 2–5 μm, and the thickness is about 100–200 nm. The diameter of the NP is about 20 ± 10 nm, and the mechanism of self-assembly process of the hexagonal sheet-like structure is shown in Figure 9.6.

Figure 9.7 shows a TEM microstructural analysis of GO, MnO_2 nanoneedle, MnO_2/rGO nanocomposite, and Mn_3O_4/rGO nanocomposite. Figure 9.7a shows the TEM image of GO. It can be clearly seen from the figure that the GO sheet structure is quite large and thick, the surface is wrinkle and disordered, and its length and width are micrometer scale. We can find that the edge of the GO is quite different

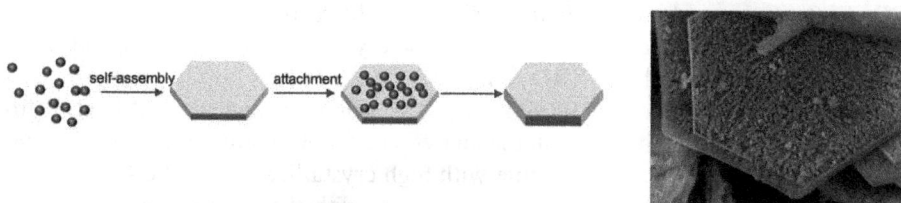

FIGURE 9.6 Schematic illustration showing the formation and growth mechanism of Mn_3O_4 hexagonal plates.

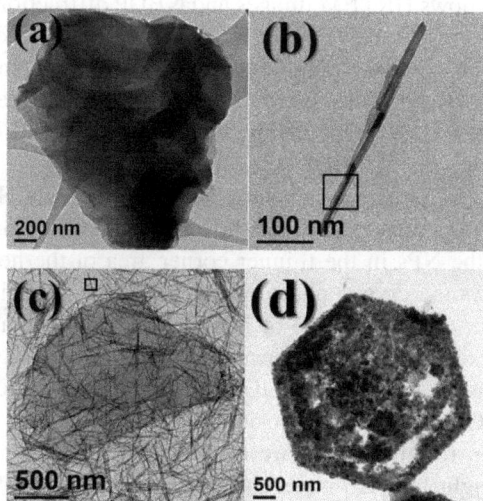

FIGURE 9.7 TEM images of (a) GO, (b) MnO_2, (c) MnO_2/rGO nanocomposite, and (d) Mn_3O_4/rGO hexagonal-flat structure nanocomposite.

from single-layer graphene with the warpage and unevenness edge described in the literature. So it is proved that the GO is not a single-layer graphite oxide after dried. Figure 9.7b shows the SEM images of a single MnO_2 nanoneedle. It can be observed that the length and the diameter of MnO_2 are within the range of 450 ± 40 nm and 20 ± 10 nm, respectively. This result is consistent with the SEM analysis. Figure 9.7c shows the results of TEM analysis of the MnO_2/rGO nanocomposite. From the results, it is shown that the MnO_2 is uniformly distributed on the surface of GO sheet, and with the one-dimensional needle-like structure. The Mn_3O_4/rGO hexagonal sheet-like nanocomposite is shown in Figure 9.7d. It can be clearly observed from the analysis image that each hexagonal sheet structure is composed of many NPs and is consistent with the SEM results.

Figure 9.8 shows a high-resolution transmission electron microscope (HRTEM) and selected area diffraction pattern (SADP) of GO, MnO_2 nanoneedle, MnO_2/rGO nanocomposite, and Mn_3O_4/rGO nanocomposite. Figure 9.8a and b shows HRTEM images and SADP of GO. From Figure 9.8a, we can observe that the edge area of the GO is wrinkled and irregularly arranged, because the oxygen-containing functional group destroys the C–C sp2 bond. The SADP analysis is performed on a block in the GO sheet structure (shown in Figure 9.8b), three diffraction rings can be observed, and the calculated d-spacing corresponds to the crystal faces (001), (002), and (100), which are consistent with the XRD results.

Figure 9.8c and d shows HRTEM images of MnO_2 nanoneedle and NBDP diffraction pattern, respectively. From the analysis result of Figure 9.8c, it can be obviously observed that MnO_2 nanoneedle with high crystallization can be synthesized, even though the synthesizing process just reacts with the temperature lower than 100°C and without high-temperature treatment. The crystal faces (211) and (110) were observed in the HRTEM image, and the d-spacing are 2.39 and 6.79 Å, respectively, which corresponded to the results of NBDP and XRD.

Figure 9.8e and f shows HRTEM images and NBDP diffraction patterns of MnO_2/rGO nanocomposite, respectively. One of the MnO_2 nanoneedles of MnO_2/rGO nanocomposite was analyzed by high-resolution TEM (shown in Figure 9.8e). As a result, two apparent crystal planes were observed, and the d-spacing of the crystal planes were 2.36 and 2.13 Å, respectively, which corresponded exactly to the (211) and (301) diffraction points of NBDP (shown in Figure 9.8f).

Finally, the Mn_3O_4/rGO nanocomposites were analyzed via HRTEM and NBDP diffraction patterns, respectively, and are shown in Figure 9.8g and h. The HRTEM analysis focuses on the NPs in the thinner corner area of the hexagonal sheet. The HRTEM image shows the crystal phase of the NPs with d-spacing = 3.035 Å, corresponding to the crystal phase (112) of Mn_3O_4. The NP was analyzed via NBDP (Figure 9.8h), and the results were also consistent with HRTEM (Figure 9.8g) and XRD.

Figure 9.9 shows a TGA diagram of GO and MnO_2 nanoneedle, MnO_2/rGO nanocomposite, and Mn_3O_4/rGO nanocomposite. The analysis was carried out by raising the temperature from room temperature to 800°C in a nitrogen-filled atmosphere, and the detailed weight loss is in the inset. Figure 9.9a shows the weight change from room temperature to 210°C. The weight loss from room temperature to 120°C is mainly from the loss of water molecules adsorbed on the surface and inside of the material. In the range from 120°C to 210°C, the weight loss occurs due to the

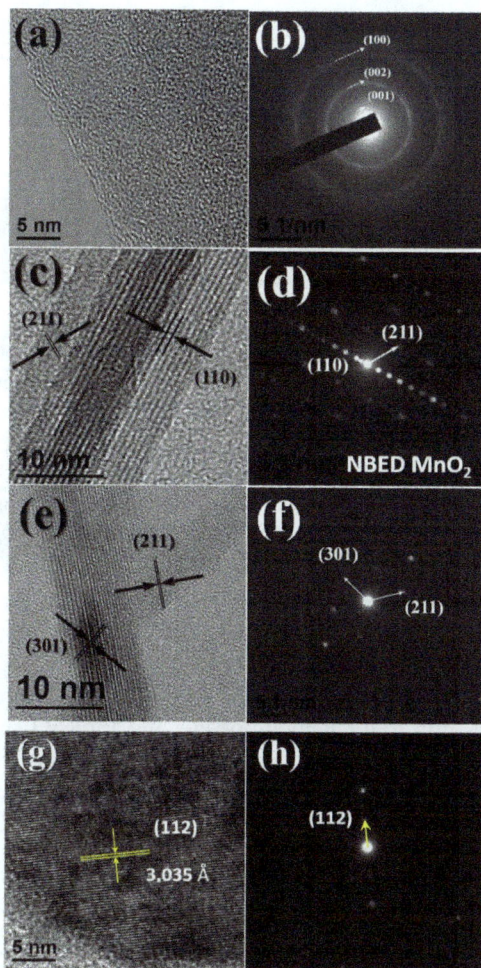

FIGURE 9.8 HRTEM images of (a) GO, (c) MnO_2, (e) MnO_2/rGO nanocomposite, and (g) Mn_3O_4/rGO hexagonal-flat structure nanocomposite. SADP of (b) GO. NBDP of (d) MnO_2, (f) MnO_2/rGO nanocomposite, and (h) Mn_3O_4/rGO hexagonal-flat structure nanocomposite.

cracking of GO as GO has more oxygen-containing functional groups, and these oxygen-containing functional groups will be cleaved and overflow via carbon monoxide (CO) or carbon dioxide (CO_2). As we know that the GO will be completely cracked before 210°C, the weight loss from 210°C to 580°C is mainly due to the oxygen dispersion inside the material. Finally, when the temperature is higher than 580°C, the manganese oxide material still has some weight change, because the manganese oxide in this study is synthesized without high-temperature heat treatment, and there may be defects in the structure. Therefore, when the TGA temperature rises to a high temperature, the structure may have a phase transformation phenomenon, resulting in a slight loss of weight. The results of TGA show that the proportion of

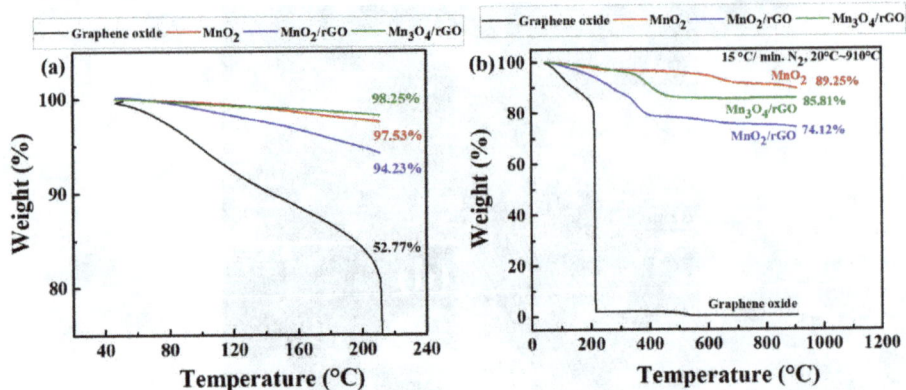

FIGURE 9.9 (a) The TGA from room temperature to 210°C of GO (black line), MnO_2 nanoneedle (red line), MnO_2/rGO nanocomposite (blue line), and Mn_3O_4/rGO hexagonal-flat structure nanocomposite (green line). (b) The TGA from room temperature to 800°C of GO (black line), MnO_2 nanoneedle (red line), MnO_2/rGO nanocomposite (blue line), and Mn_3O_4/rGO hexagonal-flat structure nanocomposite (green line).

GO in MnO_2/rGO nanocomposites and Mn_3O_4/rGO nanocomposites is 5.77% and 1.75%, respectively.

Figure 9.10 shows charge/discharge curve of GO, MnO_2 nanoneedle, MnO_2/rGO nanocomposite, and Mn_3O_4/rGO nanocomposite, and Table 9.1 shows the charge/discharge capacity and coulombic efficiency of first three cycles. The charge/discharge tests of this study were analyzed by a coin cell device. The charge and discharge potential range was from 0.002 to 3.0 V (vs. Li/Li+), and the charge and discharge current rate was 0.1 C. The charge and discharge capacities of the four materials in the first cycle were 1,204.42/311.85 mAh g^{-1}, 506.8/331.3 mAh g^{-1}, 667.1/430.8 mAh g^{-1}, and 1,086.5/804.1 mAh g^{-1}, respectively. The irreversible capacities of the four materials were 74.2%, 34.6%, 35.4%, and 26%, respectively, and the coulombic efficiencies were 25.98%, 65.4%, 64.6%, and 74.0%, respectively. Among the four materials, only the capacity of GO is higher than 1,200 mAh g^{-1}, and it is speculated that the oxygen-containing functional groups to break the van der Waals force of GO and the destruction of the original sp2 bond provide more activation points for reaction with lithium ions, thus increasing the capacitance.

Comparing MnO_2 nanoneedle and MnO_2/rGO nanocomposite, the conductivity of the GO is improved because the manganese dioxide nanoneedle is successfully combined with the carbon layer, and in the process of forming composite material, the GO is indirectly reduced. Therefore, the capacitance is increased. Comparing MnO_2/rGO nanocomposite with Mn_3O_4/rGO nanocomposite, the capacity of the latter exceeds 1,000 mAh g^{-1}. From the images of SEM and TEM, the analysis results show that the structure of Mn_3O_4/rGO is relatively loose, and since Mn_3O_4/rGO is the only product obtained by reduction treatment of MnO_2/rGO, the loose structure (providing a shorter lithium-ion diffusion route) and the less content of oxygen-containing functional group allow the Mn_3O_4/rGO nanocomposite to have a higher charge capacity. From the observation of the irreversible capacitance of

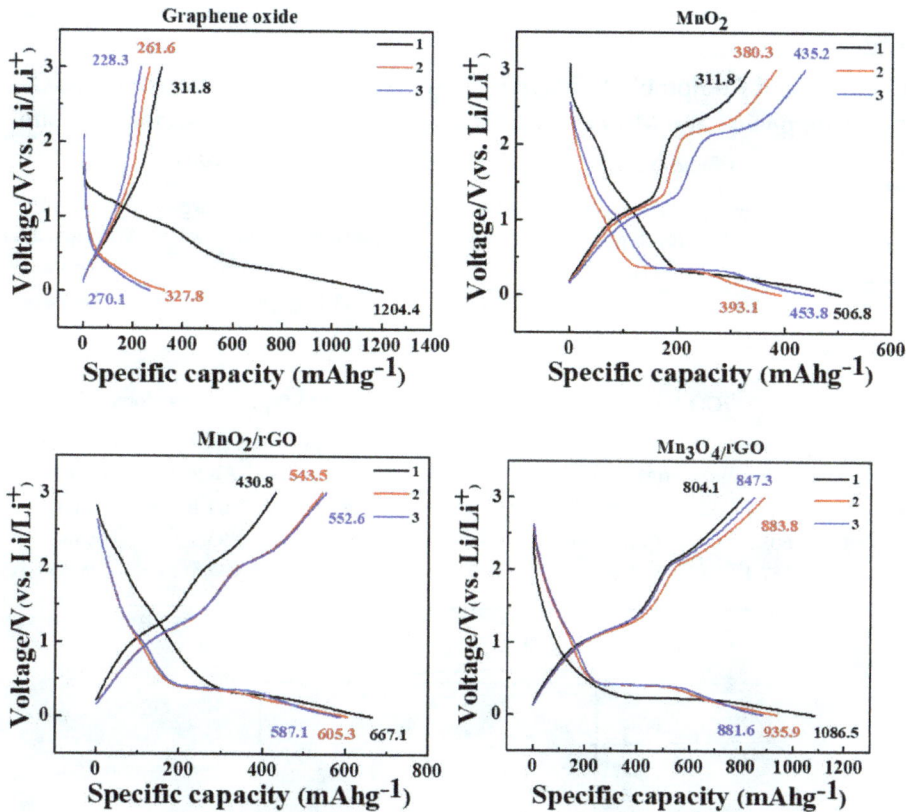

FIGURE 9.10 Charge/discharge curve of (a) GO, (b) MnO_2, (c) MnO_2/rGO nanocomposite, and (d) Mn_3O_4/rGO hexagonal-flat structure nanocomposite.

the first cycle, the irreversible capacitance of GO is as high as 74.2%; on the other hand, the irreversible capacitances of MnO_2 nanoneedle, MnO_2/rGO nanocomposite, and Mn_3O_4/rGO nanocomposite are only 34.6%, 35.4%, and 26%, respectively. MnO_2 nanoneedle and composites have a high reversible capacity due to the indirect reduction of GO during the combination of manganese ions and GO, thereby improving the conductivity and electrical properties of the overall material. When lithium ions react with oxygen-containing functional groups of GO and form lithium oxide (Li_2O), the lithium ions were trapped in the material and cannot be re-released back to the positive-electrode material causing capacity decreasing.

Many literatures indicate that GO (reduced) is combined with MO (transitional) to form a composite material, and the efficiency or property is improved due to the synergistic effect. There shows the reasons why synergistic effect can improve the performance of LIBs: (1) graphene as a 2D carrier for uniformly anchoring or dispersing MOs having well-defined shape and crystallinity; (2) MOs can suppress the restacking phenomenon of graphene; (3) graphene can be used as a 2D conductive template of pure oxide or 3D porous network to provide a good charge transfer pathway for oxides; (4) graphene can suppress volume change and agglomeration of MOs;

TABLE 9.1
Capacity and Coulombic Efficiency of GO, MnO_2 Nanoneedle, MnO_2/rGO Nanocomposite, and Mn_3O_4/rGO Hexagonal-Flat Structure Nanocomposite

	GO			MnO_2		
Cycle	Charge (mAh g⁻¹)	Discharge (mAh g⁻¹)	Coulombic Efficiency (%)	Charge (mAh g⁻¹)	Discharge (mAh g⁻¹)	Coulombic Efficiency (%)
First	1,204.4	311.8	25.9%	506.8	331.3	65.4%
Second	327.8	261.5	79.8%	393.1	380.3	96.7%
Third	270.1	228.3	84.5%	453.8	435.2	95.9%

	MnO_2/rGO Nanocomposite			Mn_3O_4/rGO Nanocomposite		
Cycle	Charge (mAh g⁻¹)	Discharge (mAh g⁻¹)	Coulombic Efficiency (%)	Charge (mAh g⁻¹)	Discharge (mAh g⁻¹)	Coulombic Efficiency (%)
First	667.1	430.8	64.6%	1,086.5	804.1	74.0%
Second	605.3	543.5	89.8%	935.9	883.8	94.4%
Third	587.1	552.6	94.1%	881.6	847.3	96.1%

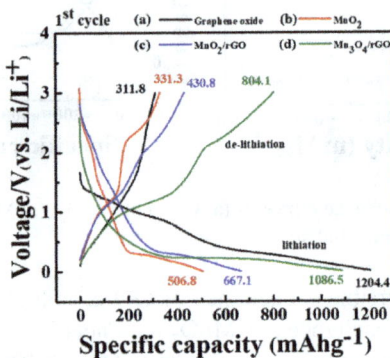

FIGURE 9.11 First-cycle lithiation and delithiation of (a) GO, (b) MnO_2, (c) MnO_2/rGO nanocomposite, and (d) Mn_3O_4/rGO hexagonal-flat structure nanocomposite.

and (5) oxygen-containing groups on GO ensure good binding, interfacial interactions, and electrical contacts between GO and MO. Such effects have been proven in many areas, such as energy storage, supercapacitor, gas sensor, water splitting, photocatalytic application, and the application of environmental hazardous substances decomposition.

Figure 9.11 shows the charge/discharge curves of the four materials. It can be observed from the figure that the charge and discharge diagram of GO has no obvious charge and discharge plateau, because the charge and discharge reaction between carbon material and lithium ion belongs to the intercalation/deintercalation reaction, so no charge and discharge plateau will be observed. On the other hand, the manganese oxide series materials have obvious charging and discharging platforms, and there is an electrochemical platform at 0.5–2.5V. The meaning of this electrochemical plateau is two: (1) irreversible decomposition of the electrolyte, and (2) solid electrolyte interphase (SEI) formation, and below 0.5 V, there is another electrochemical platform, which represents lithium ions enter the inside of the material to form LixMnO$_2$ and further form Li$_2$O and metal Mn (metallic Mn). In addition to the electrochemical platform of the oxide series, the material of the manganese oxide series is observed. The charge and discharge curves of the first to third circles of the three (Figure 9.10) can be found in the discharge of the second and third circles. The capacitance is higher than the first circle because the MnO$_2$ nanoneedle, MnO$_2$/rGO nanocomposite, and Mn$_3$O$_4$/rGO nanocomposite have the following three characteristics: (1) 3D pores or voids provide a large specific surface area and more electrolyte-filled pore channels, allowing electrons and lithium ions to be transported quickly; (2) these holes or voids are like flexible nanocontainers, and it can adapt to the volume change of manganese oxide during charge and discharge; and (3) the larger specific surface area of the three materials, providing more active sites for reacting with lithium ions to increase the capacity.

Figure 9.12 shows the results of capacity versus cycle number of GO, MnO$_2$ nanoneedle, MnO$_2$/rGO nanocomposite, and Mn$_3$O$_4$/rGO nanocomposite applied to an anode electrode of a LIB. Table 9.2 shows a consolidation table of discharge capacity of different cycle and retention for the four materials. Figure 9.12a shows the results of GO for 50 cycles of charge/discharge test, it can be observed that the first-cycle charge capacity can be higher than 1,200 mAh g^{-1}, and it is speculated that the van der Waals force between the carbon layers and the original C–C sp2 bond was destroyed by the oxygen-containing functional group, providing more activation points for reaction with lithium ions, thus increasing the capacitance. From the observation of the irreversible capacitance of the first cycle, the irreversible capacitance of GO is as high as 74.2%, the coulombic efficiency is only 25.98%, and the lithium ion is trapped in the material due to the reaction of the lithium ion with the oxygen-containing functional group to form lithium oxide (Li$_2$O) and to cause lithium ion cannot be released back to the positive-electrode material. This

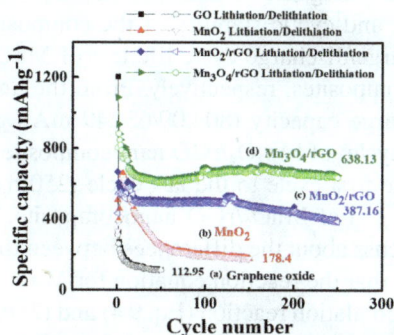

FIGURE 9.12 Capacity versus cycle number plots of (a) GO, (b) MnO$_2$ nanoneedle, (c) MnO$_2$/rGO nanocomposite, and (d) Mn$_3$O$_4$/rGO hexagonal-flat structure nanocomposite.

TABLE 9.2

Capacity and Coulombic Efficiency of GO, MnO$_2$ Nanoneedle, MnO$_2$/rGO Nanocomposite, and Mn$_3$O$_4$/rGO Hexagonal-Flat Structure Nanocomposite

	1st Discharge (mAh g^{-1})	50th Discharge (mAh g^{-1})	150th Discharge (mAh g^{-1})	250th Discharge (mAh g^{-1})	Retention (=final/first)
GO	311.85	133.82	–	–	42.91%
MnO$_2$ Nanoneedle	331.32	248.00	178.4	–	53.85%
MnO$_2$/rGO Nanocomposite	430.78	487.13	472.59	387.15	89.87%
Mn$_3$O$_4$/rGO Nanocomposite	804.09	610.27	676.72	631.49	78.53%

situation caused capacity fading, so that the capacity retention rate is only 36.22%. Figure 9.12b–d shows the results of charge/discharge cycle life test analysis for 150, 250, and 250 cycles of MnO$_2$ nanoneedle, MnO$_2$/rGO nanocomposite, and Mn$_3$O$_4$/rGO nanocomposite, respectively. The first/last cycle discharge capacities of MnO$_2$ nanoneedle, MnO$_2$/rGO nanocomposite, and Mn$_3$O$_4$/rGO nanocomposite are 331.32/178.4, 430.78/387.15, and 804.09/631.49 mAh g^{-1}, respectively. The capacity retention rates of the three materials were 53.85%, 89.87%, and 78.53%, respectively. It can be observed from the above data that MnO$_2$/rGO nanocomposites and Mn$_3$O$_4$/rGO nanocomposites are higher in capacity and retention than MnO$_2$ nanoneedle, mainly due to the fact that composites have higher conductivity because the reduced GO exists in the material. The reduction of GO suppresses the volume expansion and contraction of MnO$_2$ and Mn$_3$O$_4$ during lithiation/delithiation, and decreases the occurrence of pulverization of the material. Reduced GO provides a high-conducting and flexible network structure. The addition of MnO$_2$/Mn$_3$O$_4$ can reduce the agglomeration problem of GO, and GO can be uniformly dispersed not to aggregate. Based on the above reasons, it is confirmed that when GO is combined with MnO$_2$/Mn$_3$O$_4$ to form composites, the synergistic effects increase capacity, conductivity, energy density (power density), and cyclic stability of the composites. Figure 9.12c and d shows the results of charge/discharge cycle life test of MnO$_2$/rGO nanocomposites and Mn$_3$O$_4$/rGO nanocomposites, respectively. From the analysis results, it can be observed that the discharge capacity (804.09/631.49 mAh g^{-1}) from the first cycle to the last cycle (250th cycle) of Mn$_3$O$_4$/rGO nanocomposite always leading the discharge capacity from the first cycle to the last cycle (250th cycle) discharge capacity (430.78/387.15 mAh g^{-1}) of MnO$_2$/rGO nanocomposite. There have been many related literatures to discuss about the differences between the two composites. Dave Andre et al. pointed out that the reaction equation for Mn$_3$O$_4$ with lithium ions consists of two steps: (1) intercalation reaction (Eq. 9.4) and (2) two-stage transformation reaction (double-step conversion reaction) (Eqs. 9.5 and 9.6). The literature indicates that because Mn$_3$O$_4$ has a two-stage conversion reaction, it can lead its theoretical capacity from 900 to 1,000 mAh g^{-1}.

$$Mn_3O_4 + xLi^+ + xe^- \rightarrow Li_xMn_3O_4 \tag{9.4}$$

$$Li_xMn_3O_4 + (2-x)Li^+ + (2-x)e^- \rightarrow 3MnO + Li_2O \tag{9.5}$$

$$3MnO + Li_2O + 6Li^+ + 6e^- \Delta 3Mn + 4Li_2O \tag{9.6}$$

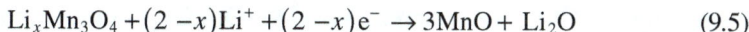

In addition, Xiangpeng Fang et al. not only analyzed MnO, MnO_2, Mn_2O_3, and Mn_3O_4 with cycle life test, but also pointed out that the reaction between MnO, MnO_2, $Mn2O_3$, Mn_3O_4, and lithium ions will produce different amounts of Li_2O, and it is ranked as follows: MnO_2 (1 • Li_2O) > Mn_2O_3 (1/2 • Li_2O) > Mn_3O_4 (1/3 • Li_2O) > MnO (0 • Li_2O). It can be known that MnO does not produce excess Li_2O during the reaction. From Eqs. (9.4)–(9.6), we can observe that eventually reversibly react with lithium ions to be MnO $\leftrightarrow$ Mn, which can avoid excessive Li_2O formation and slow down the rate of capacity decay. The results of the above literatures can explain that the capacity retention of Mn_3O_4/rGO nanocomposites is higher than that of MnO_2/rGO nanocomposites after 250 cycles of charge and discharge cycles, and the decay rate of Mn_3O_4/rGO nanocomposites is lower than MnO_2/rGO nanocomposites. Therefore, Mn_3O_4/rGO nanocomposites are among the four materials with the best cyclic stability and electrochemical properties.

Figure 9.13 shows the capacitance variation with charge/discharge rate of GO, MnO_2 nanoneedle, MnO_2/rGO nanocomposite, and Mn_3O_4/rGO nanocomposite. Table 9.3 shows the analysis results of four materials with different C-rates. The data statistics table records the discharge capacity of the last cycle of each C-rate. The 0.1 C charge/discharge rates of GO and manganese-oxide-based material are 74.4 and 123.2 mA g^{-1}, respectively. First of all, the first three cycles were tested at 0.1 C to ensure forming a stable SEI film on the surface between the electrode and the electrolyte, then using five cycles as the test frequency to increase the charge/discharge

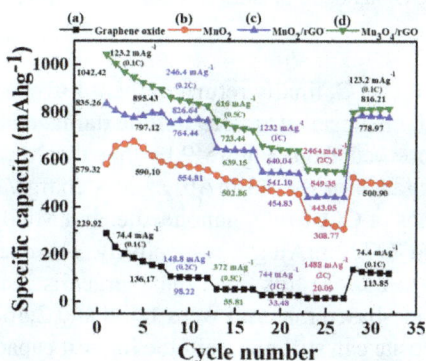

FIGURE 9.13   The cycle life curve of 32 cycles with different C-rates of (a) GO, (b) MnO_2 nanoneedle, (c) MnO_2/rGO nanocomposite, and (d) Mn_3O_4/rGO hexagonal-flat structure nanocomposite. The cell was first cycled at 0.1 C for 7 cycles with the voltage ranging between 0.002 and 3.0 V, after which the rate was increased in stages to 0.2 C for 5 cycles, 0.5 C for 5 cycles, 1 C for 5cycles, 2 and 0.1 C for final 5 cycles.

TABLE 9.3

C-Rate Test of GO, MnO$_2$ Nanoneedle, MnO$_2$/rGO Nanocomposite, and Mn$_3$O$_4$/rGO Hexagonal-Flat Structure Nanocomposite

Materials	GO	MnO$_2$	MnO$_2$/rGO	Mn$_3$O$_4$/rGO
1st Discharge Capacity	229.92	579.32	835.26	1,042.42
0.1 C	136.17	590.10	797.12	895.43
0.2 C	98.22	554.81	764.44	826.64
0.5 C	55.81	502.86	639.15	723.44
1 C	33.48	454.83	541.10	640.04
2 C	20.09	308.77	443.06	549.35
0.1 C	113.85	500.90	778.97	814.13
Retention	16.17%	79.39%	93.26%	78.10%

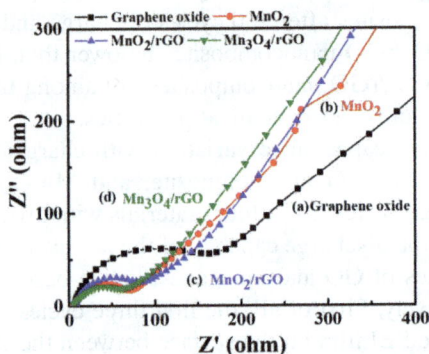

FIGURE 9.14 The Nernst curve of EIS analysis of four materials: (a) GO, (b) MnO$_2$ nanoneedle, (c) MnO$_2$/rGO nanocomposite, and (d) Mn$_3$O$_4$/rGO hexagonal-flat structure nanocomposite.

rate from 0.1, 0.2, 0.5, 1, to 2 C, finally return to the 0.1 C charge/discharge rate, and then observe whether the material structure will be damaged by the large current.

It can be clearly observed from Figure 9.13 that the Mn$_3$O$_4$/rGO nanocomposite can provide a capacitance of 895.43 mAh g^{-1} at a charge/discharge rate of 0.1 C (7th cycle). The capacities of GO, MnO$_2$ nanoneedle, and MnO$_2$/rGO nanocomposites were 136.17, 590.10, and 797.12 mAh g^{-1}, respectively. The charge/discharge rate was continuously increased to 2 C. At this time, the charge/discharge current densities of GO and manganese-oxide-based material were 1,488 and 2,464 mA g^{-1}, respectively. Mn$_3$O$_4$/rGO nanocomposite can still maintain the highest capacity of 549.35 mAh g^{-1}. On the other hand, the capacities of GO, MnO$_2$ nanoneedle, and MnO$_2$/rGO nanocomposite in 2 C current density are 20.09, 308.77, and 443.05 mAh g^{-1}, respectively. When the current density returns to 0.1 C, the retention capacities of GO, MnO$_2$ nanoneedle, MnO$_2$/rGO nanocomposites, and Mn$_3$O$_4$/rGO nanocomposites are 14.75%, 52.31%, 55.58%, and 61.35%, respectively. The Mn$_3$O$_4$/rGO nanocomposites still maintain high retention of more than 60%, representing their structure can overcome high-current charge/discharge test, and lithium ion can quickly diffuse in/out of the material without hindrance. Finally, when the charge and discharge rate returns to 0.1 C, the retention capacities of GO, MnO_2 nanoneedle, MnO_2/rGO nanocomposite, and Mn_3O_4/rGO nanocomposite were 113.85 (16.17%), 500.90 mAh g^{-1} (79.39%), 778.97 mAh g^{-1} (93.26%), and 816.21 mAh g^{-1} (78.10%). From the analysis results, it can be found that the two nanocomposites have been tested by high-current charge and discharge. It can maintain the capacitance of nearly 800 mAh g^{-1}, which implies that the structure with nanosize has more lithium-ion diffusion channels, which not only provides high capacitance, but also makes the structure strong enough to withstand the rapid charging and discharging process with large current without being destroyed.

Figure 9.14 shows the EIS curves of GO, MnO_2 nanoneedle, MnO_2/rGO nanocomposite, and Mn_3O_4/rGO nanocomposite. For all of them, a semicircle in high-frequency region and a line in low-frequency region were observed. The start point of the EIS curve, which is called the solution resistance (Rs) and contributed by electrolyte, separator, and electrodes, is small. The first semicircle at high-frequency region is mainly contributed by the charge/discharge reaction at the SEI, and the intercept with the real part of impedance is the resistance generated at the SEI region. The diameter of the semicircle Rct (charge transfer resistance), which is contributed from the electron transfer during the redox reaction, occurs between lithium ion and anode materials. The resistance value corresponds to the diameter of the semicircle. It can be obviously observed that Mn_3O_4/rGO nanocomposite has the smallest semicircle diameter, which means that the composite material has a stable and perfect SEI film formation and lower resistance, let electrons can go through the SEI film smoothly without lost.

Figure 9.15 shows the oblique line spectrum observed in the low-frequency region. When the slope of the oblique line is inclined, it represents lithium. The oblique line at the low frequency corresponds to the resistance generated due to the diffusion of Li ions between the active material and the electrolyte. The faster the diffusion rate of ions in the material, the more obvious the slope of the manganese oxide series material which is steeper than that of the GO, and the line of the Mn_3O_4/rGO nanocomposite pattern is steepest. Therefore, it can be known that the structure of the composite material is favorable for the diffusion behavior of lithium ions in the material.

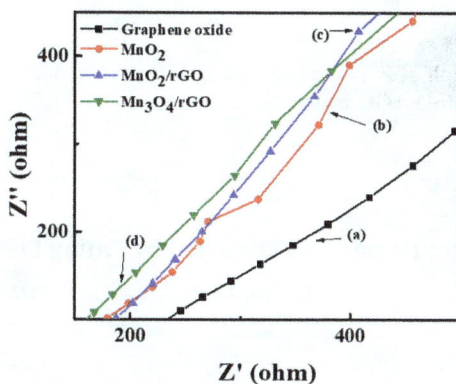

FIGURE 9.15   The slope line of EIS analysis of four materials: (a) GO, (b) MnO_2 nanoneedle, (c) MnO_2/rGO nanocomposite, and (d) Mn_3O_4/rGO hexagonal-flat structure nanocomposite.

The lithium-ion diffusion coefficient can be calculated from the formula as follows:

$$D_{Li^+} = \frac{R^2 T^2}{2 A^2 n^4 F^4 C^2 \sigma^2}$$

where A is the surface area of the electrode, n is the number of the electrons per molecule attending the electronic transfer reaction, F is the Faraday constant, C is the concentration of lithium ion in electrolyte, R is the gas constant, T is the test temperature in our analysis, and σ is the slope of the line $Z'{\sim}\omega^{-1/2}$ which can be obtained from the line of $Z'{\sim}\omega^{-1/2}$ (shown in Figure 9.16), respectively, where the constant values of F and R are 96,500 C mol^{-1} and 8.314 JK^{-1}, respectively. A is the electrode area which is 1.33×10^{-4} m^2, n is 1, T is 295 K, and C is 1 mol m^{-3}. The lithium diffusion coefficients D (cm^2 s^{-1}) of GO, MnO$_2$ nanoneedle, MnO$_2$/rGO nanocomposite, and Mn$_3$O$_4$/rGO nanocomposite are 7.32×10^{-12}, 1.3×10^{-10}, 4.7×10^{-11}, and 2.4×10^{-10} (shown in Table 9.4), respectively. From the results of D value calculation, it is proved that the GO combined with transition MO and forming MnO$_2$/rGO and Mn$_3$O$_4$/rGO composites can improve the conductivity and further enhance irreversible capacity and the performance of charge/discharge with different C-rates.

Table 9.5 provides a brief review of the performance of MnO$_2$, MnO$_2$/graphene (rGO) composites, and Mn$_3$O$_4$/rGO composite when used as anode materials in

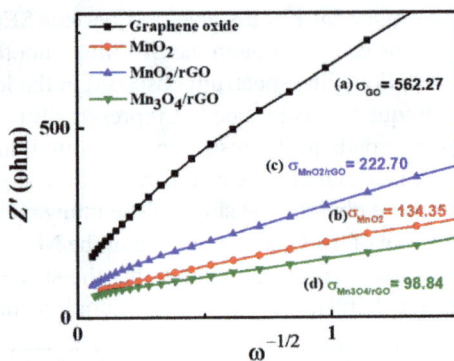

FIGURE 9.16 The graph of Z' plotted against $\omega^{-1/2}$ at low-frequency region of (a) GO, (b) MnO$_2$ nanoneedle, (c) MnO$_2$/rGO nanocomposite, and (d) Mn$_3$O$_4$/rGO hexagonal-flat structure nanocomposite.

TABLE 9.4
The Element Parameters Obtained by Fitting EIS Data

Materials	σ (Ω cm^{-2} s$^{-0.5}$)	D_{Li^+} (cm^{-2} s^{-1})
GO	562.27	7.32×10^{-12}
MnO$_2$/rGO	222.70	4.7×10^{-11}
MnO$_2$	134.35	1.3×10^{-10}
Mn$_3$O$_4$/rGO	98.84	2.4×10^{-10}

TABLE 9.5
Physical Properties and Electrochemical Li Cycling Data of MnO_2 and MnO_2/Graphene Nanocomposites and Mn_3O_4/rGO Hexagonal-Flat Structure Nanocomposite

Authors	Morphology	MnO_2 Morphology/size	Current Rate	Reversible Capacity of 1st Cycle (mAh g^{-1})	Voltage Range	Capacity Retention After n Cycles (Cycling Range)	References
Kim et al.	MnO_2 nanowires	Diameters: 10–30 nm	123 mA g^{-1}	1,573 mAh g^{-1}	0.01–3.0 V	300 mAh g^{-1} (n = 2–100)	[24]
Chen et al.	MnO_2 nanorods	Diameter: 100 nm Length: 10 μm	100 mA g^{-1}	1,119.2 mAh g^{-1}	0.01–3.0 V	1,074.8 mAh g^{-1} (n = 2–100)	[25]
Zhao et al.	Nanoporous γ-MnO_2 hollow microspheres	MnO_2 hollow. Diameter: 2.5 μm and a 500-nm-thick shell	100 mA g^{-1}	1,300 mAh g^{-1}	0.02–3.3 V	602.1 mAh g^{-1} (n = 2–20)	[26]
Li et al.	Mesoporous γ-MnO_2 microcrystal	Pore size: 4–50 nm	100 mA g^{-1}	1,271 mAh g^{-1}	0.01–3.3 V	626 mAh g^{-1} (n = 2–100)	[27]
Li et al.	α-MnO_2 hollow urchins	MnO_2 nanorods Diameters: 30 nm Lengths: 200–300 nm.	270 mA g^{-1}	746.0 mAh g^{-1}	0.01–2.0 V	481 mAh g^{-1} (n = 2–40)	[28]
Wu et al.	γ-MnO_2	Diameter: 12–18 nm	85 mA g^{-1}	600 mAh g^{-1} (100°C annealing)	0.01–2.0 V	170 mAh g^{-1} (n = 2–50)	[29]
Li et al.	MnO_2	Needle-shaped structure	100 mA g^{-1}	265 mAh g^{-1}	0.01 and 3.0 V	105 mAh g^{-1} (n = 2–6)	[30]
This work	α-MnO_2 nanoneedles	α-MnO_2 Diameter: 15–20 nm Length: 450–550 nm	123 mA g^{-1}	331.32 mAh g^{-1}	0.002–3.0 V	178.4 mAh g^{-1} (n = 2–150)	-
Kim et al.	MnO_2/rGO nanocomposites	Diameters: 10–30 nm	123 mA g^{-1}	1,215 mAh g^{-1}	0.01–3.0 V	1,100 mAh g^{-1} (n = 2–100)	[24]
Jiang et al.	MnO_2 nanorods/rGO nanocomposites	-	1.0 A g^{-1}	1,945.8 mAh g^{-1}	0.01–3.0 V	1,635.3 mAh g^{-1} (n = 2–450)	[31]
Zhang et al.	Graphene/α-MnO_2 nanocomposites	α-MnO_2 nanowire Diameter: 40–50 nm, Length: 5–10 μm	60 mA g^{-1}	~1,150 mAh g^{-1}	0.01–3.0 V	998 mAh g^{-1} (n = 2–30)	[32]

(Continued)

TABLE 9.5 (*Continued*)

Physical Properties and Electrochemical Li Cycling Data of MnO_2 and MnO_2/Graphene Nanocomposites and Mn_3O_4/rGO Hexagonal-Flat Structure Nanocomposite

Authors	Morphology	MnO_2 Morphology/size	Current Rate	Reversible Capacity of 1st Cycle (mAh g⁻¹)	Voltage Range	Capacity Retention After n Cycles (Cycling Range)	References
Wen et al.	MnO_2/graphene composite	MnO_2 NPs	100 mA g⁻¹	746 mAh g⁻¹	0.01–3.0 V	752 mAh g⁻¹ (n = 2–65)	[33]
Xing et al.	α-MnO_2/graphene nanocomposites	α-MnO_2 nanosheets	0.1 C	726.5 mAh g⁻¹	0.01–3.0 V,	575 mAh g⁻¹ (n = 2–20)	[34]
Yu et al.	Graphene/MnO_2 nanotube (NT) thin film composites	MnO_2 NTs Diameters: 70–80 nm. Lengths: 1 μm	100 mA g⁻¹	686 mA g⁻¹	0.01–3.0 V	495 mAh g⁻¹ (n = 2–40)	[35]
This work	α-MnO_2/rGO nanocomposites	α- MnO_2 nanoneedles Diameter: 15–20 nm Length: 450–550 nm	123 mA g⁻¹	430.78 mAh g⁻¹	0.002–3.0 V	387.15 mAh g⁻¹ (n = 2–250)	-
Inho Nam et al.	Mn_3O_4/graphene composites	Mn_3O_4 NP: 15 nm	60 mA g⁻¹	700 mAh g⁻¹	0.01–3.0 V	660.9 mAh g⁻¹ (n = 2–40)	[36]
Nathalie Lavoie et al.	Mn_3O_4/RGO composite	Needle-like Mn_3O_4 particles	75 mA g⁻¹	1,175 mAh g⁻¹	0.002–3.0 V	675 mAh g⁻¹ (n = 2–100)	[37]
Zhao-Hui Wang et al.	Mn_3O_4/MWCNTs nanocomposite	Mn_3O_4 NP: 15 nm	100 mA g⁻¹	710 mAh g⁻¹	0.01–3.0 V	592 mAh g⁻¹ (n = 2–50)	[38]
Lu-Lu Wu et al.	Nanorod Mn_3O_4@GNS	Mn_3O_4 nanorod Diameters: range 200 nm Length: 1.5 μm.	100 mA g⁻¹	1,155.2 mAh g⁻¹	0.01–3.0 V	561.7–1,119.4 mAh g⁻¹ (n = 2–100)	[8]
Jiayuan Chen et al.	Mn_3O_4/N-doped graphene (Mn_3O_4/NG) hybrids	Mn_3O_4 NP: 20–60 nm	88 mA g⁻¹	794.1 mAh g⁻¹	0.01–3.0 V	1,208.4 mAh·g⁻¹ (n = 2–150)	[39]
Changbin Wang et al.	Mn_3O_4@C nanocomposite	Mn_3O_4 nanorod Width: 200–300 nm Thickness: 15–20 nm	40 mA g⁻¹	723 mAh g⁻¹	0.01–3.0 V	473 mAh g⁻¹ (n = 2–50)	[40]
This work	Mn_3O_4/rGO hexagonal-flat structure nanocomposite	Mn_3O_4 NP: 15–20 nm	123 mA g⁻¹	804.09	0.002–3.0 V	638.13 (n = 2–250)	-

LIBs. Comparing the structure, morphology, charge and discharge current density, first-cycle reversible capacitance, charge/discharge test range, and capacity retention rate of literature with this study, the results confirmed that Mn_3O_4/rGO nanocomposites synthesized in this study have high capacitance, low irreversible capacity, high lithium-ion diffusion rate, good conductivity, low resistance, and high cycle stability.

9.4 CONCLUSIONS

The nanocomposite of MnO_2, MnO_2/rGO, Mn_3O_4, and Mn_3O_4/rGO could be prepared by an easy chemical reduction procedure. MnO_2/rGO and Mn_3O_4/rGO had shown significant electrochemical property and are suitable anode materials for LIBs. The enhanced electrochemical performance strongly depends on the electrical transport property of the composite having rGO as conductive buffer, size and distribution of the nanostructure over the rGO, and available surface active sites.

ACKNOWLEDGMENT

This work was financially supported by the Hierarchical Green-Energy Materials (Hi-GEM) Research Center, from The Featured Areas Research Center Program within the framework of the Higher Education Sprout Project by the Ministry of Education (MOE) and the Ministry of Science and Technology (MOST 110–2634-F-006–017) in Taiwan.

REFERENCES

1. Poizot P, Laruelle S, Grugeon S, Dupont L and Tarascon J 2000 Nano-sized transition-metal oxides as negative-electrode materials for lithium-ion batteries *Nature* **407** 496–9.
2. Li H, Liu Z, Yang S, Zhao Y, Feng Y, Bakenov Z, Zhang C and Yin F 2017 Facile synthesis of ZnO nanoparticles on nitrogen-doped carbon nanotubes as high-performance anode material for lithium-ion batteries *Materials* **10** 1102.
3. Balaya P, Li H, Kienle L and Maier J 2003 Fully reversible homogeneous and heterogeneous Li storage in RuO_2 with high capacity *Adv Funct Mater* **13** 621–5.
4. Varghese B, Reddy M, Yanwu Z, Lit C S, Hoong T C, Subba Rao G, Chowdari B, Wee A T S, Lim C T and Sow C-H 2008 Fabrication of NiO nanowall electrodes for high performance lithium ion battery *Chem Mater* **20** 3360–7.
5. Hou C-C, Brahma S, Weng S-C, Chang C-C and Huang J-L 2017 Facile, low temperature synthesis of SnO_2/reduced graphene oxide nanocomposite as anode material for lithium-ion batteries *Appl Surf Sci* **413** 160–8.
6. Weng S-C, Brahma S, Chang C-C and Huang J-L 2017 Synthesis of MnOx/reduced graphene oxide nanocomposite as an anode electrode for lithium-ion batteries *Ceram Int* **43** 4873–9.
7. Wang Z-H, Yuan L-X, Shao Q-G, Huang F and Huang Y-H 2012 Mn_3O_4 nanocrystals anchored on multi-walled carbon nanotubes as high-performance anode materials for lithium-ion batteries *Mater Lett* **80** 110–3.
8. Wu L-L, Zhao D-L, Cheng X-W, Ding Z-W, Hu T and Meng S 2017 Nanorod Mn_3O_4 anchored on graphene nanosheet as anode of lithium ion batteries with enhanced reversible capacity and cyclic performance *J Alloy Compd* **728** 383–90.

9. Nam I, Kim N D, Kim G-P, Park J and Yi J 2013 One step preparation of Mn_3O_4/graphene composites for use as an anode in Li ion batteries *J Power Sources* **244** 56–62.

10. Lavoie N, Malenfant P R, Courtel F M, Abu-Lebdeh Y and Davidson I J 2012 High gravimetric capacity and long cycle life in Mn_3O_4/graphene platelet/LiCMC composite lithium-ion battery anodes *J Power Sources* **213** 249–54.

11. Wang C, Yin L, Xiang D and Qi Y 2012 Uniform carbon layer coated Mn3O4 nanorod anodes with improved reversible capacity and cyclic stability for lithium ion batteries *ACS Appl Mater Inter* **4** 1636–42.

12. Geim A K and Novoselov K S 2010 *Nanoscience and Technology: A Collection of Reviews from Nature Journals*: World Scientific. pp 11–9.

13. Novoselov K S, Geim A K, Morozov S V, Jiang D, Zhang Y, Dubonos S V, Grigorieva I V and Firsov A A 2004 Electric field effect in atomically thin carbon films *Science* **306** 666–9.

14. Wang G, Shen X, Yao J and Park J 2009 Graphene nanosheets for enhanced lithium storage in lithium ion batteries *Carbon* **47** 2049–53.

15. Yoo E, Kim J, Hosono E, Zhou H-S, Kudo T and Honma I 2008 Large reversible Li storage of graphene nanosheet families for use in rechargeable lithium ion batteries *Nano Lett* **8** 2277–82.

16. Ma X, Ye K, Wang G, Duan M, Cheng K, Wang G and Cao D 2017 Facile fabrication of gold coated nickel nanoarrays and its excellent catalytic performance towards sodium borohydride electro-oxidation *Appl Surf Sci* **414** 353–60.

17. Sarioğlan Ş 2013 Recovery of palladium from spent activated carbon-supported palladium catalysts *Platin Met Rev* **57** 289–96.

18. Shin H J, Kim K K, Benayad A, Yoon S M, Park H K, Jung I S, Jin M H, Jeong H K, Kim J M and Choi J Y 2009 Efficient reduction of graphite oxide by sodium borohydride and its effect on electrical conductance *Adv Funct Mater* **19** 1987–92.

19. Liu H, Hu Z, Su Y, Ruan H, Hu R and Zhang L 2017 MnO_2 nanorods/3D-rGO composite as high performance anode materials for Li-ion batteries *Appl Surf Sci* **392** 777–84.

20. Majid S R 2015 Green synthesis of in situ electrodeposited rGO/MnO_2 nanocomposite for high energy density supercapacitors *SCI REP-UK* **5** 1–13.

21. Zhao G, Huang X, Wang X, Connor P, Li J, Zhang S and Irvine J T 2015 Synthesis and lithium-storage properties of MnO/reduced graphene oxide composites derived from graphene oxide plus the transformation of Mn (VI) to Mn (II) by the reducing power of graphene oxide *J Mater Chem A* **3** 297–303.

22. Yuan C, Zhang Y, Pan Y, Liu X, Wang G and Cao D 2014 Investigation of the intercalation of polyvalent cations (Mg2+, Zn2+) into λ-MnO_2 for rechargeable aqueous battery *Electrochim Acta* **116** 404–12.

23. Longo R, Kong F, Santosh K, Park M, Yoon J, Yeon D-H, Park J-H, Doo S-G and Cho K 2014 Phase stability of Li–Mn–O oxides as cathode materials for Li-ion batteries: Insights from ab initio calculations *Phys Chem Chem Phys* **16** 11233–42.

24. Li B, Rong G, Xie Y, Huang L and Fen C 2006 Low-temperature synthesis of α-MnO_2 hollow urchins and their application in rechargeable Li+ batteries *Inorg Chem* **45** 6404–10.

25. Chen J, Wang Y, He X, Xu S, Fang M, Zhao X and Shan Y 2014 Electrochemical properties of MnO_2 nanorods as anode materials for lithium ion batteries *Electrochim Acta* **142** 152–6.

26. Li L, Raji A R and Tour J M 2013 Graphene-wrapped MnO_2-graphene nanoribbons as anode materials for high-performance lithium ion batteries *Adv Mater* **25** 6298–302.

27. Jiang Y, Jiang Z-J, Chen B, Jiang Z, Cheng S, Rong H, Huang J and Liu M 2016 Morphology and crystal phase evolution induced performance enhancement of MnO_2 grown on reduced graphene oxide for lithium ion batteries *J Mater Chem A* **4** 2643–50.

28. Zhanga Y, Liu H, Zhu Z, Wong K-W, Mi R, Mei J and Lau W-M 2013 A green hydro-thermal approach for the preparation of graphene α-MnO_2 3D network as anode for lithium ion battery *Electrochim Acta* **108** 465–71.
29. Zhao J, Tao Z, Liang J and Chen J 2008 Facile synthesis of nanoporous γ-MnO_2 structures and their application in rechargeable li-ion batteries *Cryst Growth Design* **8** 2799–805.
30. Wen K, Chen G, Jiang F, Zhou X and Yang J 2015 A facile approach for preparing mno2-graphene composite as anode material for lithium-ion batteries *Int J Electrochem Sci* **10** 3859–66.
31. Xing L, Cui C, Ma C and Xue X 2011 Facile synthesis of α-MnO_2 graphene nanocomposites and their high performance as lithium-ion battery anode *Mater Lett* **65** 2104–6.
32. Alfaruqi M H, Islam S, Gim J, Song J, Kim S, Pham D T, Jo J, Xiu Z, Mathew V and Kim J 2016 A high surface area tunnel-type α-MnO2nanorod cathode by a simple solvent-free synthesis for rechargeable aqueous zinc-ion batteries *Chem Phys Lett* **650** 64–8.
33. Wang X and Li Y 2002 Rational synthesis of α-MnO_2 single-crystal nanorods *Chem Commun* **2002** 764–5.
34. Devaraj S and Munichandraiah N 2008 Effect of crystallographic structure of MnO_2 on its electrochemical capacitance properties *J Phys Chem C* **112** 4406–17.
35. Li J, Xi B, Zhu Y, Li Q, Yan Y and Qian Y 2011 A precursor route to synthesize mesoporous γ-MnO_2 microcrystals and their applications in lithium battery and water treatment *J Alloys Comp* **509** 9542– 8.
36. Nam I, Kim N D, Kim G-P, Park J and Yi J 2013 One step preparation of Mn_3O_4_graphene composites for use as an anode in Li ion batteries *J Power Sources* **244** 56–62.
37. Lavoie N, Malenfan P R L, Courtel F M, Abu-Lebdeh Y and Davidson I J 2012 High gravimetric capacity and long cycle life in Mn_3O_4_graphene platelet_LiCMC composite lithium-ion battery anodes *J Power Sources* **213** 249–54.
38. Wang Z-H, Yuan L-X, Shao Q-G, Huang F and Huang Y-H 2012 Mn_3O_4 nanocrystals anchored on multi-walled carbon nanotubes as high-performance anode materials for lithium-ion batteries *Mater Lett* **80** 110–3.
39. Chen J, Wu X, Gong Y, Wang P, Li W, Tan Q and Chen Y 2017 Synthesis of Mn_3O_4_N-doped graphene hybrid and its improved electrochemical performance for lithium-ion batteries *Ceram Int* **43** 4655–62.
40. Wang C, Yin L, Xiang D and Qi Y 2012 Uniform carbon layer coated Mn_3O_4 nanorod anodes with improved reversible capacity and cyclic stability for lithium ion batteries *ACS Appl Mater Interfaces* **4** 1636–42.

10 In-situ Synthesis of Solid-State Polymer Electrolytes for Lithium-Ion Batteries

*Yu-Chao Tseng, Ting-Yuan Lee, Yuan-Shuo Hsu,
Febriana Intan, and Jeng-Shiung Jan*
National Cheng Kung University

CONTENTS

10.1 INTRODUCTION

Lithium-ion batteries (LIBs) possess advantages such as high-energy density, fast-charging rate, great cycle stability, long cycle life, and shape variability. Currently, LIBs have become well-known power components of portable mobile devices such as laptops and smart phones. Meanwhile, LIBs will play a much more important role in our future life owing to their flourishing development in the field of household appliances and electric/hybrid vehicles [1–5]. LIBs employing lithium metal

DOI: 10.1201/9781003263807-10

batteries (LMBs) as the anode material further refer to ideal candidates to the next-generation energy storage systems due to their low redox potential of -3.04 V and ultrahigh specific capacity of 3,860 mAh g^{-1} [6]. However, dendrite formation and lithium corrosion during the processes of charge/discharge remain problems causing that LMBs have the risks of short circuit and thermal runaway [7,8]. Meanwhile, most commercial electrolyte systems are composed of mixtures of ethylene carbonate/diethyl carbonate and lithium salts, and volatile and flammable are exactly the most fatal flaw to the kinds of liquid electrolytes [9–11]. These safety issues posted by lithium anode and liquid electrolytes have limited the extensive production of LMBs and driven the researchers to pursue an electrolyte that is flame-resistant and robust enough to mechanically prevent dendrite formation.

One strategy to realize the purpose is to replace liquid electrolytes with other materials. Recently, ionic liquids (ILs) have been recommended as the suitable alternatives due to their broad electrochemical window, negligible flammability, and vapor pressure [12,13], which are especially in demand of safe and stable electrolytes [14–19]. However, although the flammability can be suppressed, issues such as the electrolyte leakage and liquid accumulation inside the batteries let the fundamental problem still unresolved. Another approach is to incorporate lithium salt and plasticizer into the polymer matrix that is also called polymer electrolytes. The properties such as high thermal, chemical, electrochemical stabilities, as well as great flame resistance, make polymer electrolytes become ideal candidates to guarantee the intrinsic security of the equipped batteries [20–23]. To date, several kinds of polymers have been introduced for electrolytes, including poly(acrylonitrile) (PAN), poly(ethylene glycol) (PEG), poly(methyl methacrylate) (PMMA), and poly(vinylidene fluoride-co-hexafluoropropylene) (PVdF-co-HFP) [24–28]. Beside these conventional materials, poly(ionic liquid)s (PILs), which are polymerized from IL molecules, have emerged as a novel topic of polymer electrolytes as their extraordinary combination of ionic properties and polymer natures [29]. According to the space charge theory put forward by Chazalviel et al., anion would loiter around the anode during the process of charge/discharge [30,31]. Meanwhile, a depletion layer caused by the accumulated anions will generate on the anode, resulting in a huge space charge and electric field. Ion transfer, mainly driven by such large electric fields around this area, would let lithium ions deposit unevenly on the anode, leading to the growth of dendrites. Chazalviel et al. also explained that cationic-type ILs can provide local reservoir to alleviate unstable electric field in the space charge region and mitigate the lithium dendrites growth. The superiority emanating from ILs has driven researchers to invest in the application of PILs for LIBs.

In this context, pyrrolidinium-based PILs were first applied to polymer electrolytes by Appetecchi et al. [32]. The assembled Li/LiFePO$_4$ cell can export discharge capacities of around 144 mAh g^{-1} at 0.1 C rate at 40°C. Later, several groups continuously exploit the similar electrolytes by selecting other additives to improve the battery performance [33–35]. Especially, the materials were developed by Yang and coworkers using succinonitrile as the additive, and the corresponding capacity can reach 150 mAh g^{-1} at 0.1 C rate [34], while Li et al. reported electrolytes based on guanidinium-based PILs, which can maintain the charge/discharge capacity values of about 130 mAh g^{-1} to at least 70 cycles at 80°C [36,37]. Subsequently, Li et al. also synthesized tetraalkylammonium-based PILs, and the results showed that the corresponding Li/LiFePO$_4$ cell can support a stable progress of charge/discharge to 50 cycles at 60°C [38,39]. Inspired by above studies, the electrolyte performance may be further optimized by increasing the ion density in PIL structures. Yin et al., for example, have confirmed that the electrolytes containing dicationic imidazolium-based PILs afford an ionic conductivity of 4.6×10^{-5} S cm^{-1} at 25°C, which is a value almost five times of magnitude higher than that of the electrolytes with monocationic PILs [40,41]. Besides, dicationic tetraalkylammonium-based PILs with multi-armed architectures were constructed by Zhou et al. [31]. The superior interfacial characteristics endow the assembled cells with the discharge capacities up to 149 mAh g^{-1} at 0.5 C rate at 60°C. Table 10.1 concludes several charge/discharge capacities of the

TABLE 10.1

Summary of the Performances of Batteries with Common PILs [32,34–39,41,45]

PIL	Discharge Capacity (cathode material: LiFePO$_4$)	References
Pyrrolidinium-based PILs	150 mAh g^{-1} after 70 cycles at 0.05 C (40°C)	[32]
	150 mAh g^{-1} after 40 cycles at 0.1 C (25°C)	[34]
	140 mAh g^{-1} after 70 cycles at 0.2 C (80°C)	[35]
Guanidinium-based PILs	130 mAh g^{-1} after 100 cycles at 0.1 C (80°C)	[36]
	115 mAh g^{-1} after 70 cycles at 0.1 C (80°C)	[37]
Tetraalkylammonium-based PILs	125 mAh g^{-1} after 130 cycles at 0.1 C (60°C)	[38]
	135 mAh g^{-1} after 50 cycles at 0.1 C (60°C)	[39]
Imidazolium-based PILs	160 mAh g^{-1} after 50 cycles at 0.1 C (40°C)	[41]
Pyridinium-based PILs	140 mAh g^{-1} after 250 cycles at 0.1 C (25°C)	[45]

batteries based on typical PILs, most of which, however, were conducted at the temperatures higher than 40°C, while their cycle performances at room temperature are still unsatisfied, which cannot be used in practical.

In this chapter, a thiol-ene click cross-linking reaction was employed to form a thiosiloxane/poly(ethylene glycol) diacrylate (PEGDA)/imidazolium-based cross-linker (C-VIm) conetwork for application as a polymer scaffold in the electrolytes. The thiol-ene click reaction is verified to be fast, which is conducive to produce a high yield and a high purity of the products without performing complicated chemical synthesis. In our experiment, firstly, the miscibility of a ternary thiosiloxane/PEGDA/C-VIm mixture was determined to ensure their complete compatibility. Secondly, the polymer membranes were formed to be solid-state electrolytes through the photopolymerization in the presence of initiator. Then, thermal, chemical, and electrical properties of the membranes were evaluated in detail. Charge/discharge performance of the Li/LiFePO$_4$ cells based on the membranes was also carried out to check the potential utilization of the materials as solid-state electrolytes for LIBs.

10.2 EXPERIMENTAL

10.2.1 MATERIALS

(Mercaptopropyl)methylsiloxane homopolymer (thiosiloxane) was purchased from Gelest. PEGDA (Mn = 700 g mol^{-1}), poly(ethylene glycol) dimethyl ether (PEGDME) (Mn = 700 g mol^{-1}), 1-vinylimidazole (VIm), and 1,12-dibromododecane were bought from Sigma-Aldrich. Lithium-bis(trifluoromethylsulfnyl) amide (LiTFSI) was obtained from Solvay. Photo initiator (2,2-dimethoxy-2-phenylacetophenone, DMPA) was supplied from Acros. Lithium metal and aluminum foil were got from UBIQ company. LiFePO$_4$ and Super P were provided by Aleees and Timcal, respectively. The solvents involved in the study were of ACS reagent grade and used without further purification.

10.2.2 SYNTHESIS OF PREPOLYMER, IMIDAZOLIUM-BASED CROSS-LINKER (C-VIM)

C-VIm was synthesized by refluxing the solution of VIm (1 mmol) and 1,12-dibromododecane (0.4 mmol) in ethyl acetate for 18 hours. The received precipitate was collected and separated from the mixture at the end of the reaction. After stirring with ethyl acetate/diethyl ether, the bromide intermediate was dissolved in H$_2$O followed by adding excessive LiTFSI into the solution. The mixture was continuously stirred for 2 hours in order to exchange the counterions. Subsequently, the white solid precipitated from H$_2$O was extracted in ethyl acetate/H$_2$O until the residual lithium bromide could not be detected by 0.1 M AgNO$_3$ aqueous solution. The upper layer was further purified by adding active carbon and magnesium sulfate, which would be removed by passing a quantitative filter paper. Finally, the solution was concentrated and dried under vacuum at 80°C to yield the desired product, C-VIm. As shown in Figure 10.1, the structure was characterized by nuclear magnetic resonance (NMR) spectra, and the result shows that a highly pure material was prepared successfully.

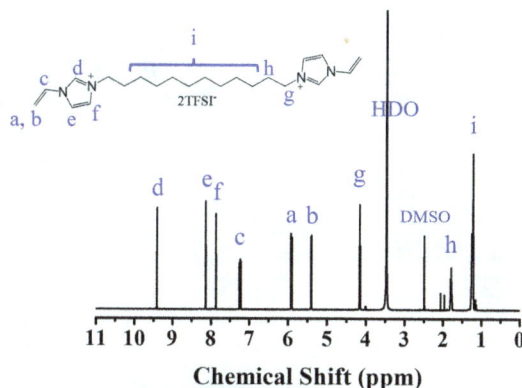

FIGURE 10.1 The ^{1}H NMR spectrum of C-VIm.

10.2.3 PREPARATION OF SOLID-STATE POLYMER ELECTROLYTES

Several polymer electrolytes were prepared via the in-situ polymerization. Thiosiloxane, C-VIm, PEGDA, PEGDME, LiTFSI, and DMPA were weighted, blended, and stirred moderately for 2 hours. The uniform mixture was cast onto lithium anode directly, and it was polymerized under ultraviolet (UV) irradiation at a wavelength of 365 nm for 5 minutes. After curing, the transparent membrane sticking to the lithium anode was thus obtained, which would be subsequently dried for 30 minutes under vacuum. It is worth noting that the step of in-situ polymerization as well as the solventless process is a novel method for the electrolyte preparation as it excludes the addition of organic solvent, which helps increase the purity of the products. For each electrolyte prepared in this study, the weight amount of polymer (thiosiloxane, C-Vim, PEGDA), PEGDME additive, and LiTFSI was kept at 1:2:1, while the percentage of thiosiloxane/C-Vim/PEGDA was adjusted to the desired ratio. The electrolytes were abbreviated as A-X. Here, X denotes the weight percentage of the C-VIm in C-VIm/PEGDA composition. The precise compositions of all prepared electrolytes were recorded in Table 10.2. The sample containing the equal content of C-VIm and PEGDA, for instance, is named as A-50%, and the

TABLE 10.2
The Composition of the Prepared Electrolyte Membranes

		Weight (mg)					
		Prepolymer					
Sample		Thiosiloxane	C-Vim	PEGDA	LITFSI	PEGDME	DMPA (initiator)
A-series	A-0%	15.0	-	85.0	100.0	200.0	2.0
	A-25%	15.0	21.2	63.8	100.0	200.0	2.0
	A-50%	15.0	42.5	42.5	100.0	200.0	2.0
	A-75%	15.0	63.8	21.2	100.0	200.0	2.0
	A-100%	15.0	85.0	-	100.0	200.0	2.0

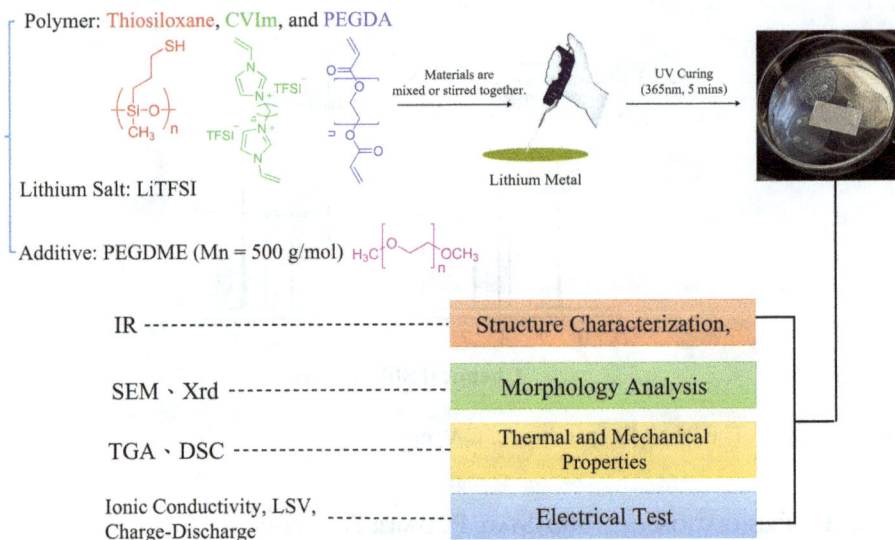

FIGURE 10.2 The experimental process of the in-situ polymerization and the optical images of electrolyte membranes.

experimental process is shown in Figure 10.2. Besides, all materials were prepared and stored in a glove box filled with argon to ensure their completeness and prevent the effect of humidity.

10.2.4 SAMPLE CHARACTERIZATION

^{1}H NMR spectra were recorded using a Bruker Avance 600NMR Spectrometer with dimethyl sulfoxide-d[6] as the solvent. Fourier transformation infrared (FT-IR) spectra were performed on a Thermo Nicolet Nexus 6700 FT-IR in the range of 400–4,000 cm[−1]. Field-emission scanning electron microscopy (Hitachi SU8010) was used to characterize the morphology of the polymer membranes with an operation voltage: 1–10 kV. X-ray diffraction (XRD) measurements were conducted using a Rigaku Ultima IV-9407F701 X-ray spectrometer at a scan rate of 10°C min[−1] from $2\theta = 5°$ to 40° with Cu Kα radiation (50 kV/250 mA). The mechanical properties were measured on a dynamic mechanical analyzer (DMA, RSA G2). The thermal stability of the samples was estimated by a thermogravimetric analyzer (TGA, TGA7 Instrument (Perkin Elmer)) with a heating rate of 15°C min[−1]. The samples were heated to 100°C and kept for 15 minutes before starting, followed by testing from 50°C to 600°C. The decomposition temperature ($T_{d5\%}$) was defined at the temperature as the samples lost the weight to 5%. Glass transition temperature (T_g) was identified by differential scanning calorimetry (DSC) using a differential scanning calorimeter (TA Instruments Q100) with a heating rate of 10°C min[−1] from −85°C to −45°C under nitrogen atmosphere.

10.2.5 Ionic Conductivity and Electrochemical Stability

The electrolytes were sandwiched between two stainless steels (SSs) to form the SS/electrolyte/SS cells. The bulk electrolyte resistance (R) was measured by AC impedance spectroscopy (CH Instruments 6116E) with the frequency set from 10^{-1} to 10^6 Hz. The measurement was conducted from 25°C to 80°C, and the value of ionic conductivity was calculated according to the following equation:

$$\sigma\left(\frac{S}{cm}\right) = \frac{d}{RA}$$

where R is the bulk resistance, d is the thickness of the sample denoted in cm, and A is the contact area of the SS and the sample (1.840 cm²). The SS/electrolyte/lithium cells were assembled to analyze the electrochemical stability of the samples via linear sweep voltammograms from 0 to 6 V at a scan rate of 5 mV s⁻¹.

10.2.6 Battery Cell Assembly

LiFePO₄ was used to prepare the cathode as the active material. In a typical process, the cathodes were prepared by pasting a mixture of LiFePO₄ powder, conductive carbon (Super P), and polyvinylidene difluoride (PVDF) (80:10:10 wt%) in N-methyl-2-pyrrolidone solvent to form a viscous slurry, which was then cast onto aluminum foil and dried at 80°C under vacuum overnight. Finally, the cathodes were roll-pressed to tighten the structures and stored in the glove box. Further, coin cell assembly (CR2032) was carried out to evaluate the charge/discharge performance using a lithium anode, and the LiFePO₄ cathode with the prepared electrolytes.

10.2.7 Charge/Discharge Long-Term Cycling Performance

The charge/discharge capacity and cycling performance were estimated on a battery testing system (BAT-700). The charge/discharge measurement of the assembled LiFePO₄/electrolyte/lithium cells was conducted under the voltage window range from 2.5 to 4.0 V at 25°C, while the cycle life of the cells was performed at a constant current rate of 0.2 C at 25°C.

10.3 RESULTS AND DISCUSSION

10.3.1 Preparation and Characterization of the Electrolytes

The electrolytes were prepared through the UV photopolymerization in this study. Thiosiloxane, lithium salt, PEGDME additive, and initiator are kept at a constant weight percentage with adjusted C-VIm/PEGDA ratios, which are 0:100, 25:75, 50:50, 75:25, and 100:0, respectively, as shown in Table 10.1. The obtained membranes are named as A-X, which X represents the weight percent of C-VIm. All the products prepared as above are found to be self-standing and translucent (Figure 10.2). To check the accuracy of the structures, FT-IR spectroscopy was

(a)

(b)

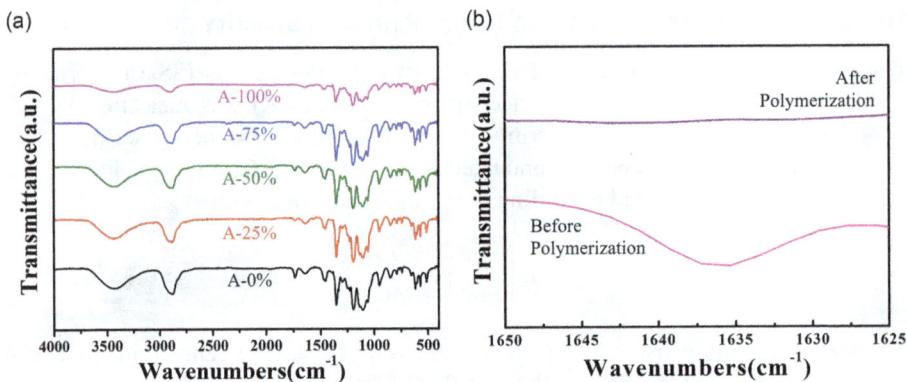

FIGURE 10.3 FT-IR spectra of A-series samples from (a) 400–4,000 cm^{-1} and partial enlarged patterns from (b) 1,625–1,650 cm^{-1}.

carried out to observe the change of functional groups before and after the reaction. As shown in Figure 10.3a, four peaks situated in 1,355, 1,195, 1,136, and 1,059 cm^{-1} are attributed to the vibration absorption of TFSI anions, ($v_a(SO_2)$, $v_a(CF_3)$, $v_s(SO_2)$, and $v_a(S-N-S)$), respectively, which indicates the existence of LiTFSI and C-VIm. The conversion of polymerization for A-50% was also characterized and is shown in Figure 10.3b. Compared with the curve measured before exposing to UV light, the characteristic peak of vinyl group (around 1638 cm^{-1}) diminishes after reacting under 365 nm UV light for 5 minutes, suggesting that most of the C=C double bonds of the prepolymers are disappear, owing to the successful reaction between thiol groups and vinyl groups. To clearly realize the nature of the materials as crystalline or amorphous, pure LiTFSI as well as the prepared samples were performed on an XRD measurement (Figure 10.4). LiTFSI possesses multiple sharp characteristic peaks ranged from $2\theta = 5°–40°$, indicating that it has a high degree of crystallinity. It is interesting that the added polymer and additive have an inhibitory action on the crystalline behavior of lithium salt since the characteristic peaks

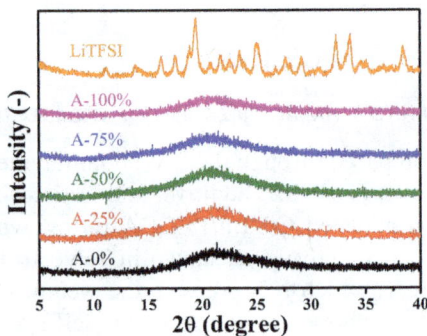

FIGURE 10.4 XRD patterns of LiTFSI and A-series samples.

of the prepared samples had completely disappeared. And it is worth noting that all the samples exhibit a broad peak at around $2\theta = 21°$, suggesting that they are mainly in an amorphous state.

10.3.2 SURFACE MORPHOLOGY

The surface morphology of the prepared membranes displays the wrinkle structures without any appearance of aggregates or micropores, indicating a good compatibility and miscibility of the constituents, as shown in Figure 10.5. It is worth noting that micropores often appear on the membrane surface during the process of solvent volatilization, which is not conducive to the maintenance of interfacial stability between the electrolyte and the electrodes. The solventless procedure adopted in this study effectively avoids the defect. There are several reasons of forming the wrinkle structures. First, it can be attributed to the truth that a thin skin of the cross-linked structures forms on an uncrosslinked soft foundation from the layer at the early stage of UV curing, and the result of micro-phase separation as a consequence of the surface tension among the individual components. Moreover, polymerization shrinkage gives rise to the formation of creases at the surface, which also promotes the creation of wrinkle structures. A complete mechanism of the wrinkle morphology occurring from a UV-cross-linkable prepolymer has been well described in the studies reported by Park et al. [42]. In our research, the wrinkle structures can be modulated by the addition of PEGDA in the formulated liquid prepolymers. The A-100% membrane, which contains 100 wt% of the hydrophobic C-VIm segments in C-VIm-PEGDA composition, shows the faint creases on the surface (Figure 10.5e). Due to the existence of hydrophilic PEGDA segments, the micro-phase separation of the other membranes (A-0%, A-25%, A-50%, and A-75%) significantly increased and therefore their surface creases become evident, as shown in Figure 10.5a–d.

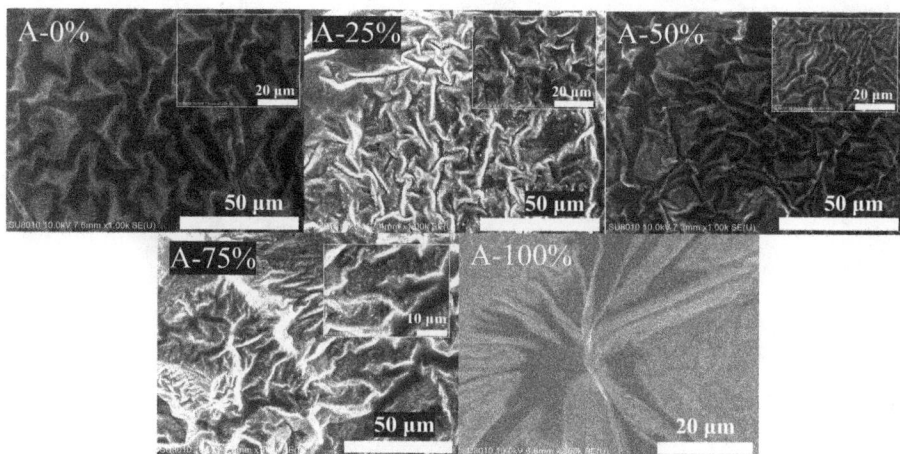

FIGURE 10.5 Surface morphology of the electrolytes: (a) A-0%, (b) A-25%, (c) A-50%, (d) A-75%, and (e) A-100%.

10.3.3 THERMAL PROPERTY AND MECHANICAL STABILITY

Thermal stabilities, which are crucial factors to polymer electrolytes in battery performance and safety issues, were evaluated by the TGA. As shown in Figure 10.6a, all the prepared samples maintained stability at a temperature over 200°C. With the further increase of temperature, A-0% began to decompose at 250°C and experienced a one-step process of the decomposition for its perfectly mixing and adequate compatibility. Finally, it was almost completely decomposed at 500°C with a total weight loss of more than 95 wt%. In contrast, for the A-100%, although the weight loss occurred at a temperature similar to that of A-0% due to the truth that the initial stage of structural decomposition is dominated by the PEG segments with weaker thermal stability for both samples, the weight loss rate was much slower than the other samples since the IL (C-VIm) segments have the extraordinary thermal stability. As a result, around 10 wt% of A-100% was remained at 500°C. Besides, the decomposition temperatures ($T_{d5\%}$) are defined as the temperature, in which the samples lose 5 wt% of the weight. The $T_{d5\%}$ of the samples are found to be increased with increasing the ratio of IL segments, from 260°C (A-0%) to 288°C (A-100%). Consequently, the addition of the IL segments can endow the materials with great thermal stability and provide stronger barrier to prevent short circuiting even in the extreme conditions. The DSC measurements were performed to observe the phase transition behavior of the used materials. As shown in Figure 10.6b, compared to the

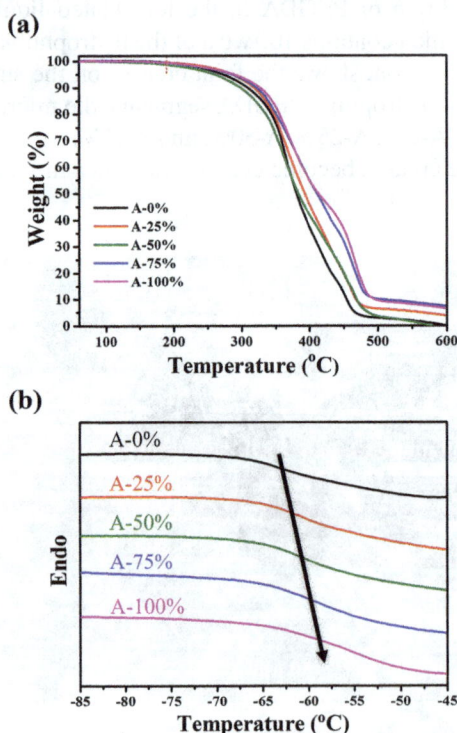

FIGURE 10.6   (a) TGA curves and (b) DSC thermograms of the A-series samples.

PEGDME additive [43], the crystalline peak in the A-series completely disappeared, which pointed out that the crystallinity of additive would be significantly reduced after incorporating polymer structure and lithium salt. The samples would thus show the amphoteric nature, and this is in agreement with the results of XRD measurement. It is worth noting that the low crystallinity of electrolytes may promote the segmental motion of polymer chains, resulting in a higher transportation ability of lithium ions. And the T_g values of A-0%, A-25%, A-50%, A-75%, and A-100% are −63.6°C, −61.8°C, −60.8°C, −57.7°C, and −55.3°C, respectively. The slightly increased trend of T_g can be regarded as the interaction among IL segments, lithium salt, and PEG additive. However, PEGDME, being a plasticizer in this study, has a significant effect on the T_g. Especially that PEGDME content is as high as 50 wt% in the system, which suggests that the T_g of A-series would be mainly dominated by such high amount of plasticizer. Herein, the differences of the T_g are thus unobvious. Also, the materials can be prepared in cylindrical shapes. All the samples have exceptional deformation ability, and the compression stress–strain curve demonstrates that A-50% does not crack under 0.44 MPa compressive stress at 37% compressive strain with the calculated compression modulus of 0.19 MPa (Figure 10.7). The character of compression resistance affords sufficient contact between the electrolyte and the electrodes, which helps to improve the performance of batteries using A-series electrolytes.

10.3.4 IONIC CONDUCTIVITY AND ELECTROCHEMICAL WINDOW

Ionic conductivity plays a crucial role in determining its applicability in lithium batteries, and for setting up appropriate operating conditions for ensuring ideal performances. Figure 10.8a shows the ionic conductivity Arrhenius plots of the A-series electrolytes from 25°C to 80°C. Relevantly, the curves present a conductivity value of higher than 10^{-4} S cm^{-1} at 25°C, that is, a temperature lower than that of the one ascribed to other reported solid-state electrolytes for LIB. The PEG-based polymer electrolytes, for instance, can provide qualified and suitable conductivity only at the temperatures higher than 40°C [44]. As indicated above, the coating of the electrolyte layer increases the continuity in surface, reduced the contact loss with electrode,

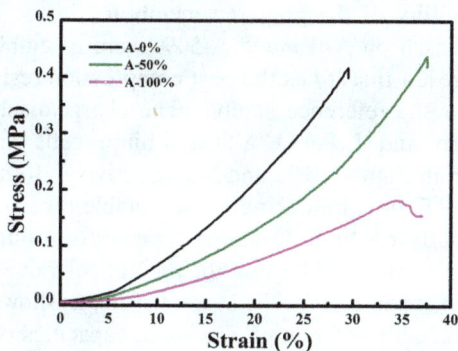

FIGURE 10.7 Compressive stress–strain curves of the A-0%, A-50%, and A-100% samples (cylindrical sample, 10.0 mm diameter × 5.0 mm thickness).

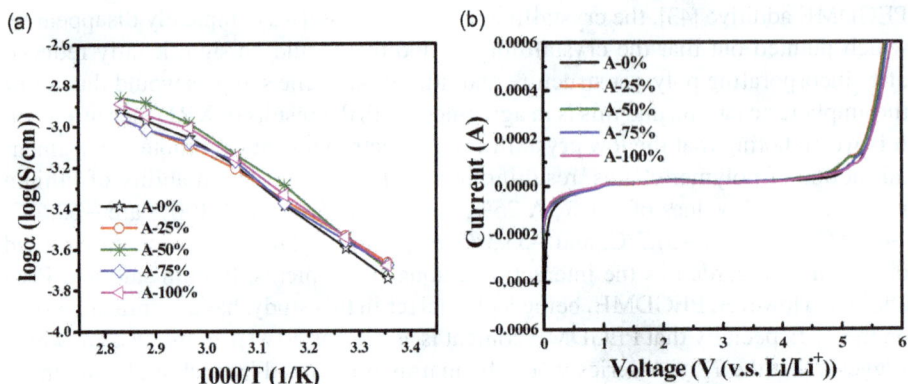

FIGURE 10.8 (a) Temperature dependence of ionic conductivity and (b) linear sweep voltammetry of the A-series samples.

and effectively decreased the interfacial resistance. The ionic conductivity, therefore, obtains a remarkable progress, which almost meets the requirement for a further application. Besides, all electrolytes exhibit a similar conductivity value within the measured temperature range, indicating that the conductivity is dominated by the ratio among polymer, PEGDME additive, and LiTFSI, which is kept at 1:2:1 in this study, while the ratio between C-VIm and PEGDA would not substantially influence the ionic conductivity. The electrochemical stability window is a crucial parameter of high-energy LIBs. As shown in Figure 10.8b, the A-series electrolytes demonstrate a stable current density up to 4.7 V, where only a slight current was detected by the CH Instrument, indicating that the materials have good tolerance to the polarization and great stability against the lithium metal. This outstanding electrochemical stability suggests that the electrolytes may be utilized for high-voltage cathode materials, such as $LiNi_{0.5}Mn_{1.5}O_4$.

10.3.5 CHARGE/DISCHARGE CAPACITY AND CYCLE PERFORMANCE

To verify the applicability of the prepared membrane in the LIBs, $LiFePO_4$/electrolyte/lithium cells based on A-0% and A-50% were assembled. The selection of A-50% is due to the result that it has the best compression resistance (Figure 10.7), and A-0% is used for the reference sample. The charge/discharge profiles of the $LiFePO_4$/A-0%/lithium and $LiFePO_4$/A-50%/lithium cells at 25°C with different C rates are shown in Figure 10.9a and b, respectively. Both cells exhibit a clear potential plateau at 0.1 C rate, indicating an acceptable reversible cycling property. Moreover, they can deliver a high discharge capacity of around 150.0 mAh g^{-1} at 0.1 C rate, which reaches almost 88.2% of the theoretical value of $LiFePO_4$, and the value reveals no obvious decrease in the second cycle, as shown in Figure 10.10. In addition, the cell with A-0% offers high discharge capacities of 143.0, 133.0, 110.0, and 20.0 mAh g^{-1} at rates of 0.2, 0.3, 0.5, and 1 C, respectively, which are superior to some of the other reported polymer electrolytes. The progress is attributed to the great ionic conductivity, and the improved interfacial compatibility resulted from

(a)

(b)

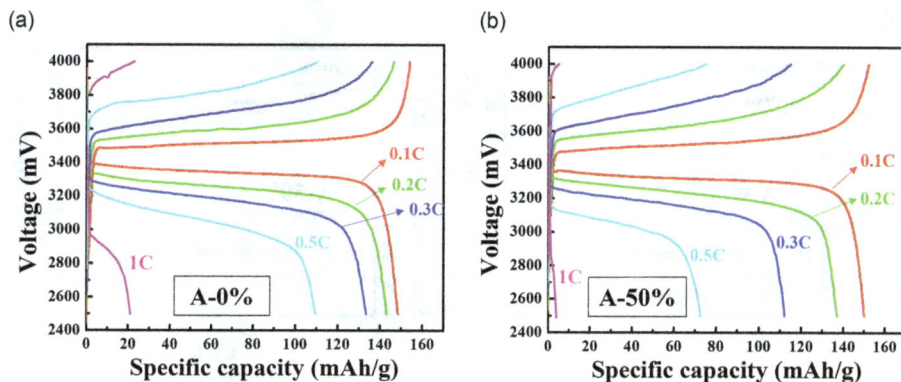

FIGURE 10.9 Charge/discharge curves of the Li/LiFePO$_4$ cells based on (a) A-0% and (b) A-50% at different C rates at 25°C.

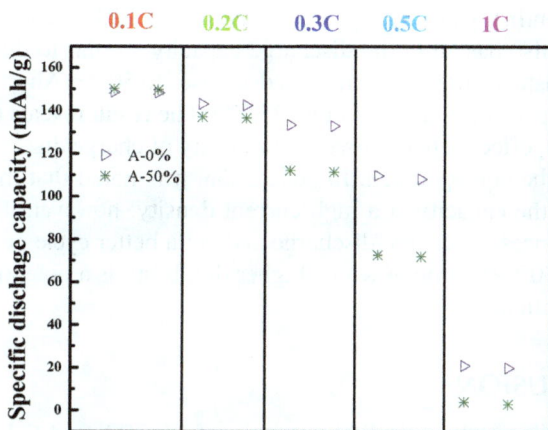

FIGURE 10.10 Discharge capacities of the Li/LiFePO$_4$ cells with A-0% and A-50% at different C rates at 25°C.

the in-situ polymerization process, which facilitates the ion transfer at the interface between electrolytes and electrodes. The capacities of the cell with A-50%, however, decrease more quickly compared to that of the cell with A-0% as increasing the rate of current. These observed rate performances indicate that the charge/discharge capacities cannot be effectively upgraded by the addition of C-VIm cross-linker, especially in the high current density.

The combination of a good ionic conductivity, a wide electrochemical window, as well as the great C rate performance makes the prepared electrolytes an excellent candidate for LIBs. Galvanostatic cycling measurements were conducted under 25°C at a specific current of 0.2 C rate to observe the cycle performance of A-0% and A-50%, as shown in Figure 10.11a and b, respectively. The cell with A-50% exhibits 140.0 mAh g^{-1} of initial discharge capacity and maintains around 125.0 mAh g^{-1} after

(a)   (b)

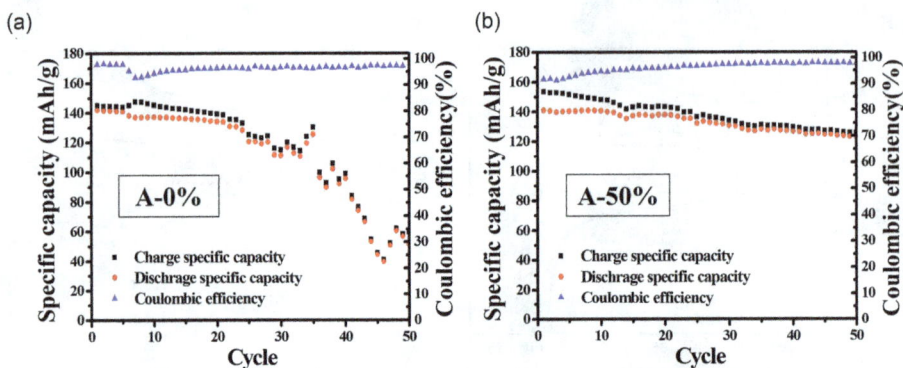

FIGURE 10.11 Cycling performance of the Li/LiFePO$_4$ cells with (a) A-0% and (b) A-50% at 0.2 C rate.

50 cycles, representing 89.3% of capacity retention and 98% of coulombic efficiency. Although A-0% also has an initial discharge capacity similar to that of A-50%, the value becomes unstable after 20 cycles and decreases to 50.0 mAh g^{-1} after 50 cycles, showing a low-capacity retention (around 35.7%). The result reveals that the addition of C-VIm, indeed, effectively improves the stability of charge/discharge and extends the cycle life of the equipped cell. In conclusion, it is noted that the use of C-VIm cannot maintain the capacity at a high current density; however, it is conducive to obtain a stable process of charge/discharge and get a better cycle performance. This indicates that A-50% electrolyte with a higher durability is a potential candidate for novel LIB applications.

10.4 CONCLUSION

In summary, the ionogels with different ratio of PEGDA and C-VIm cross-linker were designed and prepared via a facile in-situ polymerization, and they were evaluated for applications in LIBs. All of the samples have high thermal stabilities and good ionic conductivities. They display the wrinkle structures without any formation of aggregates or micropores, indicating a good compatibility and miscibility of the constituents. And the electrochemical stability of the materials was evidenced by the high onset oxidation voltages (4.7 V) and wide electrochemical windows. Furthermore, the discharge capacity of the LiFePO$_4$/A-0%/lithium cell can arrive to 110.0 mAh g^{-1} even at 0.5 C rate at 25°C, while the LiFePO$_4$/A-50%/lithium cell can maintain a high-capacity retention (89.3%) to 50 cycles. We believe that the above advantages would render the materials prepared in this study to be promising candidates as novel polymer electrolytes for LIBs.

REFERENCES

1. Armand M and Tarascon J M 2008 Building better batteries *Nature* **451** 652–7.
2. Kang B and Ceder G 2009 Battery materials for ultrafast charging and discharging *Nature* **458** 190–3.

3. Sun Y K, Myung S T, Park B C, Prakash J, Belharouak I and Amine K 2009 High-energy cathode material for long-life and safe lithium batteries *Nat Mater* **8** 320–4.

4. Scrosati B and Garche J 2010 Lithium batteries: Status, prospects and future *J Power Sources* **195** 2419–30.

5. Goodenough J B and Park K S 2013 The Li-ion rechargeable battery: A perspective *J Am Chem Soc* **135** 1167–76.

6. Cui Q, Zhong Y, Pan L, Zhang H, Yang Y, Liu D, Teng F, Bando Y, Yao J and Wang X 2018 Recent advances in designing high-capacity anode nanomaterials for li-ion batteries and their atomic-scale storage mechanism studies *Adv Sci (Weinh)* **5** 1700902.

7. Perea A, Dontigny M and Zaghib K 2017 Safety of solid-state Li metal battery: Solid polymer versus liquid electrolyte *J Power Sources* **359** 182–5.

8. Niu C Q, Liu J, Chen G P, Liu C, Qian T, Zhang J J, Cao B K, Shang W Y, Chen Y B, Han J L, Du J and Chen Y 2019 Anion-regulated solid polymer electrolyte enhances the stable deposition of lithium ion for lithium metal batteries *J Power Sources* **417** 70–5.

9. Balakrishnan P G, Ramesh R and Kumar T P 2006 Safety mechanisms in lithium-ion batteries *J Power Sources* **155** 401–14.

10. Goodenough J B and Kim Y 2010 Challenges for rechargeable li batteries *Chem Mater* **22** 587–603.

11. Lux S F, Lucas I T, Pollak E, Passerini S, Winter M and Kostecki R 2012 The mechanism of HF formation in LiPF6 based organic carbonate electrolytes *Electrochem Commun* **14** 47–50.

12. Smiglak M, Reichert W M, Holbrey J D, Wilkes J S, Sun L Y, Thrasher J S, Kirichenko K, Singh S, Katritzky A R and Rogers R D 2006 Combustible ionic liquids by design: is laboratory safety another ionic liquid myth? *Chem Commun* 2554–6.

13. Earle M J, Esperanca J M S S, Gilea M A, Lopes J N C, Rebelo L P N, Magee J W, Seddon K R and Widegren J A 2006 The distillation and volatility of ionic liquids *Nature* **439** 831–4.

14. Galinski M, Lewandowski A and Stepniak I 2006 Ionic liquids as electrolytes *Electrochim Acta* **51** 5567–80.

15. Fernicola A, Croce F, Scrosati B, Watanabe T and Ohno H 2007 LiTFSI-BEPyTFSI as an improved ionic liquid electrolyte for rechargeable lithium batteries *J Power Sources* **174** 342–8.

16. Nakagawa H, Fujino Y, Kozono S, Katayama Y, Nukuda T, Sakaebe H, Matsumoto H and Tatsumi K 2007 Application of nonflammable electrolyte with room temperature ionic liquids (RTILs) for lithium-ion cells *J Power Sources* **174** 1021–6.

17. Sakaebe H, Matsumoto H and Tatsumi K 2007 Application of room temperature ionic liquids to Li batteries *Electrochim Acta* **53** 1048–54.

18. Jin Y D, Fang S H, Chai M, Yang L, Tachibana K and Hirano S 2013 Properties and application of ether-functionalized trialkylimidazolium ionic liquid electrolytes for lithium battery *J Power Sources* **226** 210–8.

19. Sun X G, Liao C, Shao N, Bell J R, Guo B K, Luo H M, Jiang D E and Dai S 2013 Bicyclic imidazolium ionic liquids as potential electrolytes for rechargeable lithium ion batteries *J Power Sources* **237** 5–12.

20. Long L, Wang S, Xiao M and Meng Y 2016 Polymer electrolytes for lithium polymer batteries *J Mater Chem A* **4** 10038–69.

21. Liang S S, Yan W Q, Wu X, Zhang Y, Zhu Y S, Wang H W and Wu Y P 2018 Gel polymer electrolytes for lithium ion batteries: Fabrication, characterization and performance *Solid Stat Ion* **318** 2–18.

22. Zhou D, Shanmukaraj D, Tkacheva A, Armand M and Wang G X 2019 Polymer electrolytes for lithium-based batteries: Advances and prospects *Chem-Us* **5** 2326–52.

23. Zhu M, Wu J X, Wang Y, Song M M, Long L, Siyal S H, Yang X P and Sui G 2019 Recent advances in gel polymer electrolyte for high-performance lithium batteries *J Energy Chem* **37** 126–42.

24. Rao M M, Geng X Y, Liao Y H, Hu S J and Li W S 2012 Preparation and performance of gel polymer electrolyte based on electrospun polymer membrane and ionic liquid for lithium ion battery *J Membrane Sci* **399** 37–42.
25. Xue Z G, He D and Xie X L 2015 Poly(ethylene oxide)-based electrolytes for lithium-ion batteries *J Mater Chem A* **3** 19218–53.
26. Liu B, Huang Y, Cao H J, Zhao L, Huang Y X, Song A, Lin Y H, Li X and Wang M S 2018 A novel porous gel polymer electrolyte based on poly(acrylonitrile-polyhedral oligomeric silsesquioxane) with high performances for lithium-ion batteries *J Membrane Sci* **545** 140–9.
27. Tseng Y C, Wu Y, Tsao C H, Teng S S, Hou S S and Jan E S 2019 Polymer electrolytes based on Poly(VdF-co-HFP)/ionic liquid/carbonate membranes for high-performance lithium-ion batteries *Polymer* **173** 110–8.
28. Hu Z Y, Chen J J, Guo Y, Zhu J J, Qu X X, Niu W W and Liu X K 2020 Fire-resistant, high-performance gel polymer electrolytes derived from poly (ionic liquid)/P(VDF-HFP) composite membranes for lithium ion batteries *J Membrane Sci* **599** 117827.
29. Yuan J Y, Mecerreyes D and Antonietti M 2013 Poly(ionic liquid)s: An update *Prog Polym Sci* **38** 1009–36.
30. Li X W, Zheng Y W, Pan Q W and Li C Y 2019 Polymerized ionic liquid-containing interpenetrating network solid polymer electrolytes for all-solid-state lithium metal batteries *ACS Appl Mater Inter* **11** 34904–12.
31. Zhou Y, Wang B N, Yang Y, Li R J, Wang Y F, Zhou N, Shen J Y and Zhou Y 2019 Dicationic tetraalkylammonium-based polymeric ionic liquid with star and four-arm topologies as advanced solid-state electrolyte for lithium metal battery *React Funct Polym* **145** 104375.
32. Appetecchi G B, Kim G T, Montanina M, Carewska M, Marcilla R, Mecerreyes D and De Meatza I 2010 Ternary polymer electrolytes containing pyrrolidinium-based polymeric ionic liquids for lithium batteries *J Power Sources* **195** 3668–75.
33. Safa M, Chamaani A, Chawla N and El-Zahab B 2016 Polymeric ionic liquid gel electrolyte for room temperature lithium battery applications *Electrochim Acta* **213** 587–93.
34. Li X W, Zhang Z X, Li S J, Yang L and Hirano S 2016 Polymeric ionic liquid-plastic crystal composite electrolytes for lithium ion batteries *J Power Sources* **307** 678–83.
35. Li S J, Zhang Z X, Yang K H and Yang L 2018 Polymeric ionic liquid-poly(ethylene glycol) composite polymer electrolytes for high-temperature lithium-ion batteries *ChemElectroChem* **5** 328–34.
36. Li M T, Yang L, Fang S H, Dong S M, Hirano S and Tachibana K 2011 Polymer electrolytes containing guanidinium-based polymeric ionic liquids for rechargeable lithium batteries *J Power Sources* **196** 8662–8.
37. Li M T, Yang L, Fang S H, Dong S M, Hirano S and Tachibana K 2012 Polymerized ionic liquids with guanidinium cations as host for gel polymer electrolytes in lithium metal batteries *Polym Int* **61** 259–64.
38. Li M T, Yang B L, Wang L, Zhang Y, Zhang Z, Fang S H and Zhang Z X 2013 New polymerized ionic liquid (PIL) gel electrolyte membranes based on tetraalkylammonium cations for lithium ion batteries *J Membrane Sci* **447** 222–7.
39. Li M T, Wang L, Yang B L, Du T T and Zhang Y 2014 Facile preparation of polymer electrolytes based on the polymerized ionic liquid poly((4-vinylbenzyl)trimethylammonium bis(trifluoromethanesulfonylimide)) for lithium secondary batteries *Electrochim Acta* **123** 296–302.
40. Yin K, Zhang Z X, Yang L and Hirano S I 2014 An imidazolium-based polymerized ionic liquid via novel synthetic strategy as polymer electrolytes for lithium ion batteries *J Power Sources* **258** 150–4.
41. Yin K, Zhang Z X, Li X W, Yang L, Tachibana K and Hirano S I 2015 Polymer electrolytes based on dicationic polymeric ionic liquids: Application in lithium metal batteries *J Mater Chem A* **3** 170–8.

42. Park S K, Kwark Y J, Moon J, Joo C W, Yu B and Lee J I 2015 Finely formed, kinetically modulated wrinkle structures in UV-crosslinkable liquid prepolymers *Macromol Rapid Comm* **36** 2006–11.
43. Costa L T, Lavall R L, Borges R S, Rieumont J, Silva G G and Ribeiro M C 2007 Polymer electrolytes based on poly(ethylene glycol) dimethyl ether and the ionic liquid 1-butyl-3-methylimidazolium hexafluorophosphate: Preparation, physico-chemical characterization, and theoretical study *Electrochim Acta* **53** 1568–74.
44. Lu Q W, Fang J H, Yang J, Yan G W, Liu S S and Wang J L 2013 A novel solid composite polymer electrolyte based on poly(ethylene oxide) segmented polysulfone copolymers for rechargeable lithium batteries *J Membrane Sci* **425** 105–12.
45. Tian X L, Yi Y K, Yang P, Liu P, Qu L, Li M T, Hu Y S and Yang B L 2019 High-charge density polymerized ionic networks boosting high ionic conductivity as quasi-solid electrolytes for high-voltage batteries *ACS Appl Mater Inter* **11** 4001–10.

11 Rich Quasiparticle Properties of Li$_2$S Electrolyte in Lithium-Sulfur Battery

Thi Dieu Hien Nguyen and Ming-Fa Lin
National Cheng Kung University

CONTENTS

11.1 INTRODUCTION

Lithium-sulfur batteries (LSBs) might become post-rechargeable batteries due to many advantages, i.e., high energy density, higher theoretical energy storage capacity, more durable, and environment friendly. LSBs come into existence as electrochemical systems, which possess a high theoretical specific capacity of 1,675 mAh g^{-1} and a theoretical energy density of 2,600 Wh kg^{-1} [1]. A LSB cell can restore electrical energy through sulfur electrodes. Typical components of this cell are a lithium metal anode, an organic electrolyte, and a sulfur composite cathode. Similar to lithium-ion batteries, the specific mechanism is based on the charging and discharging process. In general, sulfur is in the charged state, and the operation of the cell will originate from the discharging process. In the discharging process, lithium ions and electrons are produced by the oxidation of lithium metal at the anode. The above lithium ions will transfer to the cathode, which internally passes through the electrolyte. Simultaneously, the electrons travel to the positive electrode through the external electrical circuit, thus leading to an electrical current in the circuit. In this case, sulfur is reduced to produce lithium sulfide by accepting the lithium ions and electrons at the positive electrode. Currently, significant developments of LSBs have been made and require further research efforts. The evidence for these improvements is increasing research novel sulfur composite cathodes, using various carbon materials, polymers, and other inorganic materials. Also, many emergent materials can be

DOI: 10.1201/9781003263807-11

served as an electrolyte that shows a high ionic conductivity up to 1.7×10^{-3} S cm^{-1} ($63Li_2S-36SiS_2-1Li_3PO_4$) [2]. However, in this study, we will focus on $Li_2S-P_2S_5$; other compounds could be investigated further.

In general, there are three types of sulfide solid-state electrolytes, mainly based on their structural characteristics: crystalline, amorphous (glass), and partially crystalline (glass-ceramic), which have fast ionic conductors. $Li_2S-P_2S_5$ belongs to the glass and glass-ceramic phase, which possesses non-metallic materials, leading to high electrochemical stability. The disappearance of the oxidized metals in the electrochemical reactions plays a critical role in determining the electrolyte material properties. Based on the literature research, depending on the different contributions of Li_2S and P_2S_5 and the preparation conditions, the ionic conductivity could be achieved accordingly. The metastable phases enhance the ionic conductivity of $Li_2S-P_2S_5$ glass-ceramic by forming $Li_7P_3S_{11}$ and Li_3PS_4. Some experimental methods are used to investigate the critical mechanism of high ionic conductivities, such as nuclear magnetic resonance (NMR) spectroscopy measurements [3]. The core materials in sulfide solid-state electrolyte will determine the performance of ion-based batteries.

An active chemical environment provides the unusual atom/molecule bondings [4,5] and thus creates all the physical/chemical/materials properties [4–6] and the highly potential applications [7,8]. All the elements in a periodic table could form the crystal structures using the single-atom and/or molecular forms, such as the Li-atom [9], S-atom [9], and H_2-molecule crystals [9]. It is well known that the alkali atoms exhibit the body-centered cubic (BCC) symmetries [10]. The half-occupied outmost ns valence orbital survives around each atom, and it becomes free conduction electrons through the unique form of the Fermi sea after the formation of a metallic crystal. The metallic bonds of ns-ns hybridizations [11,12], which greatly reduce the ground-state energy, are responsible for the creation of identical Fermions and the other essential properties. As for sulfur atoms, two dangling bonds, which are associated with the half-occupied orbitals of $[3p_y, 3p_z]$, could form the various molecule-gas states [9]. Furthermore, the most stable 3D crystal configuration belongs to an orthorhombic lattice with the different lattice constants under the perpendicular vectors [4,6,13]. The $sp^3 - sp^3$ bondings of $[3s, 3p_x, 3p_y, 3p_z]$ orbitals in S-S bonds should be responsible for the critical roles in the natural [12,14] and artificial world [12,14], such as biological reactions [14] and LSBs [1,7,8,15,16]. The four orbitals need to be taken into account simultaneously for crystal structures.

The chemical modifications are very powerful methods in creating the diversified compounds, in which the experimental strategy covers the direct syntheses of various elements/subsystems [17,18], the reduction or enhancement of different atom concentrations [19], the chemical adsorptions [20], and the host-atom substitutions [20]. Such enhanced manners are frequently utilized to discover the emergent materials. For example, a lot of lithium-, phosphorus-, and sulfur-related condensed-matter systems have been successfully synthesized in experimental laboratories. The up-to-date X-ray patterns show that there exist the 3D crystal structures of the binary Li_2S [5] and P_2S_5 [18], the ternary $Li_2P_2S_6$ [13,21], $Li_4P_2S_6$ [13,21], $Li_7P_3S_{11}$ [13,21], and $Li_3P_1S_4$ [13,21], LiPSCl [8], and so on [details in Table 1.2]. It should be noticed that many 3D ternary compounds could be prepared from the different ratios of two binary subsystems. Generally speaking, three kinds of synthesis techniques are developed to generate the functional materials through the solid-state route [22,23], mechano-chemical synthesis [22,23], and wet chemistry [22,23]. The various Li$_x$P$_y$S$_z$ -related compounds are combined with the different cathode and anode materials to achieve optimal battery performances [13]. The form might belong to the existence of the ternary, quaternary, pentanary, and even hexagonary materials (Table 1.2). Very interestingly, the multicomponent systems and the significant differences of the relative concentrations will exhibit a lot of crystal symmetries and result in rich and unique quasiparticle properties. There are many experimental measurements about the enhanced efficiencies of LSBs, such as the close relations among specific capacity, voltage, and cycle number [1,15,16]. However, a series of high-resolution examinations on the essential properties in the core materials are hardly done up to now. Concerning 3D binary Li$_2$S, the full observations and analyses, which mainly arise from the angle-resolved photoemission spectra [24,25], van Hove singularities [4], and optical absorption structures [26], are almost absent. On the theoretical side, the first-principles methods [4,5] are principally focused on the optimal geometry and band structure. The critical mechanisms, the multi-orbital hybridizations of the quasiparticle framework [4,5,17,27,28], are not explored thoroughly in the previous simulation papers. The concise physical and chemical pictures will be achieved from the relevant quantities under the current investigations. In addition, the phenomenological models might be useless in these respects, such as the tight-binding model [29], Kubo formula [30], and random-phase approximation [31], respectively, corresponding to electronic properties, Fermion transports/optical transitions, and Coulomb excitations.

In this work, the Vienna *ab initio* simulation package (VASP) simulations are available for determining the critical quasiparticle mechanisms of lithium-sulfur bonds [the active orbital hybridization in each bond; [12]]. Their accurate results and analyses are mainly focused on the 3D position-dependent chemical bondings [the highly nonuniform chemical environments; [5]], lithium- and sulfur-dominated band structure and wave functions within the specific energy ranges [4,5], the spatial charge density distribution [4,32], and the atom- and orbital-decomposed density of states [4]. The emerged special structures, which arise from the orbital-dependent van Hove singularities, will be utilized to identify the active orbital hybridizations. Furthermore, they are assisted by geometric and electronic properties. The theoretical predictions are consistent with part of the high-resolution measurements, such as the crystal structures by X-ray patterns [6] and the bandgap by angle-resolved photoemission spectroscopy (ARPES) [33,34] and optical measurements [6]. Also, whether the tight-binding model [29] is worthy of a full development will be discussed in detail.

11.2 3D CRYSTAL STRUCTURE OF LITHIUM-SULFUR COMPOUNDS

Our calculation uses the first-principles density functional theory (DFT) [35] to examine the optimal geometric structures, charge distribution, and electronic Li$_2$S compound. In this work, the VASP is utilized to identify an approximate solution, which directly uses DFT by solving the Kohn–Sham equations. The Perdew–Burke–Ernzerhof formula [36] is chosen in this simulated calculation, which depends on the local electron density and includes many-particle Coulomb effects. The projector-augmented wave pseudopotentials describe periodic electron-crystal scatterings. The electron Bloch wave functions are solved from the linear superposition of plane waves, with a maximum kinetic energy of 520 eV. The current study on the Li_2S compound presents that the first Brillouin zone is sampled by $9 \times 9 \times 9$ and $50 \times 50 \times 50$ k-point meshes within the Monkhorst–Pack scheme, respectively, for the optimal geometry and band structure. Moreover, the convergence condition of the ground-state energy is set to be $\sim 10^{-6}$ eV between two consecutive evaluation steps, where the maximum Hellmann–Feynman force for each ion is below 0.01 eV Å^{-1} during the atom relaxations.

After an optimal relaxation of the first-principles simulations [35], a 3D lithium-sulfur compound of Li_8S_4 presents an unusual crystal symmetry, as clearly illustrated in Table 11.1 and Figure 11.1a. This cubic system has three perpendicular lattice vectors with identical lengths along x-, y-, and z-directions, and the first Brillouin zone presents the high-symmetry along the Γ XM Γ RX/MR direction. The current calculations show $a = b = c$ of 5.710 Å, being fully consistent with the high-precision X-ray measurements of 5.708 Å [37] and the previously theoretical predictions [4,17]. Furthermore, there exist eight lithium and four sulfur atoms in a primitive unit cell, in which all the Li-S chemical bonds have identical lengths. Of course, such chemical bondings are highly position dependent; therefore, they are responsible for the anisotropic quasiparticle phenomena. It is well known that each lithium/sulfur atom provides one active 2s orbital/four significant [$3s$, $3p_x$, $3p_y$, $3p_z$] orbitals in Li-S chemical bonds. These rich multi-orbital hybridizations are able to induce unusual electronic properties, as shown in Figures 11.2 and 11.4 [discussed later].

TABLE 11.1

The Geometries Parameters and Lattice Symmetries of 3D Binary Lithium-Sulfur Compound [Li_2S]

Lattice Constant

This Work	Experiment	Other Work
• $a = b = c = 5.71060$ Å	• $a = b = c = 5.708$ Å	• $a = b = c = 5.645$ Å
• $\alpha = \beta = \gamma = 90°$		• $a = b = c = 5.715$ Å
		• $a = b = c = 5.572$ Å

Structure symmetry

Hermann Mauguin	Fd3m
Hall	F 4d 2 3 $\bar{1}$d
Point group	m $\bar{3}$ m
Crystal system	Cubic

16 Li-S chemical bonds (2.47 Å)

Unit cell: 12 atoms (8 Li- 4S)

(a)

(b)

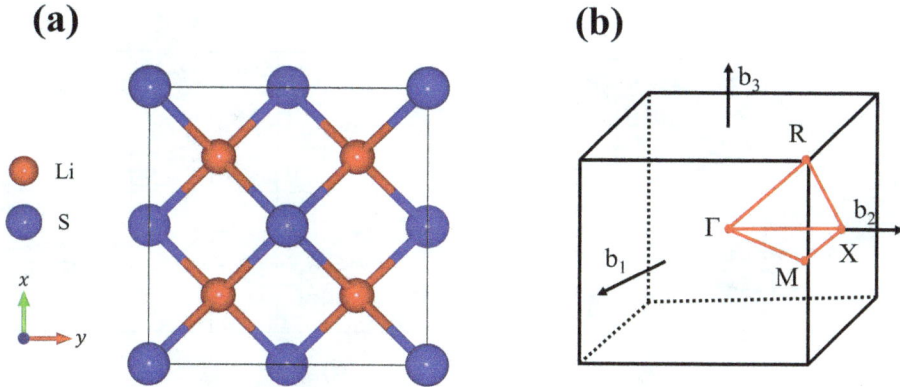

FIGURE 11.1   (a) The crystal structure and (b) the first Brillouin zone of 3D binary lithium-sulfur compound with eight Lis and four Ss in a primitive unit cell.

The 3D binary lithium sulfide compounds are frequently utilized as the solid-state electrolyte materials of LSBs [13]. During the charging and discharging processes, the lithium-related concentration and distribution will present an obvious variation. The crystal symmetries must experience the dramatic transformation among a lot of continuous intermediate configurations [4,5,17,27]. The initiated and saturated chemical reactions, which correspond to the maximum and minimum lithium-concentration cases, could be explored thoroughly for the diversified essential properties, e.g., the calculated results in this chapter. The diverse quasiparticle phenomena, being purely induced by the dynamic chemical modifications during battery operations [28], are very interesting under the current investigations [28]. On the experimental side, the time-resolved X-ray scatterings need to be greatly developed for the clear verifications of the transient crystal structures. The direct combination of theoretical predictions and experimental measurements might be very useful in clarifying the critical mechanisms of the stationary ion transports [28]. In addition, the nano-scaled techniques of scanning tunneling spectroscopy (STS) [34,38] and tunneling electron microscopy (TEM) are useless in such respect.

11.3   ELECTRONIC ENERGY SPECTRUM AND ATOM DOMINANCES

After the delicate simulations on the wave-vector-dependent energy spectrum and wave function, the featured quasiparticle properties are clearly revealed in Figure 11.2a–d along the Γ XM Γ RX/MR path. The unusual behaviors cover a direct bandgap of 3.365 eV at the Γ point [4,5], the high asymmetry of occupied hole and unoccupied electron spectra about the Fermi level [indication of multi-orbital chemical bondings in [28]], the anisotropic and oscillatory dependences [4,5], the parabolic, linear, and partial flat energy dispersions near the high-symmetry points [rich dispersion relations; [4,5]], a lot of band-edge states [the critical points in energy-wave-vector space; [4,5]], the frequent subband crossings/non-crossings [without anti-crossings; [4,5]], and the multi-degenerate states at the high-symmetry points [from the double to six-fold degeneracies; [4]]. Such characteristics are expected to

(a)

(b)

(c)

(d)

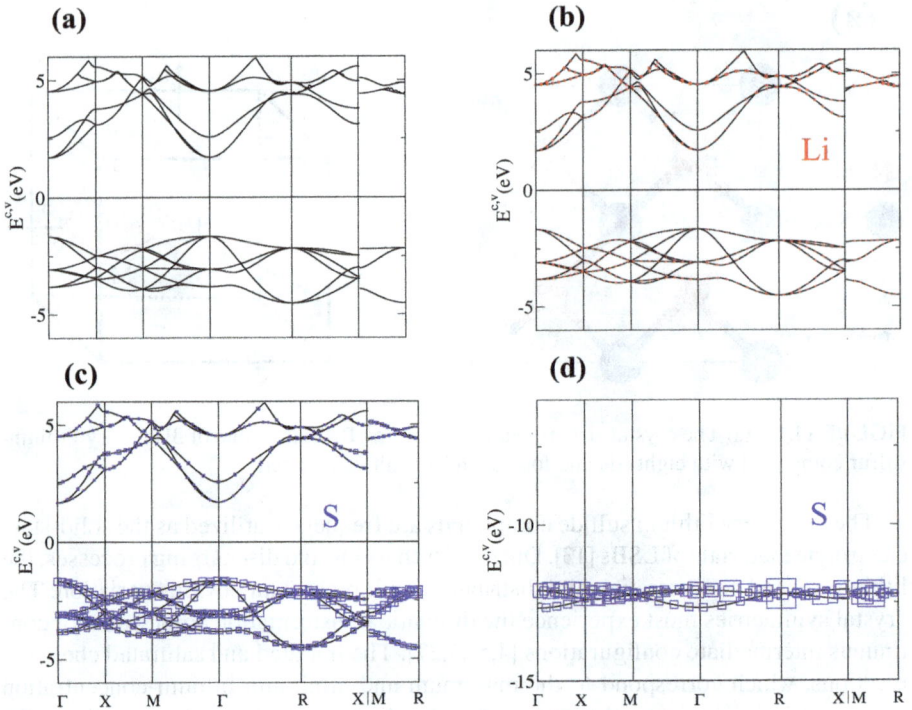

FIGURE 11.2 The Li- and S-dominated electronic energy spectra and wave functions along the high-symmetry points in the first Brillouin zone, being, respectively, indicated by red circles and blue squares: (a) the total band structure, with valence and conduction states within a specific range of $-6\,\text{eV} \le, E^{c,v} \le 6\,\text{eV}$ across the Fermi level [$E_F = 0$], the significant contributions from (b) the former and (c) the latter. (d) Similarly for the deeper valence energy subbands of $-15\,\text{eV} \le E^v \le -6$ eV, mainly owing to the S-3s orbitals.

have strong effects on the other essential properties, e.g., density of states [discussed later in Figure 11.4], transverse optical transitions [26], and longitudinal Coulomb excitations [39].

The lithium- and sulfur-dominated band structure is further achieved from the atom-decomposed contributions of the VASP wave packets. It is capable of comprehending the significant chemical bondings of the Li-S bond under the full assistance of the spatial charge density distribution (Figure 11.2a–d) and the merged van Hove singularities (Figure 11.4a–c). Concerning the former [the red solid circles], their contribution is relatively weak, but very important within a whole energy range of $-15\,\text{eV} \le E^{c,v} \le 6\,\text{eV}$, as clearly illustrated in Figure 11.2a–d. Similar results are revealed in the latter with the dominating contribution. As a result, both atoms co-dominate all the valence and conduction energy subbands in the first Brillouin zone. Very interestingly, this quasiparticle phenomenon is totally different from those of the ternary and quaternary batter materials [40–45]. For example, LiFeO [44,45] and LiFePO [40,42,46] exhibit the distinct atom dominances at the specific energy ranges. Most of the battery materials belong to such chemical bonding forms, thus leading to the giant barriers in the development of phenomenological models/quasi-particle frameworks [20,47–49].

The high-resolution measurements of ARPES [; [24,33,34]] are available in examining the wave-vector dependence of the occupied electronic states. Such examinations are very suitable for 2D layered crystal structures [details in Chapter 17], e.g., the distinct energy dispersions for few-layer graphene systems with AB/ABC stacking configurations [20], being consistent with the theoretical predictions [20]. However, the 3D condensed-matter systems present an intrinsic limit because of the specific surface confinement, leading to the mom-conservation of the perpendicular momentum transfers during the photon-electron scattering processes [50]. How to greatly reduce this measurement uncertainty has become an open issue [50], especially for the main-stream materials with Moire superlattices. For example, the various electrolyte/cathode/anode materials [details discussed earlier in Tables 1.1–1.5] have a lot of atoms and active orbitals in a unit cell and thus display many valence and conduction energy subbands. The current band structure of the 3D lithium sulfide compound exhibits the strong anisotropy, the multifold degeneracies at the high-symmetry points, and the highly non-monotonous wave-vector dependence (Figure 11.2a–d). These features can make the experimental analyses almost mission impossible. This should be the critical reason for the absence of ARPES measurements. But for successful cases, 3D ARPES measurements are done for a simple unit cell, such as the k_z-dependent band structure of the AB-stacked graphite [20]. That there exist free electrons and holes neat the K and H points [$k_z = 0$ and π; [20]], respectively, is available for the high-precision analyses. It should be noticed that Bernal graphite is a semi-metal with a low free carrier density due to a weak valence and conduction overlay through the interlayer van der Waal interactions [20,51].

11.4 ACTIVE ORBITAL HYBRIDIZATION

The spatial charge density distributions might directly provide the most important information about the active multi-orbital hybridizations of chemical bonds, as clearly revealed in the electrolyte/cathode/anode materials of ion-based batteries [52,53]. The 3D lithium-sulfur compound shows an obvious charge density variation (Figure 11.3c) after the formation of Li-S bonds, being clearly indicated from a detailed comparison with the separated cases (Figure 11.3a and c). It is well known that a uniform and isotropic carrier density appears near an isolated atom, such as the sphere-like 1s and 2s orbitals of lithium atom [the heavy and light red colors in Figure 11.3a] and the similar distribution of [1s, 2s, $2p_x,2p_y,2p_z$], 3s and [$3p_x,3p_y,3p_z$] orbitals from sulfur atom [the heavy blue, red, and yellow/green colors in Figure 11.3b]. However, the chemical bonding between lithium and sulfur atoms leads to the nonuniform and anisotropic features, especially for the close regions. The multi-orbital hybridizations, 2s-[3s, $3p_x,3p_y,3p_z$], are identified from the delicate charge variation.

The periodical charge density (Figure 11.3c) is very useful in understanding the X-ray pattern, and the experimental measurements can examine the theoretical predictions. It is well known that the X-ray diffraction spectra belong to the elastic scatterings associated with an external perturbation of electromagnetic wave [the

(a) Li   **(b) S**

(c) Li-S bond

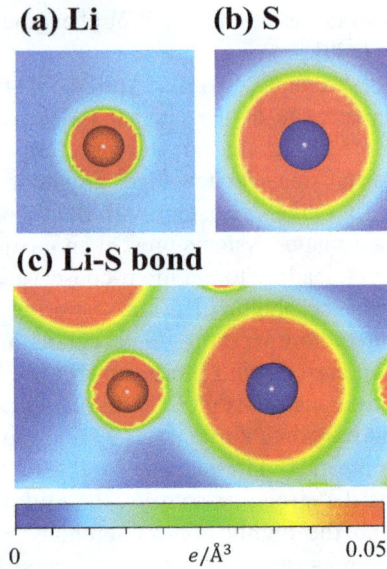

0                    $e/Å^3$ 0.05

FIGURE 11.3 The spatial charge density distribution near the lithium-sulfur bond in clearly illustrating the significant multi-orbital hybridization of 2s-[3s, $3p_x,3p_y,3p_z$]: the isolated cased of (a) lithium and (b) sulfur atoms; (c) the chemical bond on the [x, z]-plane [check].

photon–electron elastic intercalations in Ref. [54]]. This directly reflects the featured periodicity of a crystal structure and thus is very suitable for exploring the dynamic processes of electrolyte/cathode/anode materials. A series of intermediate configurations come to exist during the charging and discharge processes of ion-based batteries, in which the dramatic transformation of crystal symmetries will be clearly revealed in the position-dependent charge density. The systematic simulation investigations, being driven by the enhancement or reduction of lithium concentration/the variation of its arrangement during the chemical reactions, are required to focus on the time dependence. The close relations between the periodical charge distribution and the diffraction pattern need to be established in the further theoretical research. On the experimental side, the time-resolved diffraction techniques could be developed for more delicate examinations [6,21]. The consistency of theories and measurement is able to provide a rather strong backup for studying the critical mechanisms of ion transports. The stationary ion currents should be related to the rapid variation of charge density distribution; that is, they are mainly determined by the Coulomb potentials. However, a unified theory framework, which is based on the quasiparticle viewpoints [20,47,55,56], will be done for the transport properties.

The merged structures of van Hove singularities are able to present the critical orbital hybridizations of chemical bonds [20,47,56], or they can clarify the unusual quasiparticle behaviors [4]. It is well known that the density of states in a 3D condensed-matter system is characterized by the inverse of group velocity on a constant-energy closed surface. The vanishing velocity of a propagating wave packet, which is closely related to an extreme electronic state in band structure [a band-edge state in Figure 11.2a–d], is responsible for a van Hove singularity. Apparently, its special structure is mainly determined by the various energy dispersion relations [4,17], as well the distinct dimensionalities [4,17]. For example, the systematic investigations, being conducted on 3D graphites [AA-, AB-, and ABC-stacked bulk systems in Ref. [20]], 2D graphene systems [distinct stacking configurations and layer numbers in Ref. [20]], 1D graphene nanoribbons [56], and 1D/0D carbon nanotubes [56], clearly illustrate the diverse quasiparticle phenomena. Furthermore, most of these unique results are consistent with the high-resolution measurements of STS [34]. Most importantly, the joint van Hove singularities, the high density of states from the initial or/and final states [26], can dominate the number of available channels in Coulomb-field excitations [26], optical absorption processes [26], and carrier transports [16]. In addition, the various optical properties, with/without excitonic effects [26], are under the current investigations.

The atom- and orbital-projected van Hove singularities, as clearly shown in Figure 11.4a–c, are capable of identifying the multi-orbital hybridizations in 3D lithium-sulfur bonds. The density of states is vanishing in a specific energy range of -1.683 eV $\leq E \leq 1.683$ eV across the Fermi level, shows a large bandgap of $E_g = 3.365$ eV. Both lithium and sulfur atoms make significant contributions to the density of states within a whole energy spectrum [the red and blue curves in Figure 11.4a, respectively]. This further reflects the fact that only the Li-S bonds survive in a unit cell (Figure 11.1a). However, the latter presents the larger weight, mainly owing to more active orbitals. It is well known that the outer 2s-orbital of the Li atom is half-occupied and has a unique chemical bonding with the active ones of other atoms [27]. As a result, the Li-related van Hove singularities are available in supporting the critical multi-orbital hybridizations of each Li-S bond. Figure 11.4b and c presents the orbital-projected density of states of S, in which they display the prominent structures due to the merged van Hove singularities, associated with the on-site Coulomb potential energies [the ionization energy of atomic orbitals] [57]. 3s and $[3p_x, 3p_y, 3p_z]$ [the pink, deep blue, green, and light blue curves], respectively, dominate, in the deep- and middle-energy ranges of $E \leq -12$ eV (Figure 11.4c) and -4.8 eV $\leq E \leq -1.6$ eV (Figure 11.4b). Furthermore, such orbitals co-dominate within the conduction-state region. The valence-state parts clearly show the strong $[3p_x, 3p_y, 3p_z]-[3p_x, 3p_y, 3p_z]$-orbital hybridizations, but the weak 3s-$[3p_x, 3p_y, 3p_z]$-orbital mixings. It should be noticed that an anisotropic feature is well characterized by an observable difference among the contributions of $3p_x, 3p_y$ and $3p_z$ orbital. These critical quasiparticle behaviors are expected to determine the concise physical pictures of the other essential properties, e.g., the threshold transition frequency [58] and prominent absorption structures closely related to the specific orbital hybridizations [59].

The direct examinations on the diverse van Hove singularities could be done for the low-dimensional materials through the high-resolution STS measurements [34]. However, such tests might be mission impossible for 3D condensed-matter systems because of the intrinsic limits [60]. The tunneling quantum currents, being rather weak through the tunneling effects between small and substrate, are in a sharp contrast with the transverse thickness. The up-to-date STS experiments, which are assisted by the simultaneous measurements on nano-scaled surface morphologies [60], are

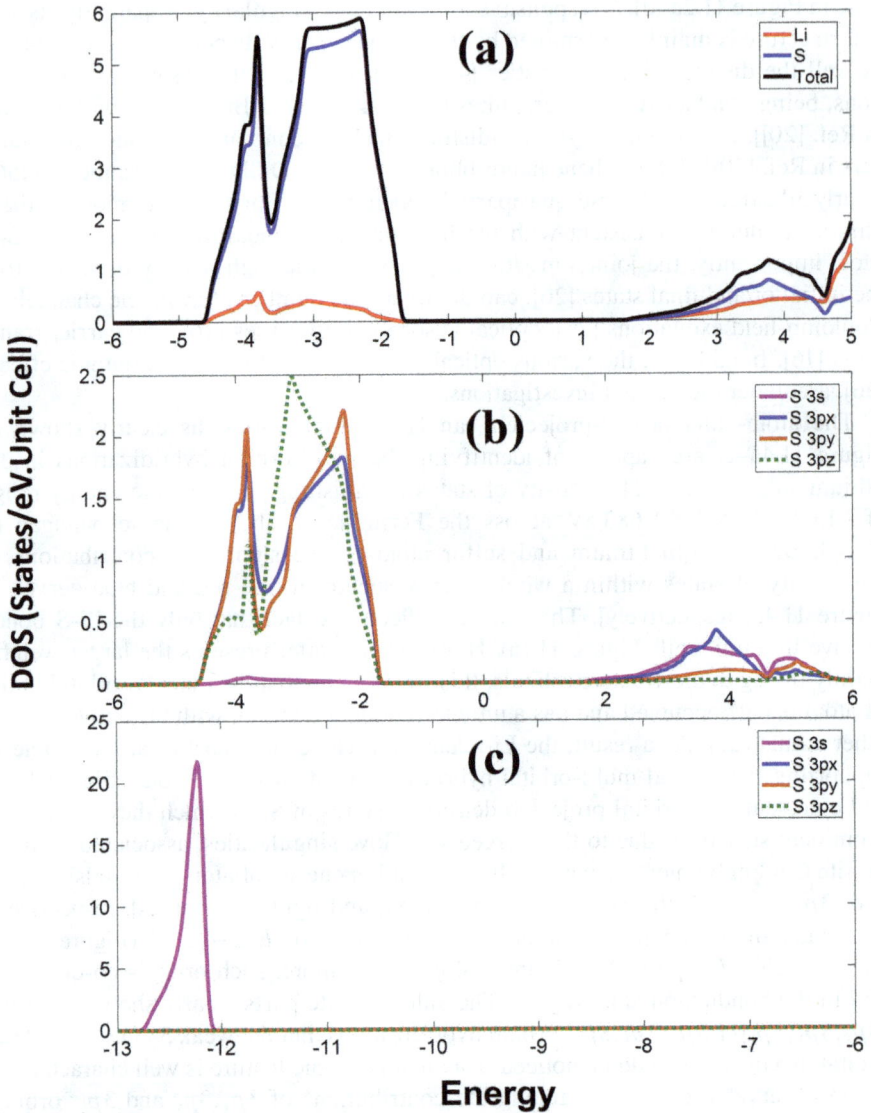

FIGURE 11.4   The valence and conduction density of states for Li$_2$S: (a) The total contribution [the black curve] separated into lithium and sulfur parts [the red and blue curves] and S-[3s, $3p_x, 3p_y, 3p_z$]-decomposed ones [the various colors] within the energy ranges of (b) -6 eV $\leq E \leq 6$ eV & (c) -15 eV $\leq E \leq -6$ eV.

very successful in identifying geometric structures and van Hove singularities. For example, the 1D carbon nanotubes [61,62], graphene nanoribbons [56,63,64], and 2D few-layer graphene systems [20], respectively, exhibit the diverse quasiparticle phenomena: (1) the radius- and chirality-enriched hexagon arrangement on a hollow cylindrical surface and three types of band property [metal, narrow- and middle-gap semiconductors due to the periodical boundary condition and the misorientation of C- $2p_z$ orbitals [20,65]], (2) the edge- and width-dependent bandgaps [only narrow and middle ones due to open periodical boundaries, nonuniform C–C bond lengths and/or spin-orbital couplings [20]], and (3) the semi-metallic and zero-gap semiconducting behaviors associated with the small band overlaps of the well-behaved stackings and the recovery of the gapless Dirac-cone structure in the asymmetric bilayer graphene systems [20]. It should be noticed that the layered Bernal graphite has been successfully examined for its density of states, such as the semi-metallic behavior with free carriers at the Fermi level. On the other hand, the STS observations cannot be finished for most electrolyte/cathode/anode materials.

There are close relations between the first-principles simulations and the phenomenological models, especially for the current band structure. How to obtain a reliable tight-binding model from from the VASP results along the high symmetry points (Figure 11.2a–d) could be conducted through the delicate analyses. All the position-dependent chemical bonds in a primitive unit cell (Figure 11.1a), the nearest-neighbor hopping integrals due to the distinct orbital hybridizations [48,66], and the different site Coulomb energies associated with the ionization energies [48,66] need to be taken into account. A successful establishment of the significant Hamiltonian simulation will be very useful in understanding the intrinsic atomic interactions [48,66], such as the different strengths of the active orbital hybridizations in $s - sp^3$ bonding of the Li-S bond. Whether similar works are available for the other electrolyte/cathode/anode materials with larger Moire superlattices need to be tested in the near-future theoretical research. In addition, the previously published works on the suitable tight-binding model seem to be absent up to now.

11.5 SUMMARY

The 3D binary lithium-sulfur compound is predicted to exhibit rich essential properties through the delicate VASP simulations and analyses. The crystal symmetry, atom-dominated electronic energy spectrum and wave packet, spatial charge density distribution, and van Hove singularities are unified under a theoretical framework of quasiparticle viewpoint [20,47,56]. Li$_8$S$_4$ has 12 atoms in a primitive unit cell, where the position-dependent chemical bonds lead to the highly anisotropic environment. The cubic crystal symmetry and the unique first Brillouin zone, with the significant $s - sp^3$ chemical bonding, are responsible for the featured electronic properties: the multi-fold state degeneracies at the high-symmetry points, a direct bandgap of $E_g = 3.365$ eV near the Γ point, the high asymmetry of the hole and electron bands about the Fermi level, the lithium and sulfur codominance within the whole energy range [the weaker weight for the former], the non-monotonous and anisotropic wave-vector dependences with the complicated dispersion relations, a lot of band-edge states, the frequent subband crossings, a giant energy spacing of -12.2 eV $\leq E^v \leq -4.7$ eV without any valence states [due to the site-energy difference between 3s and 3p orbitals in the sulfur atom; [4]], the obvious distortion of charge density distribution around two bound atoms, the vanishing density of states within the gap and vacant regions, and the merged special structures from the 2s-[$3s, 3p_x, 3p_y, 3p_z$]-orbital-projected van Hove singularities. Such characteristics are attributed to the active orbital hybridizations. This quasiparticle framework is available in studying the other properties of emergent materials. For example, the main features optical absorption spectra, with/without excitonic effects, are very suitable in establishing the close relations between the specific orbital hybridizations and the vertical transition channels [26]. Only the predicted crystal structure agrees with X-ray measurements [21]. However, the experimental examinations about band structure and van Hove singularities are absent up to now. The systematic investigations on the intermediate configurations and the stationary currents are urgently required for clarifying the critical mechanisms of ion transports and greatly enlarging the quasiparticle framework [20,47,56].

REFERENCES

1. Li G, Chen Z and Lu J 2018 Lithium-sulfur batteries for commercial applications *Chem* **4** 3–7.
2. Aotani N, Iwamoto K, Takada K and Kondo S 1994 Synthesis and electrochemical properties of lithium ion conductive glass, Li3PO4Li2SSiS2 *Solid State Ionics* **68** 35–9.
3. Hayamizu K and Aihara Y 2013 Lithium ion diffusion in solid electrolyte (Li2S) 7 (P2S5) 3 measured by pulsed-gradient spin-echo 7Li NMR spectroscopy *Solid State Ionics* **238** 7–14.
4. Eithiraj R, Jaiganesh G, Kalpana G and Rajagopalan M 2007 First-principles study of electronic structure and ground-state properties of alkali-metal sulfides–Li2S, Na2S, K2S and Rb2S *Physica Status Solidi (b)* **244** 1337–46.
5. Yi Z, Su F, Huo L, Cui G, Zhang C, Han P, Dong N and Chen C 2020 New insights into Li2S2/Li2S adsorption on the graphene bearing single vacancy: A DFT study *Applied Surface Science* **503** 144446.
6. Zhang L, Sun D, Feng J, Cairns E J and Guo J 2017 Revealing the electrochemical charging mechanism of nanosized Li2S by in situ and operando X-ray absorption spectroscopy *Nano Letters* **17** 5084–91.
7. Manthiram A, Fu Y, Chung S-H, Zu C and Su Y-S 2014 Rechargeable lithium–sulfur batteries *Chemical Reviews* **114** 11751–87.
8. Ding B, Wang J, Fan Z, Chen S, Lin Q, Lu X, Dou H, Nanjundan A K, Gleb Y and Zhang X 2020 Solid-state lithium–sulfur batteries: Advances, challenges and perspectives *Materials Today*.
9. Anastas P T and Zimmerman J B 2019 The periodic table of the elements of green and sustainable chemistry *Green Chemistry* **21** 6545–66.
10. Xie Y, Ma Y M, Cui T, Li Y, Qiu J and Zou G 2008 Origin of bcc to fcc phase transition under pressure in alkali metals *New Journal of Physics* **10** 063022.
11. Lang P F and Smith B C 2015 Metallic structure and bonding *World* **3** 30–5.
12. Lamoureux G and Ogilvie J F 2019 Hybrid atomic orbitals in organic chemistry. Part 2: Critique of practical aspects *Química Nova* **42** 817–22.
13. Kudu Ö U, Famprikis T, Fleutot B, Braida M-D, Le Mercier T, Islam M S and Masquelier C 2018 A review of structural properties and synthesis methods of solid electrolyte materials in the Li2S– P2S5 binary system *Journal of Power Sources* **407** 31–43.
14. Geist E, Kirschning A and Schmidt T 2014 sp 3-sp 3 Coupling reactions in the synthesis of natural products and biologically active molecules *Natural Product Reports* **31** 441–8.
15. Kang W, Deng N, Ju J, Li Q, Wu D, Ma X, Li L, Naebe M and Cheng B 2016 A review of recent developments in rechargeable lithium–sulfur batteries *Nanoscale* **8** 16541–88.

16. Ould Ely T, Kamzabek D, Chakraborty D and Doherty M F 2018 Lithium–sulfur batteries: State of the art and future directions *ACS Applied Energy Materials* **1** 1783–814.

17. Luo G, Zhao J and Wang B 2012 First-principles study of transition metal doped Li2S as cathode materials in lithium batteries *Journal of Renewable and Sustainable Energy* **4** 063128.

18. Ozturk T, Ertas E and Mert O 2010 A Berzelius reagent, phosphorus decasulfide (P4S10), in organic syntheses *Chemical Reviews* **110** 3419–78.

19. Pham H D, Nguyen T D H, Vo K D, Huynh T M D and Lin M-F 2020 Rich essential properties of boron, carbon, and nitrogen substituted germanenes *Applied Physics Express* **13** 085502.

20. Tran N T T, Lin S-Y, Lin C-Y and Lin M-F 2017 *Geometric and Electronic Properties of Graphene-Related Systems: Chemical Bonding Schemes*: CRC Press.

21. Camacho-Forero L E and Balbuena P B 2018 Exploring interfacial stability of solid-state electrolytes at the lithium-metal anode surface *Journal of Power Sources* **396** 782–90.

22. Rao C, Vivekchand S, Biswas K and Govindaraj A 2007 Synthesis of inorganic nanomaterials *Dalton Transactions* 3728–49.

23. Suslick K S 2001 Encyclopedia of physical science and technology *Sonoluminescence and Sonochemistry*, 3rd edn. Elsevier Science Ltd, Massachusetts 1–20.

24. Lv B, Qian T and Ding H 2019 Angle-resolved photoemission spectroscopy and its application to topological materials *Nature Reviews Physics* **1** 609–26.

25. Zhang C, Li Y, Pei D, Liu Z and Chen Y 2020 Angle-resolved photoemission spectroscopy study of topological quantum materials *Annual Review of Materials Research* **50**.

26. Fox M 2001 *Optical Properties of Solids*: Oxford University Press.

27. Liang P, Zhang L, Wang D, Man X, Shu H, Wang L, Wan H, Du X and Wang H 2019 First-principles explorations of Li2S@ V2CTx hybrid structure as cathode material for lithium-sulfur battery *Applied Surface Science* **489** 677–83.

28. Liu Z, Deng H, Hu W, Gao F, Zhang S, Balbuena P B and Mukherjee P P 2018 Revealing reaction mechanisms of nanoconfined Li 2 S: Implications for lithium–sulfur batteries *Physical Chemistry Chemical Physics* **20** 11713–21.

29. Foulkes W M C and Haydock R 1989 Tight-binding models and density-functional theory *Physical Review B* **39** 12520.

30. Nakayama T and Shima H 1998 Computing the Kubo formula for large systems *Physical Review E* **58** 3984.

31. Ren X, Rinke P, Joas C and Scheffler M 2012 Random-phase approximation and its applications in computational chemistry and materials science *Journal of Materials Science* **47** 7447–71.

32. Liu Z, Hubble D, Balbuena P B and Mukherjee P P 2015 Adsorption of insoluble polysulfides Li 2 S x (x= 1, 2) on Li 2 S surfaces *Physical Chemistry Chemical Physics* **17** 9032–9.

33. Damascelli A 2004 Probing the electronic structure of complex systems by ARPES *Physica Scripta* **2004** 61.

34. Bussolotti F, Chi D, Goh K J, Huang Y L and Wee A T 2020 *2D Semiconductor Materials and Devices*: Elsevier. pp 199–220.

35. Hafner J 2008 Ab-initio simulations of materials using VASP: Density-functional theory and beyond *Journal of Computational Chemistry* **29** 2044–78.

36. Rangel T, Caliste D, Genovese L and Torrent M 2016 A wavelet-based projector augmented-wave (PAW) method: Reaching frozen-core all-electron precision with a systematic, adaptive and localized wavelet basis set *Computer Physics Communications* **208** 1–8.

37. Chen Y-X and Kaghazchi P 2014 Metalization of Li 2 S particle surfaces in Li–S batteries *Nanoscale* **6** 13391–5.

38. Feenstra R M 1994 Scanning tunneling spectroscopy *Surface Science* **299** 965–79.
39. Albareda G, Suñé J and Oriols X 2009 Many-particle hamiltonian for open systems with full coulomb interaction: Application to classical and quantum time-dependent simulations of nanoscale electron devices *Physical Review B* **79** 075315.
40. Ouyang X, Lei M, Shi S, Luo C, Liu D, Jiang D, Ye Z and Lei M 2009 First-principles studies on surface electronic structure and stability of LiFePO4 *Journal of Alloys and Compounds* **476** 462–5.
41. Yuan L-X, Wang Z-H, Zhang W-X, Hu X-L, Chen J-T, Huang Y-H and Goodenough J B 2011 Development and challenges of LiFePO 4 cathode material for lithium-ion batteries *Energy & Environmental Science* **4** 269–84.
42. Xu J and Chen G 2010 Effects of doping on the electronic properties of LiFePO4: A first-principles investigation *Physica B: Condensed Matter* **405** 803–7.
43. Yamada A, Chung S and Hinokuma K Optimized LiFePO [sub 4] for lithium battery cathodes *Journal of the Electrochemical Society* **148** A224.
44. Boufelfel A 2013 Electronic structure and magnetism in the layered LiFeO2: DFT+ U calculations *Journal of Magnetism and Magnetic Materials* **343** 92–8.
45. Li J, Li J, Luo J, Wang L and He X 2011 Recent advances in the LiFeO2-based materials for Li-ion batteries *International Journal of Electrochemical Science* **6** 1550–61.
46. Nagpure S C, Babu S, Bhushan B, Kumar A, Mishra R, Windl W, Kovarik L and Mills M 2011 Local electronic structure of LiFePO4 nanoparticles in aged Li-ion batteries *Acta Materialia* **59** 6917–26.
47. Shih-Yang Lin H-Y L, Nguyen D K, Tran N T T, Pham H D, Chang S-L, Lin C-Y and Lin M-F 2020 *Silicene-Based Layered Materials: Essential Properties*: IOP Publishing.
48. Chiun-Yan Lin J-Y W, Chiu C-W and Lin M-F 2019 *Coulomb Excitations and Decays in Graphene-Related Systems*: CRC Press.
49. Ho J-H, Lu C, Hwang C, Chang C and Lin M-F 2006 Coulomb excitations in AA-and AB-stacked bilayer graphites *Physical Review B* **74** 085406.
50. Kittel C 1996 *Introduction to Solid State Physics*: John Wiley & Sons New York 402.
51. Chen X, Tian F, Persson C, Duan W and Chen N-X 2013 Interlayer interactions in graphites *Scientific Reports* **3** 1–5.
52. Scrosati B, Hassoun J and Sun Y-K 2011 Lithium-ion batteries. A look into the future *Energy & Environmental Science* **4** 3287–95.
53. Pistoia G 2013 *Lithium-Ion Batteries: Advances and Applications*: Newnes.
54. Speakman S A 2013 Introduction to x-ray powder diffraction data analysis. *Center for Materials Science and Engineering at MIT*.
55. Li W-B, Lin S-Y, Tran N T T, Lin M-F and Lin K-I 2020 Essential geometric and electronic properties in stage-n graphite alkali-metal-intercalation compounds *RSC Advances* **10** 23573–81.
56. Lin S Y T, N. T. T.; Chang, S. L.; Su, W. P.; Lin, M. F. 2018 *Structure- and Adatom-Enriched Essential Properties of Graphene Nanoribbons*: CRC Press.
57. Mahan G D 2013 *Many-Particle Physics*: Springer Science & Business Media.
58. Chen C J 1993 *Introduction to Scanning Tunneling Microscopy* vol 4: Oxford University Press on Demand.
59. Dien V K, Pham H D, Tran N T T, Han N T, Huynh T M D, Nguyen T D H and Fa-Lin M 2020 Orbital-hybridization-created optical excitations in Li8Ge4O12 *arXiv preprint arXiv:2009.02160*.
60. Hipps K 2006 *Handbook of Applied Solid State Spectroscopy*: Springer. pp 305–50.
61. Dai H 2002 Carbon nanotubes: Synthesis, integration, and properties *Accounts of Chemical Research* **35** 1035–44.
62. Du A J, Chen Y, Lu G and Smith S C 2008 Half metallicty in finite-length zigzag single walled carbon nanotube: A first-principle prediction *Applied Physics Letters* **93** 073101.

63. Yin W-J, Xie Y-E, Liu L-M, Chen Y-P, Wang R-Z, Wei X-L, Zhong J-X and Lau L 2013 Atomic structure and electronic properties of folded graphene nanoribbons: A first-principles study *Journal of Applied Physics* **113** 173506.
64. Yang L, Cohen M L and Louie S G 2007 Excitonic effects in the optical spectra of graphene nanoribbons *Nano Letters* **7** 3112–5.
65. Tran N T T, Lin S-Y, Glukhova O E and Lin M-F 2016 π-Bonding-dominated energy gaps in graphene oxide *RSC Advances* **6** 24458–63.
66. Lin C-Y, Ho C-H, Wu J-Y, Do T-N, Shih P-H, Lin S-Y and Lin M-F 2019 *Diverse Quantization Phenomena in Layered Materials*: CRC Press.

12 Diversified Quasiparticle Phenomena of P_2S_5

Electrolyte in Lithium-Sulfur Battery

Thi Dieu Hien Nguyen and Ming-Fa Lin
National Cheng Kung University

CONTENTS

12.1 INTRODUCTION

Lithium-sulfur batteries are holding many advantages such as high theoretical energy density up to 2,600 Wh kg^{-1}. Lots of efforts to increase the cycle life and efficiency of rechargeable batteries are the current attention, which can be employed in applied technologies. The most important mechanism is related to the conversion of active sulfur and lithium sulfide in the discharge and charge process. Principally, sulfur exists in the charged process, in which the discharging process plays the role in operating the system. It is noted that lithium ions and electrons are formed in the discharge state due to the oxidation of lithium metal at the anode. In the next step, lithium ions will travel through the electrolyte to the cathode. At the same time, in the external circuit, the electrons travel to the positive electrode to create the current. In this case, sulfur will directly take part in the chemical reaction to produce lithium sulfide. Very interestingly, lithium-sulfur battery principally consists of an anode, a cathode, and an electrolyte, mainly from lithium metal, a sulfur composite, and organic sulfur electrolyte, respectively. Currently, the sulfur electrolyte has an excellent touch with cathode and anode in batteries, and sulfur is one of the most abundant elements, leading to an efficient price. The binary compound Li_2S-P_2S_5 is the main focus of sulfur electrolyte, which can increase ionic conductivity by creating

DOI: 10.1201/9781003263807-12

diversified metastable forms. There are some identical anions that could be achieved, such as PS_4^{3-} tetrahedral (>75 mol% Li_2S) [1–3], $P_2S_7^{4-}$ (<75 mol% Li_2S) [4–6].

Both phosphorus and sulfur atoms, which possess 15 and 16 occupied orbitals in a periodic table [7], are very active elements in nature. A plenty of diversified compounds are successfully synthesized by the various chemical reactions [8,9]. The up-to-date experimental syntheses clearly show that all the active isomers can be characterized by a formula of P_4S_x of $x \leq 10$, such as P_4S_2 [10], P_4S_3 [10], P_4S_4 [10], P_4S_5 [10,11], P_4S_6 [10,14], P_4S_7 [10,11], P_4S_8 [10,12], P_4S_9 [10], and P_4S_{10} [10,11,13]. The unusual crystal symmetries and the unique phonon modes of the coherent atomic vibrations have been thoroughly examined by X-ray patterns [14] and Raman scatterings [15], respectively. These high-precision measurements clearly indicate the rich geometric properties. However, the very complicated multi-orbital hybridizations of phosphorus-sulfur bonds, which are responsible for the fundament properties, remain unsolved in the previous works, e.g., the close relations between chemical bondings and electronic properties [16]. Among these 3D binary compounds, phosphorus pentasulfide and sesquisulfide [P_4S_{10} and P_4S_3] are frequently utilized in the commercial products. The former will be chosen for a model study in this chapter. Such condensed-matter systems are outstanding candidates for developing a unified theoretical framework of quasiparticle viewpoints through a series of first-principles simulations [17] and/or the systematic investigations of phenomenological models [18–20]. In addition, the ternary/quaternary/pentanary lithium phosphorus sulfur compounds, which are generated from the different ratios of Li-S- and P-S-related subsystems [8,11,13,21,22], could serve as the high-efficient solid-state electrolytes. Their diversified quasiparticle phenomena will be very interesting research topics.

Very obviously, the complicated and active chemical bondings are indicated from the successful syntheses of the diversified phosphorus sulfide compounds. The previously described theoretical works show that only few first-principles simulations are focused on their geometric structures and electronic energy spectra [11,13,16,23]. However, the delicate analyses about the important features of electronic states are almost absent, such as the different atom dominances and their relations with the multi-orbital hybridizations. This work will directly link all the relevant physical quantities to clearly identify the significant multi-orbital hybridizations in P-S bonds, in which they cover the highly nonuniform chemical environment [16], the atom-dominated band structure [24], the non-spherical charge density distribution [17], and the merged van Hove singularities [17]. It is also noticed that there are no high-resolution measurements about electronic structures [angle-resolved photoemission spectroscopy (ARPES) in [25,26]], optical absorption spectra [27], and Coulomb excitations [electron-energy-loss spectroscopy in [18–20]].

The 3D binary phosphorus sulfur compound, P_2S_5 [28], is one of a part of an outstanding electrolyte candidate in fully exploring the diversified quasiparticle phenomena of lithium-ion-based batteries. Furthermore, the first-principles simulations, but not the phenomenological models [e.g., the tight-binding model; [29]], are very reliable in examining the critical multi-orbital hybridizations of phosphorus-sulfur bonds, which are responsible for the unusual physical/chemical/materials [23]. The Vienna *ab initio* simulation package (VASP) simulations, which cover the many-particle electron-electron Coulomb interactions [exchange and correlation ones in

[17]] and the periodical crystal potential [7], are operated through the minimum net force on each atom [~10^{-6} eV] [17]. By the delicate analyses [17], the optimal crystal structure is expected to present a Moire superlattice [13] with an observable fluctuation of phosphorus-sulfur bonds. The close relations between the highly nonuniform chemical environments and the electronic properties will be thoroughly explored in this work. Specifically, the detailed examinations are conducted on the phosphorous- and sulfur-dominated energy spectra and wave functions within the distinct energy ranges [30], the 3D spatial charge density [30], and the atom- and orbital-projected density of states [17]. The main features of band structure, the band property near the Fermi level [30], the asymmetry of hole and electron spectra [30], the various energy dispersions [30], the band-edge states [the critical points in energy-wave-vector space, van Hove singularities; [30]], and the non-crossing, crossing, or anti-crossing behaviors need to be checked in detail. Moreover, a plenty of prominent structures, which appear in the energy-dependent density of states, will be clearly identified from the composite contributions of different atoms and orbitals. Under the consistency of all calculate results, the active multi-orbital hybridizations could be achieved for the phosphorus-sulfur bonds. The theoretical predictions need to be thoroughly examined through the high-precision experimental measurements, e.g., the crystal symmetry through X-ray technique/Raman scattering [14,15], the band gap by ARPES [25,26], and the threshold absorption frequency using optical spectroscopies [27]. The significant differences of phosphorus and lithium sulfide compounds are fully discussed. In addition, whether the reliable tight-binding model [29] is worthy of a closer development for this complicated system will be done in the near-future investigations [29]. Also, the significant differences between the phosphorus and lithium sulfide compounds/among the P- and S-related isomers are discussed later [7].

12.2 REAL- AND WAVE-VECTOR-SPACE LATTICE SYMMETRIES

The optimal geometric structures, charge distribution, and electronic state of P_2S_5 are calculated by the first-principles density functional theory (DFT) calculations [17,31]. In this chapter, we performed the calculations with the VASP, which evaluates an approximate solution within DFT by solving the Kohn–Sham equations. The Perdew–Burke–Ernzerhof formula used in VASP, which depends on the local electron density, is utilized to deal with many-particle Coulomb effects. The frequent electron-crystal scatterings are characterized by the projector-augmented wave pseudopotentials. The electron Bloch wave functions are solved from the linear superposition of plane waves, with a maximum kinetic energy of 520 eV. The current study on P_2S_5 compounds shows that the first Brillouin zone is sampled by $9 \times 9 \times 9$ and $30 \times 30 \times 30$ k-point meshes within the Monkhorst–Pack scheme, respectively, for the optimal geometry and band structure. Moreover, the convergence condition of the ground-state energy is set to be ~10^{-6} eV between two consecutive evaluation steps, where the maximum Hellmann–Feynman force for each ion is set below 0.01 eV Å^{-1} during the atom relaxations.

The 3D binary phosphorus sulfide of P_2S_5, as clearly illustrated in Figure 12.1a and b, exhibits the unusual geometries in the real- and wave-vector spaces. This condensed-matter system belongs to a triclinic material [32], in which the lattice

(a)   (b)

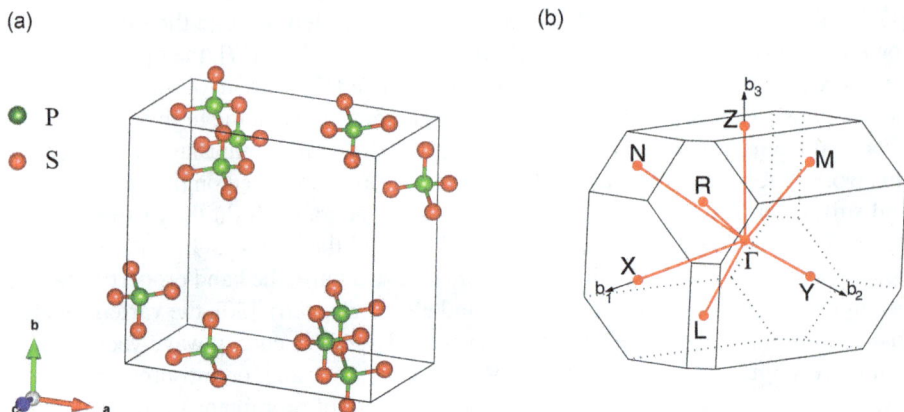

FIGURE 12.1 The unusual crystal structures of P_2S_5 within the (a) real and (b) wave-vector spaces, with the green and red balls for phosphorus and sulfur atoms, respectively. Also shown in (b) are those of the high-symmetry points.

constants are slightly different from one another [$a = 9.679$ Å, $b = 9.824$ Å and $c = 9.825$ Å; [33]] through the non-orthogonal angles of $\alpha = 92.894°$, $\beta = 109.712°$, and $\lambda = 90°$ [33]. There exist 28 atoms in a primitive unit cell using eight phosphorus and 20 sulfur atoms, apparently generating a large Moire superlattice [8]. An observable fluctuation of the P-S chemical bond lengths is revealed within the range of ~1.712–2.519 Å. This obviously indicates the position-dependent multi-orbital hybridizations; that is, the highly nonuniform and anisotropic chemical environment have resulted in the spatial modulation of the chemical bonding strengths [34]. More important, the hopping integrals, which are characterized by the Coulomb interaction of the neighboring atomic orbitals [29], are expected to be greatly diversified by the strong dependences on the different electronic states and local positions. As a result, it will be very difficult to get a set of reliable parameters in the tight-binding model because of too complicated multi-orbital hybridizations [29]. On the wave-vector-space side, the first Brillouin zone displays an unusual structure (Figure 12.1b). The electronic states, which is defined on the specific paths of high-symmetry points, should be rather unique. The experimental verifications by the ARPES [26] might become mission impossible.

Very interestingly, the 3D binary phosphorus sulfide compounds have shown the isomer cases of P_4S_x of $x \leq 10$ by modulating their relative concentrations and distributions. The diverse crystal structures are directly verified by the elastic X-ray scatterings [14,35], which are assisted by the inelastic Raman scatterings [23] and scanning electron microscopy [SEM in [34]]. It should be noticed that the latter are only focused on the atomic components. There also exist the Li-, P, and S-related multi-component main-stream materials closely associated with the solid-state electrolytes of lithium-sulfur batteries [8,36–40]. The 3D ternary [8,41], quaternary [40,42] and pentanary [42,43] lithium phosphorus sulfides possess larger Moire superlattices and more complex chemical environments. All the main-stream materials can provide the various lattice symmetries in clarifying the close relations between the active

orbital hybridizations/the spin configurations and the featured fundamental properties. The unified framework of quasiparticle viewpoints could be thoroughly from
them [24,44,45].

12.3 ATOM-DETERMINED ELECTRONIC ENERGY SPECTRUM AND WAVE FUNCTION

The electronic energy spectrum and atomic contributions, which are thoroughly conducted on the 3D ternary P_2S_5 systems, are predicted to exhibit the unusual quasiparticle phenomena. There are a plenty of valence and conduction energy subbands
along the high-symmetry points of the first Brillouin zone (Figure 12.2a), directly
reflecting a large Moire superlattice of 28 atoms and the active [$3s$, $3p_x$, $3p_y$, $3p_z$] orbitals. The occupied hole bands are highly asymmetric to the occupied electron ones
about the Fermi level [$E_F = 0$]; [24,44,45]. All the electronic states strongly depend
on the variation of wave vector: the main features of k-dependent energy spectrum

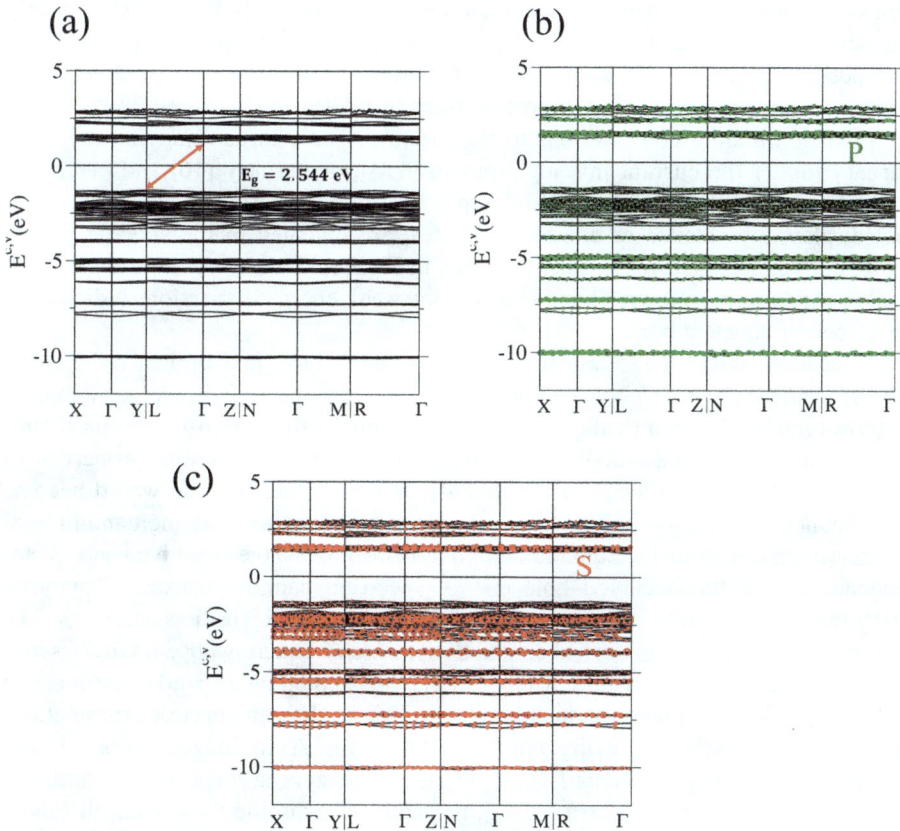

FIGURE 12.2 The 3D P- and S-created band structures and wave functions within the first
Brillouin zone along the high-symmetry points, which are illustrated by green circles and red
circles, respectively.

are anisotropic, dispersive, non-monotonous, and degenerate/non-degenerate. The different energy dispersions [parabolic/linear/oscillator/partially flat forms] might reveal at the specific wave vectors. The extreme states belong to the critical points in the energy-wave-vector space; furthermore, they can induce the unique van Hove singularities due to the vanishing group velocities [30]. The crossing and non-crossing behaviors come to exist frequently, but only few for the subband anti-crossings. An indirect band gap, which is characterized by the highest occupied state at Y/L and the lowest unoccupied state at Γ, is revealed as $E_g \sim 2.544$ eV. This result clearly presents the significant feature of covalent bonding between the neighboring phosphorus and sulfur atoms. However, the rich and unique electronic structure cannot be utilized to identify the specific chemical bondings through certain full subband widths. For example, π and σ chemical bondings are very meaningful in graphite-related systems, while they become vanishing for Li-, P-, and S-combined compounds [8].

The phosphorus- and sulfur-dominated electronic states, as clearly illustrated in Figure 12.2b and c by the green and red balls, respectively, are capable of providing their roles in chemical bonds. Both atoms make the non-negligible contributions within a wide valence and conduction energy range [-12 eV $\leq E^{c,v} \leq 5$ eV]. However, the relative weight is larger for the latter, especially in the significant valence-energy range of -8 eV $\leq E^v$. Sulfur atoms should be active in condensed-matter systems and thus play important roles in a lot of main-stream materials, e.g., their dominating contributions to the lithium-, phosphorus-, and sulfur-related ternary [under the current investigations in [8,41], quaternary [46] and pentanary compounds [46]. In addition to band gap, certain wide energy spacings, which do not display any valence or conduction states, are revealed in the specific energy ranges [also discussed for density of states in Figure 12.4]. Most importantly, several energy spacings, a wide band gap, and the weak energy dispersions indicate the existence of covalent bonds.

In general, ARPES [25,26], transport equipment [46], and optical reflectance/absorption/transmission/photoluminescence spectroscopies [27] are available in determining band gap/optical gap [25,27]. For example, these high-resolution examinations are very successful in clarifying a lot of unusual essential properties of few-layer emergent graphene materials [24]. However, the first one would become nonconductivity, mainly owing to many energy subbands and the momentum non-conservation due to surface quantum confinement. That is, the sensitive wave-vector dependence for the occupied hole energy spectrum cannot be exactly analyzed from the angle-resolved measurements of photon-electron inelastic scatterings. The temperature-dependent electrical conductivity, which frequently appears in the semiconductor using the exponential decay form [28,46], could be utilized to estimate an intrinsic property, the magnitude of band gap. This method and optical examinations cannot directly distinguish a direct or indirect band gap. As to transverse optical transitions under the long wavelength limit [47], the reflectance/absorption/transmittance and photoluminescence spectra [27,47], respectively, examine the optical threshold absorption [the optical gap in [27]] and the bound-state energy of exciton. Their combination could evaluate the many-body red shift [48] and thus band gap [27,48]. The high-precision transport and optical measurements are urgently necessary in testing this critical quasiparticle property.

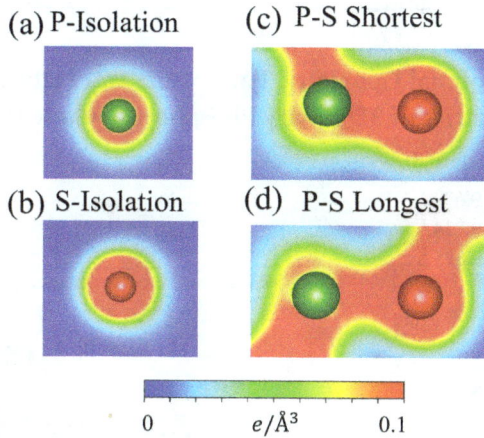

(a) P-Isolation (c) P-S Shortest

(b) S-Isolation (d) P-S Longest

0    $e/Å^3$ 0.1

FIGURE 12.3 The spatial charge density distributions for (a) the isolated phosphorus and (b) sulfur atoms, as well as the 3D P_2S_5 compound in the (c) shortest and (d) longest P-S bonds through the active multi-orbital hybridization of [3s, $3p_x$, $3p_y$, $3p_z$]–[3s, $3p_x$, $3p_y$, $3p_z$] [the (x, z)-plane projection]. The different plane decompositions are comparable to one another in the carrier density.

12.4 SIGNIFICANT MULTI-ORBITAL HYBRIDIZATIONS: CHARGE DENSITY DISTRIBUTION AND VAN HOVE SINGULARITIES

The spatial charge density distributions, which correspond to the isolated phosphorus and sulfur atoms (Figure 12.3 a and b) of a condensed-matter system (Figure 12.1a and b), are very useful in identifying the unusual sp^3–sp^3 chemical bonding of [3s, $3p_x$, $3p_y$, $3p_z$]–[3s, $3p_x$, $3p_y$, $3p_z$] multi-orbital hybridizations. The former cases present the almost spherical forms, in which there exist the heavy green/red region [the inner orbitals in [13]], the light red region [3s orbitals in [13]], and the yellow/light green region [($3p_x$, $3p_y$, $3p_z$) orbitals in [13]]. This is drastically changed by the very strong mixings of sp^3–sp^3 orbitals in P-S bonds (Figure 12.3 c and d); that is, carrier densities are bound together except for the inner region. The highly anisotropic feature will be directly reflected in all the fundamental properties, such as a rather large fluctuation of bond length in a Moire superlattice (Table 12.1), optical transitions [27], and Coulomb-field excitations [48]. Such behavior is quite different from of lithium-sulfur bonds [a uniform in bond length Table 11.1]. Their important difference might account for the experimental observations of the various isomer cases for the phosphorus sulfide compounds [P_4S_x of $x \leq 10$ in [12]]. The above-mentioned quasiparticle viewpoint is further supported by the merged van Hove singularities due to the sp^3–sp^3 eight orbitals.

The atom- and orbital-decomposed van Hove singularities, which are characterized by the zero group velocities on the constant-energy surfaces [30], are available in identifying the multi-orbital hybridizations of the active chemical bonds. Very interesting, Figure 12.4a–c present rich and unique quasiparticle behaviors. The vanishing density of states appears in band gap across the Fermi level, as well as certain energy spacings with the sufficiently wide widths. In addition, an indirect

TABLE 12.1

The Optimal Geometric Parameters and Lattice Symmetries of 3D Binary Phosphorus Sulfide [P_2S_5]

Lattice Constant

• $a = 9.679 Å$	• $\alpha = 92.894°$
• $b = 9.824 Å$	• $\beta = 109.712°$
• $c = 9.825 Å$	• $\gamma = 90°$

Structure symmetry

Hermann-Mauguin	$P\bar{1}$ [2]
Hall	-P 1
Point group	$\bar{1}$
Crystal system	triclinic

26 Li-S chemical bonds (1.7269–2.2979 Å)

Unit cell: 28 atoms (8P- 20S)

gap could be examined from the optical measurements of the threshold absorption frequency [27]. According to the characteristics of atom and orbital dominances, the total energy spectrum is classified into four ranges: (I) $E < -8.0\,eV$ for P- & S-3s codominance, (II) $-8.0\ eV \leq E \leq -3.50\,eV$ for S-[3s, $3p_x$, $3p_y$, $3p_z$] codominance, (III) $-8.0\ eV \leq E \leq -3.50\,eV$ for S-[$3p_x$, $3p_y$, $3p_z$] codominance, and (IV) 0 $eV \leq E \leq 5.0\,eV$ for P- and S-[$3p_x$, $3p_y$, $3p_z$] codominance. Only the sp^3–sp^3 bonding can illustrate the rich mergences of prominent structures, in which their specific energies are mainly determined by the ionization energies of 3s orbitals [the on-site Coulomb potentials in Hamiltonian of the tight-binding model; [29]] and the bonding strengths of the neighboring sp^3 orbitals [the orbital and length-dependent hopping integrals; [29]]. It will play a critical role in establishing the phenomeno-logical models under a unified quasiparticle framework [24,44,45]. The similar investigations could be generalized to the isomer cases of P_4S_x [$x \leq 10$], as well as the Li-, P-, and S-related ternary/quaternary/pentanary compounds [8], which are expected to display the diversified multi-orbital hybridizations within the specific energy ranges.

The prediction of van Hove singularities, which is done for a 3D binary P_2S_5 compound, cannot be directly verified by the high-resolution STS measurements [49,50]. The main reason is a negligible quantum tunneling current due to a thick sample. The similar results are expected to occur in the L-, P-, and S-related multi-component materials. Such limits will induce some troubles for fully understanding the quasiparticles of lithium-sulfur batteries. However, the direct STS examina-tions are very successful for the low-dimensional materials, such as the layered graphene systems [24], 1D carbon nanotubes [51], planar/cured graphene nanorib-bons [44], and C_{60}-related fullerenes [52]. Very interestingly, the main features of lattice symmetries and van Hove singularities are identified simultaneously in the delicate analyses [17,31]. Specifically, the joint van Hove singularities, which are associated with the occupied initial state and the unoccupied final state through

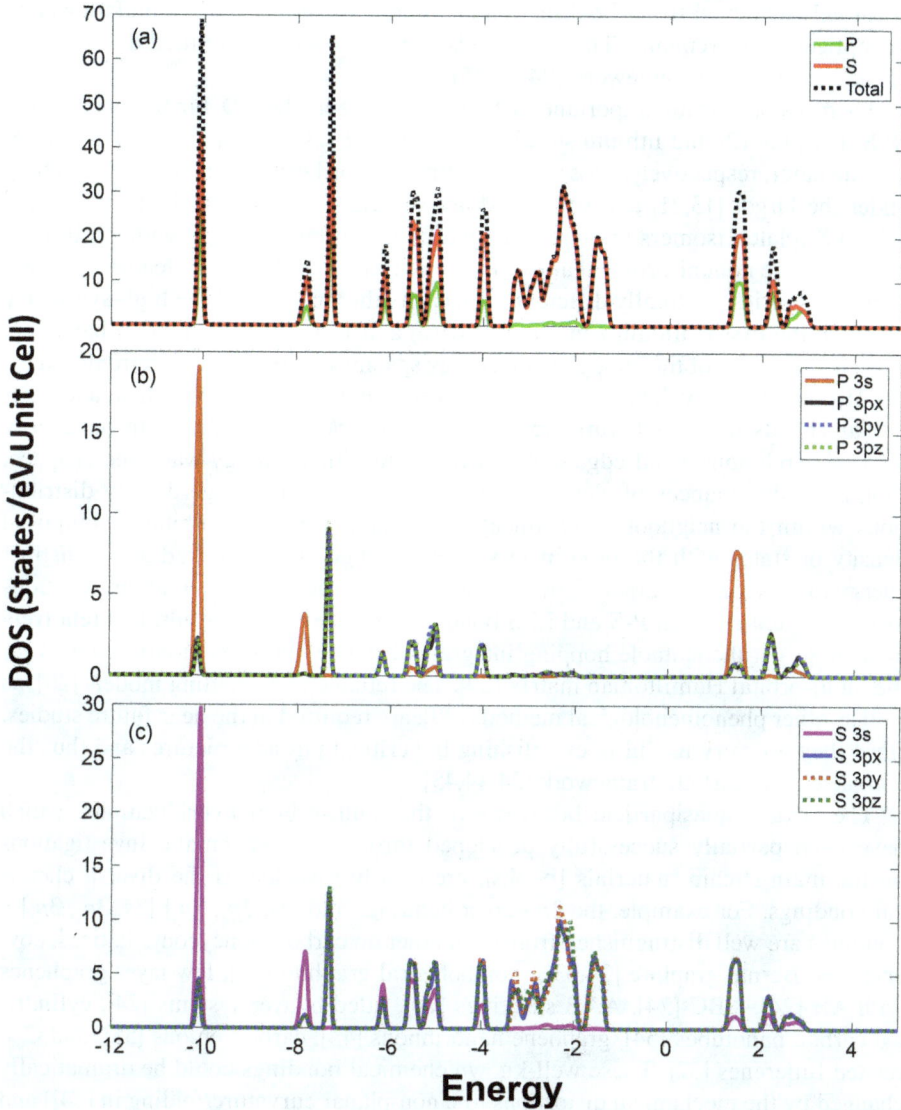

FIGURE 12.4   The energy-dependent density of states for P_2S_5: (a) the total contribution [the black curve], with the separation parts of phosphorus and sulfur parts [the red and blue curves], (b) P-[3s, $3p_x$, $3p_y$, $3p_z$]-projected parts [the distinct colors], and (c) S-[3s, $3p_x$, $3p_y$, $3p_z$]-decomposed ones [the different colors].

an external perturbation of an electromagnetic wave, could be verified from the high-precision measurements of optical reflectance/absorption/transmission spectroscopies [27]. The experimental analyses should be finished by taking very careful steps, since there are a lot of prominent absorption structures, as revealed in other electrolyte/cathode/anode materials [36]. The significant physical pictures, the active orbital hybridizations versus the strong optical excitations, would be accurately achieved from the direction combinations of the VASP simulations and the optical measurements. This strategy is consistent with the unified development of the quasiparticle framework [24,44,45].

There exist certain important differences between the 3D binary phosphorus [P_2S_5 Chapter 12] and lithium sulfide compounds [Li_2S in Chapter 11]. The former and the latter, respectively, possess the triclinic and orthorhombic lattice symmetries under the larger [13,21] and smaller Moire superlattices [21]. Furthermore, all the P- and S-related isomers are characterized by the formula of P_4S_x with $x \leq 10$ [10], creating more nonuniform and anisotropic environments. The main features of electronic properties are totally different from each other in terms of the high-symmetry points in the first Brillouin zone [Figs. 11.1(b) and 12.1(b)], magnitude of band gap, existence/absence of the enough wide energy spacings [without any electronic states; [13,21]], number of valence and conduction subbands, asymmetry of hole and electron and bands about the Fermi level, parabolic/linear/oscillatory/partially flat energy dispersion relations, band-edge states [critical points in an energy-wave-vector space; [30], atom dominances of electronic states, non-spherical charge density distributions within the neighboring chemical bonds, and atom- and orbital-decomposed density of states with the prominent van Hove singularities merged at the distinct energy ranges. Such characteristics are attributed to the multi-orbital hybridizations of sp^3–sp^3 and s–sp^3 in P-S and Li-S bonds, respectively. As a result, it is relatively difficult to get the suitable hopping integrals of the neighboring atomic orbitals for the multi-orbital Hamiltonian matrix [29]. The reliable tight-binding models [29], as well as other phenomenological methods [29], are required in the near-future studies, since they are very useful in establishing the critical physical pictures and thus the complete quasiparticle framework [24,44,45].

The critical quasiparticle behaviors of the multi-orbital hybridizations, which have been partially successfully developed through the systematic investigations on the main-stream materials [books], are clearly revealed as the diverse chemical bondings. For example, the $2p_z$–$2p_z$ π bondings and [3s, $3p_x$, $3p_y$]-[3s, $3p_x$, $3p_y$] σ bondings are well distinguished from each other on carbon-honeycomb lattices, covering the Bernal graphite [24,53], rhombohedral graphite [24], few-layer graphenes with AB [24], ABC [24], AAB stackings [24], titled bilayer systems [24], cylindrical carbon nanotubes [54], graphene nanoribbons [44], carbon onions [55], and C_{60}-related fullerenes [52]. These well-known chemical bondings could be dramatically changed by the mechanical distortions [the non-planar curvature/folding in [44]] and chemical modifications [adsorption in [56] and substitutions in [57]], i.e., the co-dominances of π and σ bondings, or the new chemical bondings [e.g., the s–sp^3 in the hydrogenated graphene systems with observable bucklings; [58]]. Concerning certain cathode/electrolyte/anode materials [38,59], their active orbital hybridizations are delicately identified in the previous works, such as the strong evidences of 2s-[2s, $2p_x$, $2p_y$, $2p_z$] and [d_{xy}, d_{yz}, d_{zx}, $d_{x^2-y^2}$, d_{z^2}]-[2s, $2p_x$, $2p_y$, $2p_z$] orbital hybridizations for LiTiO/LiFeO [Li- & Ti-/Fe-O bonds in [60–62]] and 2s-[2s, $2p_x$, $2p_y$, $2p_z$] & [3s, $3p_x$, $3p_y$, $3p_z$]-[2s, $2p_x$, $2p_y$, $2p_z$] orbital mixings for LiGeO/LiSiO [Li- and Ge- and Si-O bonds in [63]]. In short, the diversified chemical bondings are response for the rich and unique quasiparticle phenomena, which is a development strategy of quasiparticle framework [24,44,45].

12.5 CONCLUSIONS AND CHALLENGES

The first-principles delicate simulations and analyses, which are thoroughly conducted on the relevant physical and chemical quantities of the 3D binary P_2S_5 compound, have identified the sp^3–sp^3 multi-orbital hybridizations of $[3s, 3p_x, 3p_y, 3p_z]$-$[3s, 3p_x, 3p_y, 3p_z]$ in P-S chemical bonds. The featured geometric and electronic properties are revealed as the optimal geometry with a large Moire superlattice/a highly nonuniform chemical environment, the complicated atom dominances of phosphorus and sulfur atoms, a lot of asymmetric valence and conduction bands without the specific orbital hybridization, an indirect band gap of ~2.544 eV at the L Γ points, many band-edge states in the various energy dispersions, obvious deformations of charge density distributions close to two nearest-neighbors atoms [associated with $[3s, 3p_x, 3p_y, 3p_z]$ orbitals], and the rich emergences of sp^3-orbital-projected van Hove singularities within the specific energy ranges. The predicted crystal structure has been verified by the high-resolution X-ray techniques [14,35,64] and Raman scatterings [15]. However, the ARPES [25,26] and STS [26] measurements are absent up to now because of the intrinsic limits, leading to open issues in the near-future studies. The similar researches could be generalized to the isomer materials of P_4S_x [$x \leq 10$]. The important differences among them will cover all the essential properties [10] and even the potential applications [10], since their crystal symmetries are greatly diversified by the distinct ratios of phosphorus and sulfur atoms. These are under the systematic investigations. On the theoretical side, whether the phenomenological models could be developed under a unified quasiparticle framework deserves the closer examinations, e.g., the tight-binding model [29], the Kubo formula [65], and the random-phase approximation [66,67].

The quasiparticle properties can be greatly diversified by changing the core components of condensed-matter systems. Very interestingly, the specific ratio combinations of P_2S_5 and Li_2S binary subsystems have successfully generated the various ternary compounds through the various experimental synthesis methods, such as the very efficient solid-state route [68,69], mechano-chemical synthesis [68,69], and wet chemistry [68,69]. From the viewpoints of basic science researches, the systematic investigations on the essential properties and the highly potential applications are urgently requested for the Li-, P-, and S-related materials. These main-stream systems cover the mono-element crystals of Li [7], P [7] and S atoms [7], the binary Li_2S and P_2S_5, the ternary $Li_2P_2S_6$ [8], $Li_4P_2S_6$ [8], $Li_7P_3S_{11}$ [8], and $Li_3P_1S_4$ [8,41], and so on. The significant differences in component-enriched lattice symmetries will be responsible for the unusual physical/chemical/material phenomena [8]. The VASP simulations, but not the phenomenological models [17], could be utilized to achieve the optimal crystal structures under the full relaxation processes [17].

The close relations, which appear between the nonuniform and anisotropic chemical environment and the main features of electronic properties, could be thoroughly examined from the multi-atom-dominated band structures, the non-spherical charge density distributions, and the very complicated mergences of the orbital-decomposed van Hove singularities at the fixed energies [also discussed in Problem 18. 11]. As a result, all the active multi-orbital hybridizations in various sulfur-related 3D compounds are expected to be thoroughly settled in the near-future studies. This strategy is capable of examining the previous viewpoints of the quasiparticle framework [24,44,45].

REFERENCES

1. Mizuno F, Hayashi A, Tadanaga K and Tatsumisago M 2006 High lithium ion conducting glass-ceramics in the system Li2S–P2S5 *Solid State Ionics* **177** 2721–5.
2. Liu Z, Fu W, Payzant E A, Yu X, Wu Z, Dudney N J, Kiggans J, Hong K, Rondinone A J and Liang C 2013 Anomalous high ionic conductivity of nanoporous β-Li3PS4 *Journal of the American Chemical Society* **135** 975–8.
3. Phuc N H H, Totani M, Morikawa K, Muto H and Matsuda A 2016 Preparation of Li3PS4 solid electrolyte using ethyl acetate as synthetic medium *Solid State Ionics* **288** 240–3.
4. Seino Y, Nakagawa M, Senga M, Higuchi H, Takada K and Sasaki T 2015 Analysis of the structure and degree of crystallisation of 70Li 2 S–30P 2 S 5 glass ceramic *Journal of Materials Chemistry A* **3** 2756–61.
5. Aoki Y, Ogawa K, Nakagawa T, Hasegawa Y, Sakiyama Y, Kojima T and Tabuchi M 2017 Chemical and structural changes of 70Li2S-30P2S5 solid electrolyte during heat treatment *Solid State Ionics* **310** 50–5.
6. Busche M R, Weber D A, Schneider Y, Dietrich C, Wenzel S, Leichtweiss T, Schröder D, Zhang W, Weigand H and Walter D 2016 In situ monitoring of fast Li-ion conductor Li7P3S11 crystallization inside a hot-press setup *Chemistry of Materials* **28** 6152–65.
7. Anastas P T and Zimmerman J B 2019 The periodic table of the elements of green and sustainable chemistry *Green Chemistry* **21** 6545–66.
8. Kudu Ö U, Famprikis T, Fleutot B, Braida M-D, Le Mercier T, Islam M S and Masquelier C 2018 A review of structural properties and synthesis methods of solid electrolyte materials in the Li2S– P2S5 binary system *Journal of Power Sources* **407** 31–43.
9. Chu W, Calise F, Duić N, Østergaard P A, Vicidomini M and Wang Q 2020 Recent advances in technology, strategy and application of sustainable energy systems *Energies* **13** 5229.
10. Von Schnering H G and Hönle W 1988 Chemistry and structural chemistry of phosphides and polyphosphides. 48. Bridging chasms with polyphosphides *Chemical Reviews* **88** 243–73.
11. Nizamov I Y S, Al'metkina L A, Garifzyanova G G, Sergeenko G N G and Batyeva E S 1995 Reactions of phosphorus sulfides (P4S5, P4S7 and P4S10) and 2, 4-BIS (alkylthio)-2, 4-dithioxo-1, 3, 2λ5, 4λ5-dithiadiphosphetanes with dialkyldisulfides and thioacetals in the presence of iodine *Phosphorus, Sulfur, and Silicon and the Related Elements* **102** 71–81.
12. Jason M E 1997 Transfer of sulfur from arsenic and antimony sulfides to phosphorus sulfides. Rational syntheses of several less-common P4S n species *Inorganic Chemistry* **36** 2641–6.
13. Ozturk T, Ertas E and Mert O 2010 A Berzelius reagent, phosphorus decasulfide (P4S10), in organic syntheses *Chemical Reviews* **110** 3419–78.
14. Speakman S A 2013 Introduction to x-ray powder diffraction data analysis *Center for Materials Science and Engineering at MIT.*
15. Cantarero A 2015 Raman scattering applied to materials science *Procedia Materials Science* **9** 113–22.
16. Tahri Y, Chermette H and Hollinger G 1991 Electronic structures of phosphorus oxide P4O10 and phosphorus sulfides P4S10 and P4S7 *Journal of Electron Spectroscopy and Related Phenomena* **56** 51–69.
17. Hafner J 2008 Ab-initio simulations of materials using VASP: Density-functional theory and beyond *Journal of Computational Chemistry* **29** 2044–78.
18. Chiun-Yan Lin J-Y W, Chiu C-W and Lin M-F 2019 *Coulomb Excitations and Decays in Graphene-Related Systems*: CRC Press..

19. Lin C-Y, Ho C-H, Wu J-Y, Do T-N, Shih P-H, Lin S-Y and Lin M-F 2019 *Diverse Quantization Phenomena in Layered Materials*: CRC Press..
20. Chen S, Wu J, Lin C and Lin M 2016 Theory of magnetoelectric properties of 2d systems *New Journal of Physics* **18** 103024.
21. Yi Z, Su F, Huo L, Cui G, Zhang C, Han P, Dong N and Chen C 2020 New insights into Li2S2/Li2S adsorption on the graphene bearing single vacancy: A DFT study *Applied Surface Science* **503** 144446.
22. Zhang L, Sun D, Feng J, Cairns E J and Guo J 2017 Revealing the electrochemical charging mechanism of nanosized Li2S by in situ and operando X-ray absorption spectroscopy *Nano letters* **17** 5084–91.
23. Mu A, Cyvin B, Cyvin S, Pohl S and Krebs B 1976 Spectroscopic studies of As4O6, Sb4O6, P4S10, Ge4S104− and organometallic compounds containing the M4X6 cage. The Raman and ir spectrum of Ge4S104− *Spectrochimica Acta Part A: Molecular Spectroscopy* **32** 67–74.
24. Tran N T T, Lin S-Y, Lin C-Y and Lin M-F 2017 *Geometric and Electronic Properties of Graphene-Related Systems: Chemical Bonding Schemes*: CRC Press..
25. Damascelli A 2004 Probing the electronic structure of complex systems by ARPES *Physica Scripta* **2004** 61.
26. Bussolotti F, Chi D, Goh K J, Huang Y L and Wee A T 2020 *2D Semiconductor Materials and Devices*: Elsevier. pp 199–220.
27. Fox M 2001 *Optical Properties of Solids*: Oxford University Press..
28. Ertaş E, Öztürk T and Ösken I 2018 Purification of phosphorus decasulfide (P4S10). Google Patents).
29. Foulkes W M C and Haydock R 1989 Tight-binding models and density-functional theory *Physical Review B* **39** 12520.
30. Kittel C 1996 *Introduction to Solid State Physics*: John Wiley & Sons New York 402.
31. Lozhnikov V E, Mamonov A V, Borzilov V O, Mamonova M V, Prudnikov P V, Sorokin A A and Baksheev G G 2019 Estimating the performance of Ab initio calculation by VASP on openpower high performance system. In: *CEUR Workshop Proceedings*, pp 24–9.
32. Auffray N, Le Quang H and He Q-C 2013 Matrix representations for 3D strain-gradient elasticity *Journal of the Mechanics and Physics of Solids* **61** 1202–23.
33. Jain A, Ong S P, Hautier G, Chen W, Richards W D, Dacek S, Cholia S, Gunter D, Skinner D and Ceder G 2013 Commentary: The materials project: A materials genome approach to accelerating materials innovation *APL Materials* **1** 011002.
34. Li M, Liu X, Li Q, Jin Z, Wang W, Wang A, Huang Y and Yang Y 2020 P4S10 modified lithium anode for enhanced performance of lithium–sulfur batteries *Journal of Energy Chemistry* **41** 27–33.
35. Li J and Sun J L 2017 Application of X-ray diffraction and electron crystallography for solving complex structure problems *Accounts of Chemical Research* **50** 2737–45.
36. Li G, Chen Z and Lu J 2018 Lithium-sulfur batteries for commercial applications *Chem* **4** 3–7.
37. Ding B, Wang J, Fan Z, Chen S, Lin Q, Lu X, Dou H, Nanjundan A K, Gleb Y and Zhang X 2020 Solid-state lithium–sulfur batteries: Advances, challenges and perspectives *Materials Today*.
38. Kang W, Deng N, Ju J, Li Q, Wu D, Ma X, Li L, Naebe M and Cheng B 2016 A review of recent developments in rechargeable lithium–sulfur batteries *Nanoscale* **8** 16541–88.
39. Ould Ely T, Kamzabek D, Chakraborty D and Doherty M F 2018 Lithium–sulfur batteries: State of the art and future directions *ACS Applied Energy Materials* **1** 1783–814.

40. Barghamadi M, Best A S, Bhatt A I, Hollenkamp A F, Musameh M, Rees R J and Rüther T 2014 Lithium–sulfur batteries—the solution is in the electrolyte, but is the electrolyte a solution? *Energy & Environmental Science* **7** 3902–20.

41. Lepley N, Holzwarth N and Du Y A 2013 Structures, Li+ mobilities, and interfacial properties of solid electrolytes Li 3 PS 4 and Li 3 PO 4 from first principles *Physical Review B* **88** 104103.

42. Yang H, Naveed A, Li Q, Guo C, Chen J, Lei J, Yang J, Nuli Y and Wang J 2018 Lithium sulfur batteries with compatible electrolyte both for stable cathode and dendrite-free anode *Energy Storage Materials* **15** 299–307.

43. Choi J-W, Cheruvally G, Kim D-S, Ahn J-H, Kim K-W and Ahn H-J 2008 Rechargeable lithium/sulfur battery with liquid electrolytes containing toluene as additive *Journal of Power Sources* **183** 441–5.

44. Lin S Y T, N. T. T.; Chang, S. L.; Su, W. P.; Lin, M. F. 2018 *Structure- and Adatom-Enriched Essential Properties of Graphene Nanoribbons*: CRC Press..

45. Shih-Yang Lin H-Y L, Nguyen D K, Tran N T T, Pham H D, Chang S-L, Lin C-Y, Lin M-F 2020 *Silicene-Based Layered Materials: Essential Properties*: IOP Publishing..

46. Li X, Liang J, Lu Y, Hou Z, Cheng Q, Zhu Y and Qian Y 2017 Sulfur-rich phosphorus sulfide molecules for use in rechargeable lithium batteries *Angewandte Chemie International Edition* **56** 2937–41.

47. Jackson J D 1999 *Classical Electrodynamics*: American Association of Physics Teachers..

48. Mahan G D 2013 *Many-Particle Physics*: Springer Science & Business Media..

49. Kano S, Tada T and Majima Y 2015 Nanoparticle characterization based on STM and STS *Chemical Society Reviews* **44** 970–87.

50. Hipps K 2006 *Handbook of Applied Solid State Spectroscopy*: Springer. pp 305–50.

51. Lin M-F and Shung K W-K 1994 Plasmons and optical properties of carbon nanotubes *Physical Review B* **50** 17744.

52. Sanchís J, Milačič R, Zuliani T, Vidmar J, Abad E, Farré M and Barceló D 2018 Occurrence of C60 and related fullerenes in the Sava River under different hydrologic conditions *Science of the total environment* **643** 1108–16.

53. Lai Y, Ho J, Chang C and Lin M-F 2008 Magnetoelectronic properties of bilayer Bernal graphene *Physical Review B* **77** 085426.

54. Chen S, Chen P and Wang Y 2011 Carbon nanotubes grown in situ on graphene nanosheets as superior anodes for Li-ion batteries *Nanoscale* **3** 4323–9.

55. Tomita S, Sakurai T, Ohta H, Fujii M and Hayashi S 2001 Structure and electronic properties of carbon onions *The Journal of Chemical Physics* **114** 7477–82.

56. Tran N T T, Nguyen D K, Glukhova O E and Lin M-F 2017 Coverage-dependent essential properties of halogenated graphene: A DFT study *Scientific Reports* **7** 1–13.

57. Lv R, Li Q, Botello-Méndez A R, Hayashi T, Wang B, Berkdemir A, Hao Q, Elías A L, Cruz-Silva R and Gutiérrez H R 2012 Nitrogen-doped graphene: Beyond single substitution and enhanced molecular sensing *Scientific Reports* **2** 586.

58. Huang H-C, Lin S-Y, Wu C-L and Lin M-F 2016 Configuration-and concentration-dependent electronic properties of hydrogenated graphene *Carbon* **103** 84–93.

59. Tarascon J-M and Armand M 2011 *Materials for Sustainable Energy: A Collection of Peer-Reviewed Research and Review Articles from Nature Publishing Group*: World Scientific. pp 171–9.

60. Nguyen T D H, Pham H D, Lin S Y and Lin M F 2020 Featured properties of Li+-based battery anode: Li4Ti5O12 *RSC Advances* **10** 14071–9.

61. Buyukyazi M and Mathur S 2015 3D nanoarchitectures of a-LiFeO2 and alpha-LiFeO2/C nanofibers for high power lithium-ion batteries *Nano Energy* **13** 28–35.

62. Zhou F, Kang K, Maxisch T, Ceder G and Morgan D 2004 The electronic structure and band gap of LiFePO4 and LiMnPO4 *Solid State Commun* **132** 181–6.

63. Khuong Dien V, Thi Han N, Nguyen T D H, Huynh T M D, Pham H D and Lin M-F 2020 Geometric and electronic properties of Li2GeO3 *Frontiers in Materials* **7**.

64. Baltes H, Yacoby Y, Pindak R, Clarke R, Pfeiffer L and Berman L 1997 Measurement of the x-ray diffraction phase in a 2D crystal *Physical Review Letters* **79** 1285–8.

65. Nakayama T and Shima H 1998 Computing the Kubo formula for large systems *Physical Review E* **58** 3984.

66. Ren X, Rinke P, Joas C and Scheffler M 2012 Random-phase approximation and its applications in computational chemistry and materials science *Journal of Materials Science* **47** 7447–71.

67. Olsen T and Thygesen K S 2013 Random phase approximation applied to solids, molecules, and graphene-metal interfaces: From van der Waals to covalent bonding *Physical Review B* **87** 075111.

68. Rao C, Vivekchand S, Biswas K and Govindaraj A 2007 Synthesis of inorganic nanomaterials *Dalton Transactions* 3728–49.

69. Suslick K S 2001 *Encyclopedia of Physical Science and Technology Sonoluminescence and Sonochemistry*, 3rd edn. Elsevier Science Ltd, Massachusetts 1–20.

13 Cathode/Electrolyte Interface in High-Voltage Lithium-Ion Batteries
A First-Principles Study

Zhe-Yun Kee, Tai-Yu Pan, and Wen-Dung Hsu
National Cheng Kung University

CONTENTS

13.1 INTRODUCTION TO LITHIUM-ION BATTERY

Before any discussion is done, the fundamental structure and working principle of Li-ion battery (LIB) is introduced briefly for a better framework and clarity of the following content.

Conventional LIB consists of four major parts, namely, the cathode (positive electrode), anode (negative electrode), separator, and electrolyte system. The cathode can be categorized into three types based on the motion of freedom of lithium ion, specifically the 3D spinel, 2D layered, and 1D olivine structure. Furthermore, the anode can also be classified into three types by how they react with lithium ions, namely, by intercalation, conversion, and alloying reaction, with intercalation reaction (graphite) being commercially most successful. The separator plays an important role in facilitating ion exchange between the cathode and the anode, and simultaneously

DOI: 10.1201/9781003263807-13

preventing electrons flowing internally via the electrolyte. Lastly, the electrolyte system, which is a mixture of various solvent, additives, and lithium salt, provides an additional lithium ion to accelerate the transportation of lithium ion between the electrodes and stabilize the electrolyte/electrode interface.

The cyclability of rechargeable LIB arises from the transportation of lithium ion and electron between the electrodes. When the battery is depleted, the lithium ions remain in the cathode. Charging of battery is achieved by applying a reversed external potential difference to force the oxidation of transition metal (TM) ion present in the cathode, which releases electrons that flow to the anode through the external circuit. This, as a result, causes the lithium ion to intercalate into the anode and maintain charge neutrality for the anode. This step harvests the work done by the external potential difference and converts it into a chemical energy that is stored in the anode. During the discharge process, the battery is connected to an external load. The chemical potential difference between the electrodes results in spontaneous oxidation of lithium atoms, and electrons flow from the anode to the cathode, reducing the TM ions in the cathode and, at the same time, driving the external load. The lithium ions again exit the anode and intercalate into the cathode to maintain charge neutrality [1,2].

13.2   CATHODE MATERIAL

LIBs were invented in the 1970s, and the first prototype of commercial LIBs was developed by SONY in the 1990s [1,2]. One of the significant breakthroughs that might have helped SONY develop them was the discovery of intercalation compounds as cathode materials. Cathode materials used in LIBs are usually composed of the guest Li ion and the host matrix such as TM oxides, metal chalcogenides, and polyanion compounds [3]. Based on the crystal structure, TM oxide cathodes can be divided into three categories, namely, layered, spinel, and olivine. The layered cathodes show relatively high theoretical capacity but severely capacity fading in high-voltage operations, while the spinel cathode excels in operation voltage but still manifests rapid capacity fading in high-voltage operation.

A typical layered cathode is $LiCoO_2$ (LCO), which is the first and the most commercially used cathode material around the world. It shows a good rate capability and stable cycling performance when cycling below 4.2 V vs. Li/Li+, which delivers a reversible capacity of around 140 mAh g^{-1}. This indicates that only 50% Li ions of LCO are utilized. When the cycling voltage is ramped up to 4.5 V, severe capacity fading problems shorten the lifetime of batteries. To reduce the cost of cathode materials, the amount of Co is precisely controlled. Layered compounds $LiNiO_2$ and $LiMnO_2$ are potential candidates to replace LCO, but their performance cannot satisfy the demand of commercial application.

Ternary layered cathode $LiNi_xMn_yCo_{1-x-y}O_2$ (NMC) is another replacement for decreasing the usage amount of Co. By adjusting the ratio of Ni to Mn to Co, the electrochemical properties and thermal stability of NMC change significantly [4]. The commonly studied NMC cathodes include $LiNi0.3Mn0.3Co0.3O_2$ (NMC111), $LiNi0.5Mn0.3Co0.2O_2$ (NMC532), and $LiNi0.8Mn0.1Co0.1O_2$ (NMC811). Experimental results show that the increment of Ni ratio could enhance the battery capacity at the expense of thermal stability, while an increased Mn ratio can lower the production cost to the detriment of cyclability and overall performance.

Besides the most commercially successful layered cathodes, spinel $LiMn_{1.5}Ni_{0.5}O_4$ (LNMO) has been a promising cathode material for realizing high-voltage power batteries for electrical vehicles (EVs) or hybrid electrical vehicles (HEVs). It has high voltage when compared to Li/Li+, is cobalt-free, and has higher safety, making it a better option over other cathode materials in the application of EVs.

Determined by the synthesis method and heat treatment temperature, LNMO can have two different arrangements of Ni ions, namely, the disordered LNMO with $Fd\bar{3}m$ space group (both nickel and manganese ions located at the 16d octahedral sites with occupancy of 1:3) and the ordered LNMO with $P4_332$ space group (nickel ions located at the 4b octahedral sites, while manganese ions located at the 12d octahedral sites). Benefitting from its nature of the two-phase reaction, the ordered LNMO shows a stable charge/discharge plateau around 4.7 V, although this also indicates the involvement of phase nucleation, growth, and grain boundary motion, which impede the lithium inter/deintercalation and the possibility of grain pulverization. The disordered LNMO shows better structural stability and electrochemical properties owing to its partial solid-solution reaction (lithium content 0.5–1) at the cost of consistent charge/discharge plateau. The formation process of disordered LNMO induces oxygen loss, which causes part of the Mn^{4+} ions to reduce to Mn^{3+} in order to maintain charge neutrality, thus showing a two-step profile consisting of a minor plateau around 4.0V and a major plateau at 4.7 V, which corresponds to the Mn^{3+}/Mn^{4+} and Ni^{3+}/Ni^{4+} redox couples, respectively [5]. The presence of oxygen vacancies in the lattice may provide better Li ion migration path with reduced migration barrier, which facilitate the intercalation of Li ion and suppress the cracking of particle, resulting in the enhanced cyclability for the disordered LNMO.

Full LNMO-LTO cell study shows comparable coulombic efficiency and capacity in early-stage cycle for both ordered and disordered LNMO cathode materials under the same morphology [6], which suggests that the surface properties are similar in both cation ordering and disordering. This is supported by both density functional theory (DFT) calculation [7] and experimental observation [8], showing the tendency of cation reordering for the ordered phase in the surface region. Recent studies have also shown that the superior cyclability of disordered LNMO compared to the ordered phase may arise from the enhanced mechanical properties and Li ion migration path induced by the presence of oxygen vacancy. The combined conclusions of following pieces of literature imply that the mechanical properties and particle morphology may be much more crucial for a major breakthrough in the cyclability of LNMO cathode material.

13.3   GENERAL ISSUES

The common issue of layered cathodes is the fast capacity fading under high-voltage operation, which restricts the breakthrough of LIBs with higher energy density. Recently, one of the most accepted comments of the capacity fading problem is the gradual phase transformation in the cathode surface region. During the cycling process, the variation of Li concentration is large in the cathode surface region. In the Li-deficient region, the TM ions tend to leave their original sites and migrate into the tetrahedral sites or the octahedral sites in Li layers, which cause the phase transformation from layered structure to spinel or rock salt structure, especially in high-voltage operation. From the high-resolution transmission electron microscopy (HR-TEM) or high-angle annular dark field (HAADF) scanning transmission electron microscopy (STEM) images, the thickness of the phase-transformed region increases obviously after cycling. Those TM ions in the phase-transformed region are usually blamed for slowing down the Li diffusion efficiency and leading to the capacity fading [9–11].

For LNMO cathode material to be incorporated productively in high-operation voltage power batteries, there are two major problems to tackle: first, the rapid capacity fading and second, finding electrolyte systems that are compatible as well as both electrochemically and thermally stable under the desired operation voltage.

The rapid capacity fading of LNMO is primarily attributed to the two-phase reaction mechanism and the destabilization of surface during charging and discharging of batteries. The two-phase reaction involves phase nucleation and growth and grain boundary movement, which induce undesirable stress concentration between lithium-rich and lithium-deficit domains that result in grain pulverization. The surface instability causes the TM ions to either migrate into sites along the lithium diffusion path or dissolve into electrolyte, leading to the elevation of lithium inter/deintercalation resistance and the loss of active material.

Hunter's disproportionation has been the go-to-answer for a long time when referring to the TM dissolution mechanism in LMO-based batteries [12]. Therefore, the high TM dissolution rate has often been ascribed to acid attack under the inevitable presence of moisture and LiPF6 in the electrolyte system. In Hunter's theory, dissolution is promoted by the disproportionation of under-coordinated surface Mn (III) into tetravalent and divalent states. However, Leung has shown that even without the presence of water, dissolution of Mn into an organic electrolyte may still occur [13]. Meanwhile, the experimental result by Jang et al shows that the Mn dissolution rate was peaked at a state close to the fully delithiated state of LMO [14], which is in contradiction to Hunter's theory. As a result, Hunter's theory was thought to be far too simple to describe the TM dissolution mechanism effectively. In recent years, extensive studies have suggested that the TM dissolution is instead a complicated mechanism that involves phase transformation near the surface and subsurface regions, emanating from the surface instability of LMO-based material, which is facilitated by the adsorption of chemical species present in the electrolyte systems.

The surface instability of LNMO originates from the complete delithiation of the surface region during the terminal phase of charging, resulting in the decomposition of the parent phase accompanied by the oxygen loss and followed by the formation of new phase such as Mn_3O_4-like and rock salt structure. The migration of TM ions into the lithium tetrahedral site forms Mn_3O_4-like structure, while the migration of TM ions into the octahedral site results in rock salt structure under fully delithiated state.

13.4 CATHODE SURFACE

Pristine LNMO surface undergoes reconstruction during the formation process, forming a thin irreversible layer of rock salt structure on the subsurface region and Mn_3O_4-like structure on the surface region.. During the charging phase, both Mn and Ni ions have the chance to migrate into the tetrahedral sites forming an Mn_3O_4-like structure. The Ni ions have relatively low solubility in commercialized electrolyte system compared to Mn ions, as Ni ions accumulate near the surface region, ultimately resulting in the irreversible Ni-rich Mn_3O_4-like surface structure. This, to some extent, stabilizes the surface as it suppresses further TM dissolution and oxygen loss [5] due to its higher stability. On the contrary, the reversible appearance/disappearance of Mn_3O_4-like structure in the case of $LiMn_2O_4$ implies severe dissolution of tetrahedral Mn^{2+}, which is in agreement with the faster capacity degradation of LMO observed experimentally [15].

For the theoretical model to be established in our study, the surface structure of LNMO must be understood. The cross-reference of the experiment and first-principles calculation shows that the predominant surface of LNMO is (001) and (111). The first-principles calculation shows that Li-terminated (001) surface has the lowest surface energy out of all possible terminations of (001) and (111) surfaces. Despite that, some pieces of literature have pointed out that the reconstructed Li-terminated (111) surface, through the exchange of subsurface Li ions with the surface Mn ions resulting in a surficial inverse-spinel structure, has the lowest surface energy [16,17]. Some research has concluded that the (111) surface has better TM dissolution resistance, while the (001) surface has better Li ion kinetics evidenced by the preferential nucleation of delithiated phase at the (001) facets on both LMO and LNMO [17–19].

The TM dissolution of cathode material and decomposition of electrolyte have great dependency on the cathode/electrolyte interfacial characteristic as reported by some pieces of literature; for example, the adsorption modes of electrolytes, surface decorations, contaminants, and side products in electrolyte systems may alter the interfacial behavior. Correspondingly, as stated by Intan *et al* in their research, the presence of protonated surface oxygen, surface oxygen vacancy, fluoride ions in the electrolyte system, and strong adsorption energy of the electrolyte lower the energy barrier of TM dissolution [20]. Furthermore, they also identify the inverse linear relation between the TM dissolution barrier and adsorption energy of electrolytes, which provide the adsorption energy as a simple yet effective descriptor to characterize the propensity of TM dissolution under the influence of specific electrolyte system. Similarly, Yukihiro *et al* have done research regarding the influence of surface properties on the decomposition of EC molecules. They calculated the DOS for EC and LNMO (001) surface and found that the fermi level of LNMO surface lies between the lowest occupied molecular orbital (LUMO) and the highest occupied molecular orbital (HOMO) of EC under any state of charge (SoC). For spontaneous electron transfers to occur, which leads to the oxidation of electrolyte, the LUMO of EC must have an energy higher than the fermi level of LNMO surface. Therefore, they concluded that it is likely that the oxidation of electrolyte exhibits a catalytic mechanism associated with the electrolyte adsorption mode on the LNMO surface [21,22].

13.5 PARAMETERS FOR ASSESSMENT

In accordance with the frontier molecular orbital theory, K. Fukui et al have pointed out that the electrochemical window of the electrolytes is defined as the interval starting from LUMO and ending with HOMO. The reactivity of an electrolyte species toward the electrode can be predicted by comparing the fermi level of electrode (electrochemical potential) with the LUMO and HOMO of the electrolyte molecule. The LUMO and HOMO are usually referred to the electron orbitals of a molecule in the neutral state. However, the propensity for a reaction to occur, i.e., whether a molecule accepts or donates electrons, is determined by its electron affinity (EA) and ionization energy (IE), which involve structural changes. Therefore, EA and IE are considered more precise in describing the redox reaction than HOMO and LUMO as they take the structural changes into account. In other words, the relative position of IE/EA directly reflects the stability of the battery system when the electrode is in contact with the electrolyte.

Meanwhile on the cathode, fermi level and work function are the metrics that are commonly used to assess the stability of the cathode surface and are sometimes used interchangeably. When referring to the vacuum level, the work function of the exposed surface is defined as the energy difference from fermi level to the vacuum energy, which can be depicted as the minimum energy required to extract an electron from the system to the vacuum immediately outside the surface.

For an ideal molecule of an electrolyte, its IE value should be higher than the work function of cathode materials to prevent oxidation by cathode materials. For high-voltage batteries, the work function of cathode materials is lower than conventional ones; thus, a high-IE electrolyte molecule is needed.

13.6 CONVENTIONAL APPROACH AND SIMULATION

Conventionally, the development of electrolyte systems for a specific battery system is tedious, onerous, and unproductive at the same time. The most common development method is trial-and-error. Experiments are inevitable when the computation power and algorithm are not sufficient to compute such sophisticated interactions between molecules, not even atoms, and the process often involves massive investment. Before the availability of high-performance computers, the timespan of developing a new material, from laboratorial prototype to commercial application, ranged from few years to several tens of years. In recent years, high-performance computers have become readily available in most of the institutions and research centers, which has sped up the development of novel material tremendously. In addition to them, the first-principles calculation enables us to perform preliminary screening on candidate materials or material designs of our interest, and if it satisfies the criteria for our application, further calculations will be done in detail. After the initial screening process, the desired material design or novel material will be validated through experiment, and the whole process is guided by theoretical calculation, which prevents deviation or provides rectification to the course of development, thus enabling pinpoint targeting of material development.

The method that we use is based on DFT, in which the interactions among atoms and electrons are reformulated in terms of electron density, solving the Kohn-Sham equation for an approximate solution to many-body Schrodinger equations.

Most of the first-principles calculation can be categorized into three major parts, namely, structural optimization, static calculation, and molecular dynamics. Before any delicate calculation, the constructed models should go through structural optimization to ensure a reasonable model. Structural optimization algorithms relax the built model and seek to relieve any factitious traces induced during the process of model construction. Meanwhile, static calculation involves most of the property calculations under 0 K. After structural optimization, material properties such as mechanical properties, magnetic properties, and optical properties can be acquired. Further processing of the data even produces simulated XRD, IR, and Raman spectra with precise peak position.

The greatest difference between static calculation and Molecular Dynamic (MD) is the consideration of kinetics and temperature in MD, which are not considered in static calculations. First-principles molecular dynamics unlike classical MD, which uses semi-empirical force fields to calculate forces, employ forces obtained from DFT to evolve the system's dynamic in time instead, which typically provide higher accuracy than classical MD at the cost of elevated computational expense. MD provides us with the possible reaction path and the products under the condition that we investigate, combined with enhanced sampling methods to accelerate the whole process, such as applying holonomic constraint to a chosen reaction coordinate (slow-growth approach) or simply increase the environment temperature to make rare events observable. Despite their convenience, enhanced sampling methods must be used wisely to avoid the overestimation of kinetic-related parameters. After identifying the reaction path, climbing image nudged elastic band (CI-NEB) can be used to determine the reaction barrier and transition states.

13.7 OUR ADAPTED MODEL AND EXPERIENCE

In our work, both the static calculation and molecular dynamics approaches were applied to investigate the cathode/electrolyte interface. All calculations were performed using the following commercialized software packages: Vienna Ab-initio Simulation Package (VASP) [23–26], in which the DFT-based method is implemented, and Gaussian03, in which hybrid functionals are adopted. Gaussian03 [27] is mainly used to calculate the EA/IE of electrolyte molecules, while VASP is used to calculate the fermi level and DOS of LNMO and NMC surface models. The projector-augmented wave (PAW) potentials are applied with Hubbard U correction (DFT+U)-incorporated generalized gradient approximation (GGA) exchange-correlation functional.

13.7.1 LNMO

When discussing the cathode part of LIBs, only the oxidation of electrolyte is considered. The fermi levels of surface models could directly compare to the HOMO of

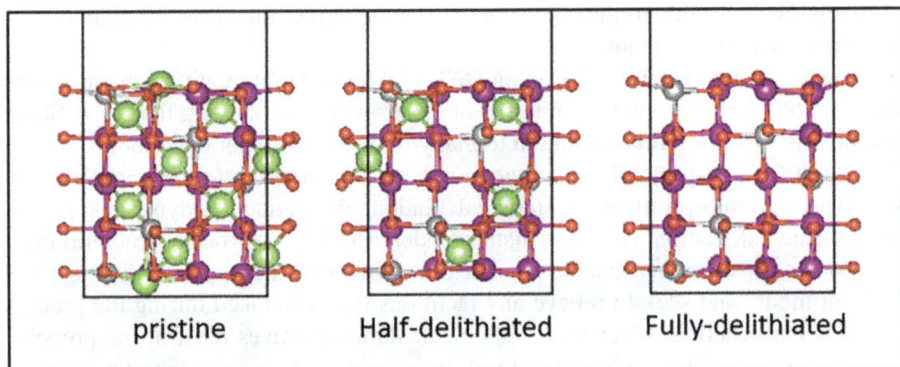

pristine   Half-delithiated   Fully-delithiated

FIGURE 13.1   Slab models corresponding to SoC 0, 50, and 100.

the electrolyte to validate whether spontaneous electron transfer from electrolyte to cathode is possible. For an electron to spontaneously transfer from the electrolyte to the cathode, the HOMO of electrolyte should have an energy level higher than the fermi level. However, there is a predicament comparing energy levels calculated using different computational software. Therefore, by subtracting the free energy of the molecule in the neutral state with the positively charged state, we get the IE of a specific electrolyte molecule – the energy required to extract an electron from the molecule. IE is a relative value rather than an absolute value, thus eliminating the potentially different reference state.

As pieces of literature have shown that the LNMO discharge profile with two distinctive plateaus corresponds to two separate "two-phase" reactions, the slab models that we built correspond to the three different possible phases within the reactions (Figure 13.1).

13.7.2 NMC

First-principles molecular dynamics simulation was applied to investigate the electrode/electrolyte interface between the (012) and (104) surface of NMC cathode and different electrolyte system EC, DMC, and ADM with 1M $LiPF_6$. NMC bulk and surface models with different Li concentrations were also introduced to mimic the charging process of cathode material. All the models were kept at 333 K for several picoseconds until they reached an equilibrium state (Figures 13.2 and 13.3).

13.8 PRELIMINARY RESULTS AND OUTLOOK

13.8.1 LNMO

The calculated IE and work functions of surface models presented show improbable spontaneous reaction upon contact of electrolyte with LNMO at any degree of lithiation (SoC). It may be inferred that the experimentally observed Cathode Electrolyte

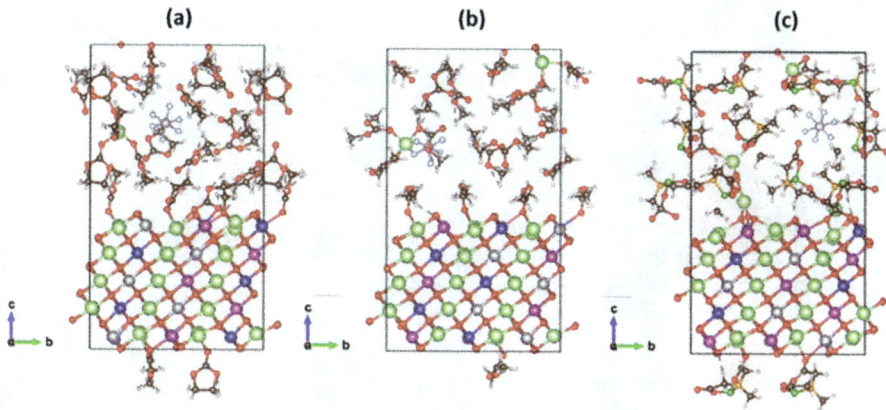

FIGURE 13.2 100%Li (012) combined models with different electrolyte systems (a) EC, (b) DMC, (c) ADM.

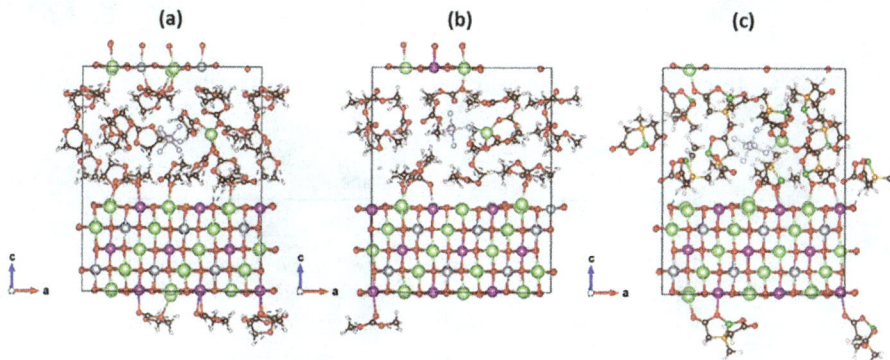

FIGURE 13.3 100%Li (104) combined models with different electrolyte systems (a) EC, (b) DMC, (c) ADM.

Interphase (CEI) formation and decomposition of electrolyte are most likely to be an adsorption-enhanced/facilitated reaction or that some prerequisite condition must be achieved for the initialization of electrolyte decomposition (Figure 13.4).

Further calculation of integrated DOS of surface models in Figure 13.5 indicate possible reaction induced by the accumulation of positive charge at the cathode. The mobility of electron flow through external circuit is very fast when compared to the inter/deintercalation of Li ion, resulting in the possibility of transient charge build-up that initiated the early-stage reaction of CEI formation. It requires about 10^{23} of positive charges that build up per cm^3 for the fermi level of cathode to be lower than the IE of EMC, which is within a reasonable range for a fast-charging device.

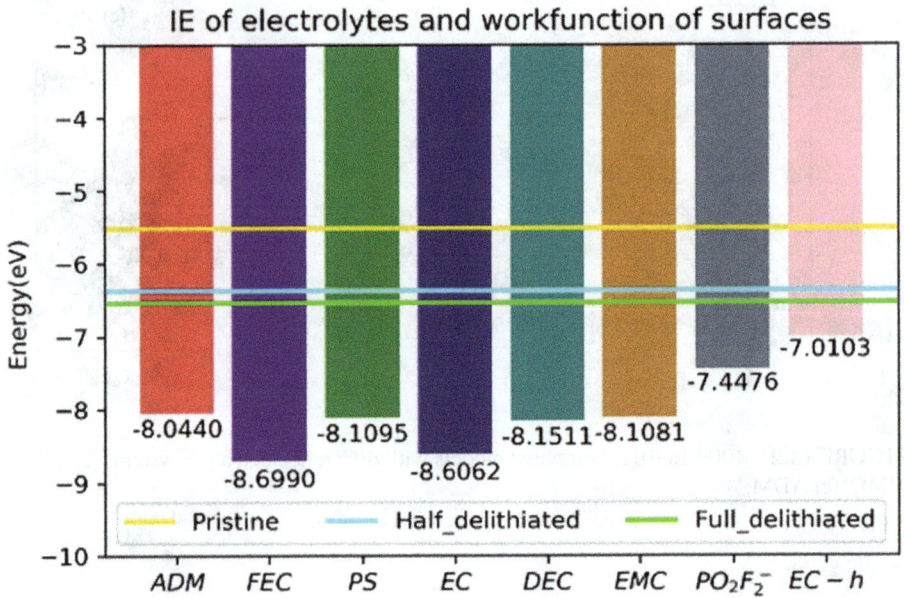

FIGURE 13.4   IE of electrolytes with work functions of surface models.

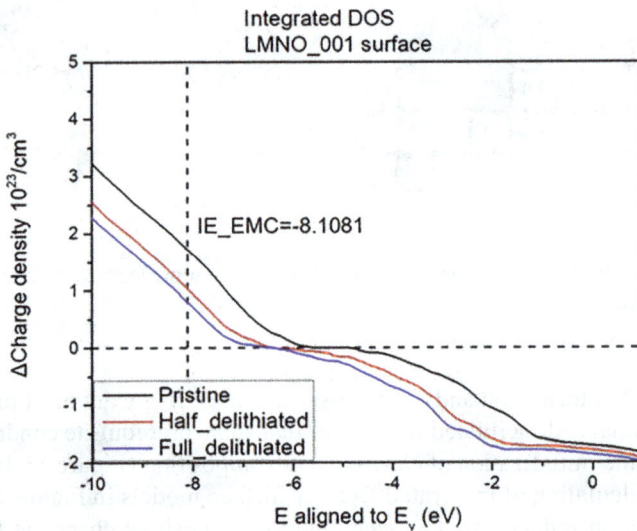

FIGURE 13.5   IE of electrolytes with fermi levels of surface models

13.8.2   NMC

Calculating the average bond length between the center metal ion and neighboring O in MO_6 octahedron (M = Ni, Co, Mn) in each model reveals that electrolyte could eliminate the bond length difference between outermost layers and the middle region of NMC surface models. Such circumstances might have the effect to stabilize the surface structure of NMC. The result reveals that the existence of electrolyte in the electrode/electrolyte interface will be a great benefit to stabilize the NMC surface.

Our calculation results indicate that the Ni-O bond length in NMC bulk is shortened when the Li concentration decreases. The Co-O bond is also shortened but to less extent. The Mn-O bond length seems to be no relation with Li concentration change. However, despite the 100% Li concentration of the surface model, the surface Ni ions on both (012) and (104) pure surfaces have much shorter Ni-O bond length than the Ni ions deep in the surface. This circumstance might be related to some structural deformation that occurs on the surface layer.

Our simulation results further show that the appearance or adsorption of electrolyte system decreases bond length difference between the surface and the middle region. It reveals that the electrolytes can help to eliminate the structural deformation and stabilize the surface structure of NMC. The construction of the electrode/electrolyte interface model between NMC surfaces and different electrolyte systems reveals that the surface deformation caused by Ni-O and Co-O bond length change tends to be alleviated by the presence of electrolytes.

ACKNOWLEDGMENTS

This work was financially supported by the Hierarchical Green-Energy Materials (Hi-GEM) Research Center, from The Featured Areas Research Center Program within the framework of the Higher Education Sprout Project by the Ministry of Education (MOE) and the Ministry of Science and Technology (MOST 109-2221-E-006-130-) in Taiwan.

REFERENCES

1. Whittinghamm S 1976 Electrical energy storage and intercalation chemistry *Science (80-.)* **192** 1126-7.
2. Mizushima K, Jones P C, Wiseman P J and GoodenoughJ B 1980 LixCoO2 (0<x<-1): A new cathode material for batteries of high energy density *Mater. Res. Bull.* **15** 783-9.
3. Nitta N, Wu F, Lee J T and Yushin G 2015 Li-ion battery materials: Present and future *Mater. Today* **18** 252-64.
4. Bak S-M, Hu E, Zhou Y, Yu X, Senanayake S D, Cho S-J, Kim K-B, Chung K Y, Yang X-Q and Nam K-W 2014 Structural changes and thermal stability of charged LiNixMnyCozO2 cathode materials studied by combined in situ time-resolved XRD and mass spectroscopy *ACS Appl. Mater. Interfaces* **6** 22594-601.
5. Liang G, Peterson V K, See K W, Guo Z and PangW K 2020 Developing high-voltage spinel LiNi 0.5 Mn 1.5 O 4 cathodes for high-energy-density lithium-ion batteries: Current achievements and future prospects *J. Mater. Chem. A* **8** 15373-98.
6. Aktekin B, Valvo M, Smith R I, Sørby M H, Lodi Marzano F, Zipprich W, Brandell D, Edström K and Brant W R 2019 Cation ordering and oxygen release in LiNi0.5-xMn1.5+ xO4-y (LNMO): In situ neutron diffraction and performance in Li ion full cells *ACS Appl. Energy Mater.* **2** 3323-35.
7. Lee E and Persson K A 2014 Erratum: First-principles study of the nano-scaling effect on the electrochemical behavior in LiNi0.5Mn1.5O4 (Nanotechnology (2013) 24 (424007)) *Nanotechnology* **25** 159501.

8. Kim J-H, Huq A, Chi M, Pieczonka N P W, Lee E, Bridges C A, Tessema M M, Manthiram A, Persson K A and Powell B R 2014 Integrated nano-domains of disordered and ordered spinel phases in LiNi 0.5 Mn 1.5 O 4 for Li-ion batteries *Chem. Mater.* **26** 4377–86.

9. Jung S-K, Gwon H, Hong J, Park K-Y, Seo D-H, Kim H, Hyun J, Yang W and Kang K 2014 Understanding the degradation mechanisms of LiNi0.5Co0.2Mn0.3O2 cathode material in lithium ion batteries *Adv. Energy Mater.* **4** 1300787.

10. Lin F, Markus I M, Nordlund D, Weng T-C, Asta M D, Xin H L and Doeff M M 2014 Surface reconstruction and chemical evolution of stoichiometric layered cathode materials for lithium-ion batteries *Nat. Commun.* **5** 3529.

11. Zheng J, Gu M, Xiao J, Zuo P, Wang C and Zhang J-G 2013 Corrosion/fragmentation of layered composite cathode and related capacity/voltage fading during cycling process *Nano Lett.* **13** 3824–30.

12. Hunter J C 1981 Preparation of a new crystal form of manganese dioxide: λ-MnO2 *J. Solid State Chem.* **39** 142–7.

13. Leung K 2017 First-principles modeling of Mn(II) migration above and dissolution from LixMn2O4 (001) surfaces *Chem. Mater.* **29** 2550–62.

14. Dong H, Jang, Shin Y J and Oh S M 1996 Dissolution of spinel oxides and capacily losses in 4 V *J. Electrochem. Soc.* **143** 2204–11.

15. Tang D, Sun Y, Yang Z, Ben L, Gu L and Huang X 2014 Surface structure evolution of LiMn2O4 cathode material upon charge/discharge *Chem. Mater.* **26** 3535–43.

16. Kim S, Aykol M and Wolverton C 2015 Surface phase diagram and stability of (001) and (111) LiM n2 O4 spinel oxides *Phys. Rev. B - Condens. Matter Mater. Phys.* **92** 115411.

17. Kim J S, Kim K, Cho W, Shin W H, Kanno R and Choi J W 2012 A truncated manganese spinel cathode for excellent power and lifetime in lithium-ion batteries *Nano Lett.* **12** 6358–65.

18. Liu H, Kloepsch R, Wang J, Winter M and Li J 2015 Truncated octahedral LiNi0.5Mn1.5O4 cathode material for ultralong-life lithium-ion battery: Positive (100) surfaces in high-voltage spinel system *J. Power Sources* **300** 430–7.

19. Li L, Chen-Wiegart Y C K, Wang J, Gao P, Ding Q, Yu Y S, Wang F, Cabana J, Wang J and Jin S 2015 Visualization of electrochemically driven solid-state phase transformations using operando hard X-ray spectro-imaging *Nat. Commun.* **6** 1–8.

20. Intan N N, Klyukin K and Alexandrov V 2019 Ab initio modeling of transition metal dissolution from the LiNi0.5Mn1.5O4 cathode *ACS Appl. Mater. Interfaces* **11** 20110–6.

21. Okuno Y, Ushirogata K, Sodeyama K, Shukri G and Tateyama Y 2019 Structures, electronic states, and reactions at interfaces between LiNi 0.5 Mn 1.5 O 4 cathode and ethylene carbonate electrolyte: A first-principles study *J. Phys. Chem. C* **123** 2267–77.

22. Leung K, Rosy and Noked M 2019 Anodic decomposition of surface films on high voltage spinel surfaces – Density function theory and experimental study *J. Chem. Phys.* **151**.

23. Kresse G and Hafner J 1993 Ab initio molecular dynamics for liquid metals *Phys. Rev. B* **47** 558–61.

24. Kresse G and Furthmüller J 1996 Efficient iterative schemes for ab initio total-energy calculations using a plane-wave basis set *Phys. Rev. B* **54** 11169–86.

25. Kresse G and Furthmüller J 1996 Efficiency of ab-initio total energy calculations for metals and semiconductors using a plane-wave basis set *Comput. Mater. Sci.* **6** 15–50.

26. Kresse G and Joubert D 1999 From ultrasoft pseudopotentials to the projector augmented-wave method *Phys. Rev. B* **59** 1758–75.

27. Frisch M J et al. 2016 G16_C01 Gaussian 16, Revision C.01, Gaussian, Inc., Wallin.

14 Electrode/Electrolyte Interfaces in Sodium-Ion Battery

Roles, Structure, and Properties

Tuyen T. T. Truong and Nhan T. Tran
Viet Nam National University - Ho
Chi Minh City (VNU HCM)
Ho Chi Minh University of Science (HCMUS)

Binh T. Tran
National Cheng Kung University

Man V. Tran and Phung M. L. Le
Viet Nam National University - Ho
Chi Minh City (VNU HCM)
Ho Chi Minh University of Science (HCMUS)

CONTENTS

DOI: 10.1201/9781003263807-14

14.1 INTRODUCTION

The increasing need for high-density energy storage capability is driving the hunt for new versatile storage systems that are alternatives to the previously successful lithium-ion technology. Among the accessible battery chemistries, sodium (Na)-ion batteries (SIBs) making use of more Earth-abundant elements and, possibly, renewable carbonaceous sources are becoming promising for "side-by-side" energy storage systems.

In recent years, significant research progress has been made on the development of electrode materials with high charge storage capacities and rate capability. However, the cycle life and safety of (SIBs) are still limited by the instability of the electrolyte when it comes to contact with both cathode and anode during the operation of sodium-ion batteries (SIBs). Indeed, like in LIBs, the formation of solid electrolyte interphase (SEI layer) originated from the decomposition of electrolyte and electrode dissolution is still poorly understood, and regulation strategies are still in their infancy in the context of SIBs [1]. For safety, such a surface film layer is essential, only allowing ionic transfer (e.g., Li^+ and Na^+) while preventing the transport of electrons and solvent molecules [2]. Unexpectedly, the interfacial reactions relating to the SEI layer mainly determine the battery function, particularly the reactivity of the electrode materials and the oxidation/reduction reactions that occur on the surface particles of the respective anode and cathode. Therefore, the SEI layer has important implications for the battery performance in terms of cycle life limitations, the capacity for reversibility, and safety. Hence, to improve Na-ion batteries (NABs), a fundamental aspect of the electrode-electrolyte interface should be explored [3,4]. Additionally, the major variations between the Na^+ and Li^+ ions in charge density and the degree of Lewis acidity lead to the large difference in solvation and desolvation capabilities that individually represent the primary induction of interphase formation and the initial activation of Na^+ ion diffusion through the interphase [5]. Therefore, the knowledge of Na-based interphase formation does not like Li SEI layer so it is essential to have an in-depth understanding of the functional of SIB interphase layer, its formation (component and structure), especially ion transport kinetic through SEI layer as the way with Li-ion SEI layer, which attracted more attention in the last decade [6,7].

In this review, from three key perspectives (Figure 14.1), we comprehensively review the current research status and the complicated issues of the as-generated interphases of the SIB. First, in Section 14.2.1, we focus on a fundamental

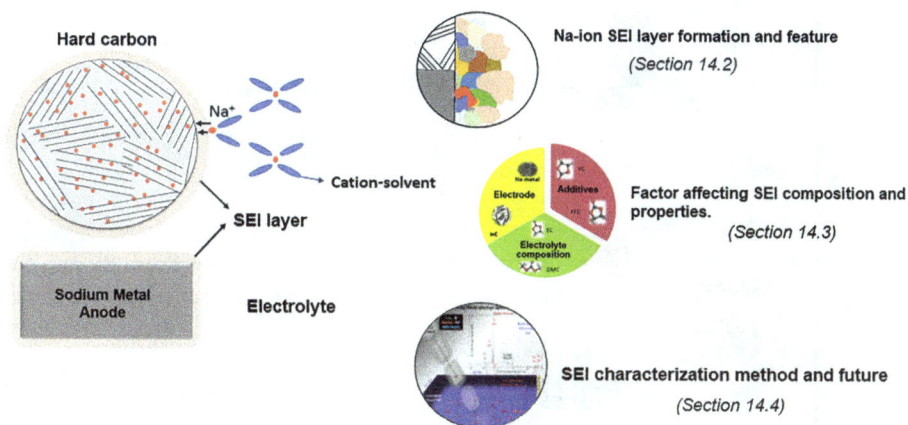

FIGURE 14.1 Schematic diagram of solid electrolyte interphase and the three main keys explored in this review.

understanding of the main body interphase layer, thermodynamic considerations of electrolyte reduction in SIBs, and their composition/structure, followed by a discussion of the current understanding of SEI and their influence on the battery performance in Section 14.2.2. In Section 14.3, we focus on finding factors affecting SEI composition (electrode types, electrolyte components, and additives) and discussed individual subsections, and the relevant components, morphology, and associated electrochemical properties are addressed. Then, we provided our understanding of the analysis techniques used to characterize the SEI-layer properties such as composition, thickness, and morphology (Section 14.4). Finally, we also provide the critical challenges for building stable Na-based interphase and the understanding of anode's SEI layer affecting cell design and SEI durability of SIBs. Additionally, we showed the perspectives of thinking about developing an artificial SEI layer to control high-performance and high-capability Na-ion rechargeable batteries in the next generation. Figure 14.1 illustrates the overall structure review about understanding the Na-ion SEI layer.

14.2 SEI FORMATION AND FEATURES

14.2.1 SEI FORMATION

In general, the presence of the SEI on the electrodes is the product of the thermodynamic balance between the energy levels of the electrode and those of the electrolyte, which provides the primary driving force for the generation of the SEI. Figure 14.2a shows the fundamental driving force derived from the energy-state differential between the electrode and the electrolyte. In the case of an anode with μ_A higher than the LUMO (Lowest Unoccupied Molecular Orbital) of the electrolyte, the electrons can spontaneously transfer from the anode to the electrolyte, leading to spontaneous reduction and the subsequent SEI generation [8]. Similarly, when the redox reaction

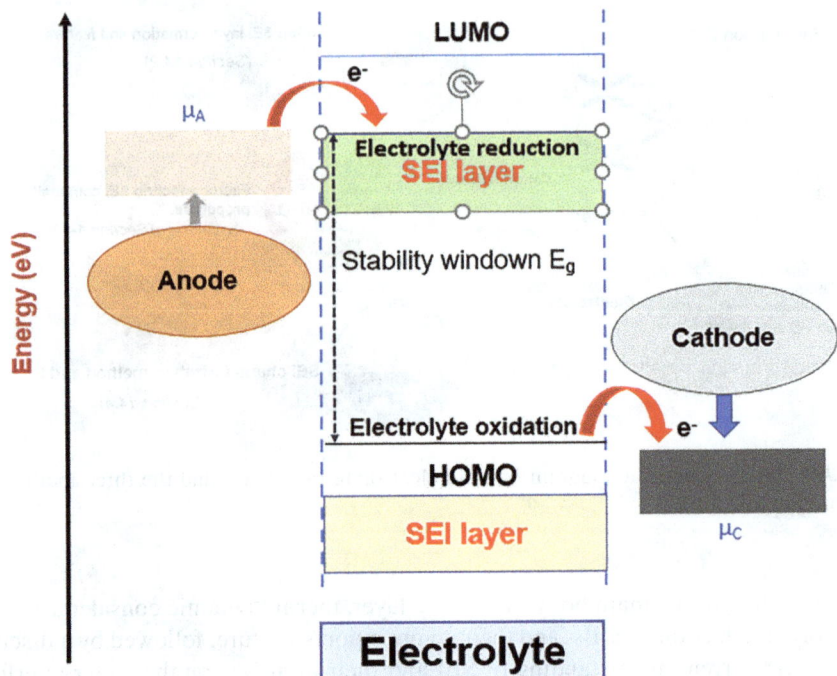

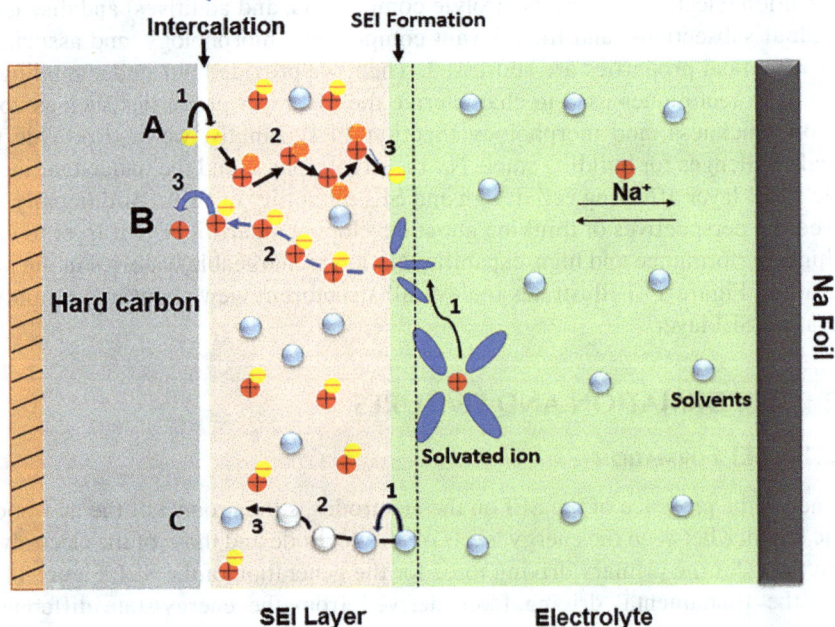

FIGURE 14.2 Formation Na-ion SEI-layer mechanism. (a) Energy of the formation of SEI layers under reduction and oxidation conditions. (b) Electrochemical kinetics of formation Na-ion SEI layers.

of the cathode has a lower μ_C than the HOMO (Highest Occupied Molecular Orbital) of the electrolyte (such as operating at a potential above 4.2 V), cathode-electrolyte interphase (CEI) occurs, which formed the solid electrolyte interphase layer on the cathode electrode. The reduction potential of organic solvents is below ~1 V vs. Na^+/Na on metal, carbon, and oxides, which means that the SEI layer was formed spontaneously below ~ 1 V vs. Na^+/Na [9]. Furthermore, the products decomposed from electrolyte oxidation retain on the cathode surface, forming a CEI passivation coating to prevent further electron transport, resulting in a reduction in the decomposition process. Specific knowledge of this field remains scarce so far, although more focus has recently been diverted to high-voltage cathode products, further stressing the importance of stable CEI-layer investigations [10].

On the other hand, the SEI-layer growth mechanism also depends on various phenomena of coupled electrodes and particle size, all of which depend on the potential and concentrations of lithium ions, solvent molecules, and electrons at a specific point in the domain [11]. Therefore, in the research of SEI-layer formation, shedding light on the influence of Na-ion transport kinetics on SEI growth plays an important role. The three-step dominant phenomena of the SEI-layer growth kinetics are represented in Figure 14.2b. In process A, the electron from carbon is integrated into the SEI layer (A-1), followed by electron conduction across the SEI layer (this stage happens after the development of the SEI layer in cycle 1) (A-2), and lastly by electron reactivity with the electrolyte to generate a new SEI layer (A-3). In this process, electrons migrate through the SEI sheet to meet the outer electrolyte interface, where they react with Na ion and electrolytes (solvent and salt) to form more SEI as a parasitic reaction that passes through the electrode. Process B depicts how ion Na^+ intercalated into carbon/electrode; before intercalation of sodium into hard carbon (B-3), the process of absorption/desolation of sodium ion from electrolyte and diffusion of Na^+ ion across the SEI layer has earlier occurred. The solvent incorporation and diffusion across the SEI layer are illustrated in Figure 14.2 C-1,2,3. The (de)intercalation reaction takes place at the electrode/SEI interface where an electron supplied from the electrode interacts with a Na^+ ion diffused from the electrolyte via the SEI. The electron transfer, Na intercalation into the electrode, and the mass transfer of reactants in the electrolyte are commonly agreed to be a fast process during early cycles, as demonstrated by the low interfacial impedance attributed to the SEI layer [12,13].

14.2.2 SEI STRUCTURE AND COMPONENTS

As the SEI components heavily depend on the electrode material, electrolyte salts, solvents, as well as the working state of the cell, no identical SEI layer can be found in two different circumstances. Therefore, an in-depth understanding of the structure and SEI composition is necessary to control the formation of the SEI layer as well as to control the performance of the SIB.

The principal components of the SEI layer on the electrode surface in organic electrolytes consist of inorganic (such as NaF and NaCl) and organic compounds. Organic SEI components are predicted to result from the degradation of the electrolyte solvents, i.e., from DEC (diethyl carbonate) and EC (ethylene carbonate) decomposition, for instance, sodium alkyl carbonates ($ROCO_3Na$), sodium carbonate

(Na_2CO_3), sodium alkoxides (RONa), ethylene oxide oligomers (such as polyethylene oxide, PEO, (CH_2–CH_2–O)$_n$), semi-carbonates, and sodium double alkyl carbonates (NEDC) [14]. Therefore, it is important to classify the solvent decomposition contributions to understand the nature and chemical composition of the SEI layer.

Figure 14.3 displays the C 1s synchrotron-based hard X-ray photoelectron spectroscopy (HAXPES) spectra of electrodes cycled 10 times in Na cells. Based on the sp^2 carbon intensity of hard carbon, all the photoelectron intensities were normalized to enable us to compare the relative intensity of each peak. The peaks assigned to the –CH_2–COO–, R-OK, –C (=O)–, and –OC(=O) O-bonds were seen by all electrodes. These bonds should be derived from alkyl carbonates and alkoxides, which are the decomposition products of the PC.

In addition, the spectra of hard carbon surfaces cycled in Na examined at a depth of ~1 nm with TOF-SIMS (time-of-flight-secondary ion mass spectroscopy) are depicted in Figure 14.3d. In the sodium half-cell, mainly inorganic fragments (e.g., Na_2O Na_2F, Na_2OH, Na_2CO_3, etc.) were confirmed in the TOF-SIMS spectra of the cycled electrode. These results emphasized that Na-alkoxides are predictably more soluble than their Li equivalents due to their lower Lewis acidity, and thus, the formation of PEO is more preferred in the case of Na-based electrolytes, leading to the deposition of more organic species on the surface of sodium-based hard carbon [15,16].

These observations are summarized in Figure 14.4 to illustrate the structure of the Na-ion SEI layer on a hard carbon anode. In the sodium cell, the inner layer was composed of both organic and inorganic species, while the outermost surface of the SEI was covered with inorganic species unlike the SEI formed in the Li cell. The result revealed that the surface chemistry was significantly different from that of the Li in Na cells.

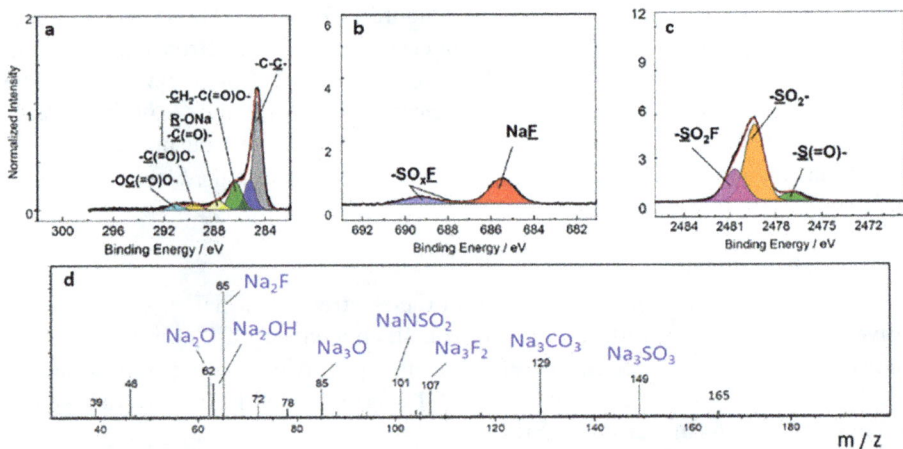

FIGURE 14.3 Na-ion SEI-layer component (a–c) HAXPES spectra of F1s for the hard carbon electrodes filled with 1M NaFSI in PC electrolyte; (d) TOF-SIMS spectra for the hard carbon after cycles in Na cells filled with1M NaFSI in PC electrolyte. (Reprinted with permission from Komaba et al. [15] Copyright 2020, American Chemical Society.)

Na cell

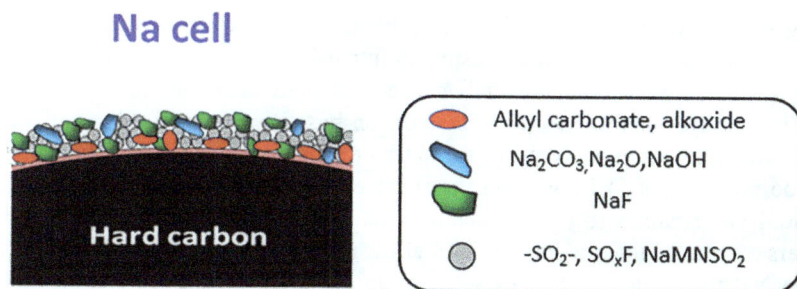

FIGURE 14.4 Schema depicts the structure and composition of the Na-ion SEI layer. (Reprinted with permission from Komaba et al. [15] Copyright 2020, American Chemical Society.)

14.2.3 INFLUENCE OF SEI ON THE BATTERY PERFORMANCE

The movement of ions is one of the key functions/properties of a layer SEI. Ions that transport through the SEI directly influence the reversibility, rate performance, and more fundamentally, the energy efficiency of the batteries. One of the main causes for battery loss is the high impedance from a dense SEI. Therefore, on the electrode surface, the SEI structure, morphology, thickness, and coverage have profound effects on the capacity of a cation to switch over. Since SEI is formed on the anode surface by electrolyte decomposition, its properties are highly dependent on the structure of the Na+ electrolyte and its chemical/electrochemical interactions with electrode materials. In this section, the influence of electrolyte selections on the Na+ transport in hard carbon SEI and hence the battery performance through regulating the SEI properties was evaluated [1,16,17].

The passivation layer composition and distribution may differ with the composition of the electrolyte (salt, solvent, and additive combination, etc.). For the best HC anode performance, there are only several studies on electrolyte optimization. For instance, early research by Komaba indicated that hard carbon EC/DEC in 1 M sodium hexafluorophosphate (NaPF$_6$) offered the best cycling stability among the typical single and binary electrolytes (consult Figure 2 from "Electrochemical Na Insertion and Solid Electrolyte Interphase for Hard-Carbon Electrodes and Application to Na-Ion Batteries" by S. Komaba et al., *Adv. Funct. Mater.* 21 3859–3867 [18]). This research shows that raising the DMC (dimethyl carbonate) content to 50% led to poor performance in Coulombic efficiency (CE) and low cycling power. Under the reduction environment, the less stable DMC could result in additional consumption of Na+. Hence, all the changes in the structure of electrolytes essentially affects the SEI properties and the related transport of ions within the SEI layer.

Furthermore, Na-ion transport and battery performance are also affected by the growth and structure of SEI. As described earlier, the persistent growth of dysfunctional SEI is responsible for growing cell impedance, decreasing ionic conductivity, absorbing electrolytes, and gradually failing cells. The Na/C half-cell fading is consistent with a synergistic effect of SEI formation and electrolyte degradation by Ji and colleagues [19]. Their further analysis found that SEI production and power fading were also linked to the kinetic problems of electrolyte depletion at the Na-metal counter electrode. This suggested an intrinsic electrolyte reactivity of the Na-metal electrode that would constantly cause electrolyte degradation.

Additionally, other injection and conversion-based anodes also witness the effect on Na^+ transport kinetics from SEI formation to HC electrodes. Early studies identified the presence of SEI layers on the surface of metal oxides and suggested their potential contribution to electrochemical behavior [20]. Munoz-Marquez and co-workers concluded that the formation of electrolyte decomposition by-products (SEI) during Na^+ insertion and their dissolution upon Na^+ extraction led to repeated SEI formation and unsustainable ion transport due to an increase in cell resistance associated with unstable SEI formation upon cycling [21].

From all of the observations mentioned above, it is inferred that the chemistry and morphology of these interfacial layers are important parameters that affect the efficiency of the battery. Hence, in addition to improving the composition of the electrolyte to enhance the bulk properties (e.g., ionic mobility, stabilization, etc.), the composition of the electrolyte influences the composition and consistency of the interphase layer(s) and hence has a severe effect on the cell performance.

14.3 FACTOR AFFECTING SEI COMPOSITION AND PROPERTIES

14.3.1 Type of Electrode (Carbonaceous Electrode, Sodium Metal, Copper Foil)

a. Sodium metal:

The sodium metal has been considered as a promising negative electrode with high specific capacity and high energy density batteries, which can be a benefit to sodium metal-derived batteries, including $Na-O_2$, Na-S, and $Na-CO_2$. However, there are drawbacks related to dendritic metal growth and unwanted parasitic reactions at the interphase of electrode/electrolyte. For enhanced cycling stability of sodium metal batteries, the SEI layer should have high ionic conductivity, high impermeability of electrolyte solvent, and high mechanical stability. In general, the SEI layer can be formed by the corrosion of the metal surface as well as the electrolyte decomposition during battery performance. Several efforts have been made in terms of the SEI formation mechanisms, chemistry, characteristics, and functions for understanding the SEI layers. Iermakova et al. [22] prepared symmetric cell Li/Li and Na/Na cell for comparison of stripping/plating processes in various electrolyte conditions, including 1 M $LiPF_6$ in $EC_{0.5}DMC_{0.5}$ (LP30) and 1 M $NaPF_6$ in $EC_{0.5}DMC_{0.5}$ and $EC_{0.45}PC_{0.45}DMC_{0.1}$. The results showed that Li/Li cell exhibits a smooth charging and discharging profile as well as low polarization at current densities of 0.1 and 1 mA cm^{-2}, whereas a large overpotential was observed for the Na/Na cells at 0.1 mA cm^{-2} without effects of the electrolytes nature. A stable SEI on the Li metal surface was confirmed by FTIR (Fourier transformation infrared spectroscopy) analysis, while a variation in terms of composition was found for the SEI on the Na electrode. In addition, the ether-based organic compounds are considered the most favored solvents for rechargeable batteries owing to their high electrochemical durability. Yi et al. [23] found that a high reversibility of the Na stripping/plating can be archived with Coulombic efficiencies of 99.9% over 300 cycles when using glymes (mono-, di-, and tetraglyme) as solvents for 1 M NaPF$_6$ electrolyte. Such high electrochemical performances were unable to achieve when using conventional carbonates solvents such as EC/DEC and EC/DMC. This could be attributed to a thinner formation of the SEI layer for the case of using diglyme as a solvent compared to those of using EC: DEC mixture in the electrolyte. The chemical composition of the SEIs was identified by X-ray photoelectron spectroscopy (XPS), and the diglyme solvent helps to generate a uniform mixture of Na$_2$O and NaF, increasing permeability toward the electrolyte, while EC: DEC shows a thicker mixture with both organic and inorganic compounds existing in the SEI, as shown in Figure 14.5a and b.

b. **Carbon-based electrode**:

For negative electrodes, hard carbon has been considered as the most practically viable material so far for SIBs. Several crucial parameters for choosing carbon as anode material are surface area, pore-size distribution, particle size and shape, the content of surface species, and impurities. Also, hard carbon contained many stacked carbon sheets, which have two kinds of edge sites, namely, armchair and zig-zag sites. Edge sites of carbon show much high reactivity compared to carbon in the basal planes, and hence, the physical and chemical properties of carbon vary from the basal plane to the edge-plane area [24]. Notably, the substrate has a much influence on SEI formation on carbonaceous materials compared to that of the electrolyte [25]. The high catalytic activity of the disturbed graphite structure in hard carbon favors the decomposition of the electrolyte. Similar to Li-ion batteries, the first lithiation capacity is generally higher than that of the first de-lithiation capacity due to the formation of the SEI, and this irreversible capacity loss (ICL) is termed as Q_{IR}. Fong et al. [26] first demonstrated a correlation of Q_{IR} with the capacity required for the SEI formation and surface area of the carbon electrode, in which a Na$_2$CO$_3$ film was formed and its thickness was calculated to be 45 ± 5 Å on the carbon particles. The component of SEI layers on the electrode surface is much dependent on the reduction reaction of salts, solvents, and impurities [27]. The products of reduction of salts anions could be NaF, NaCl, and Na$_2$O precipitating on the electrode surface, while reduction of solvent could form insoluble SEI components like Na$_2$CO$_3$, semi-carbonate, and polymers. Every parameter and property of the SEI significantly affects battery performance. ICL in the first cycle occurs as a result of salt and solvent reduction and hence is a characteristic of the SEI.

14.3.2 ELECTROLYTE COMPOSITION

The components of SEI are always contributed from the reduction and decomposition of the electrolyte. Therefore, various electrolytes induce different SEI layers. Among the Na salts in Na batteries, NaClO$_4$ has a strong oxidation ability for Na

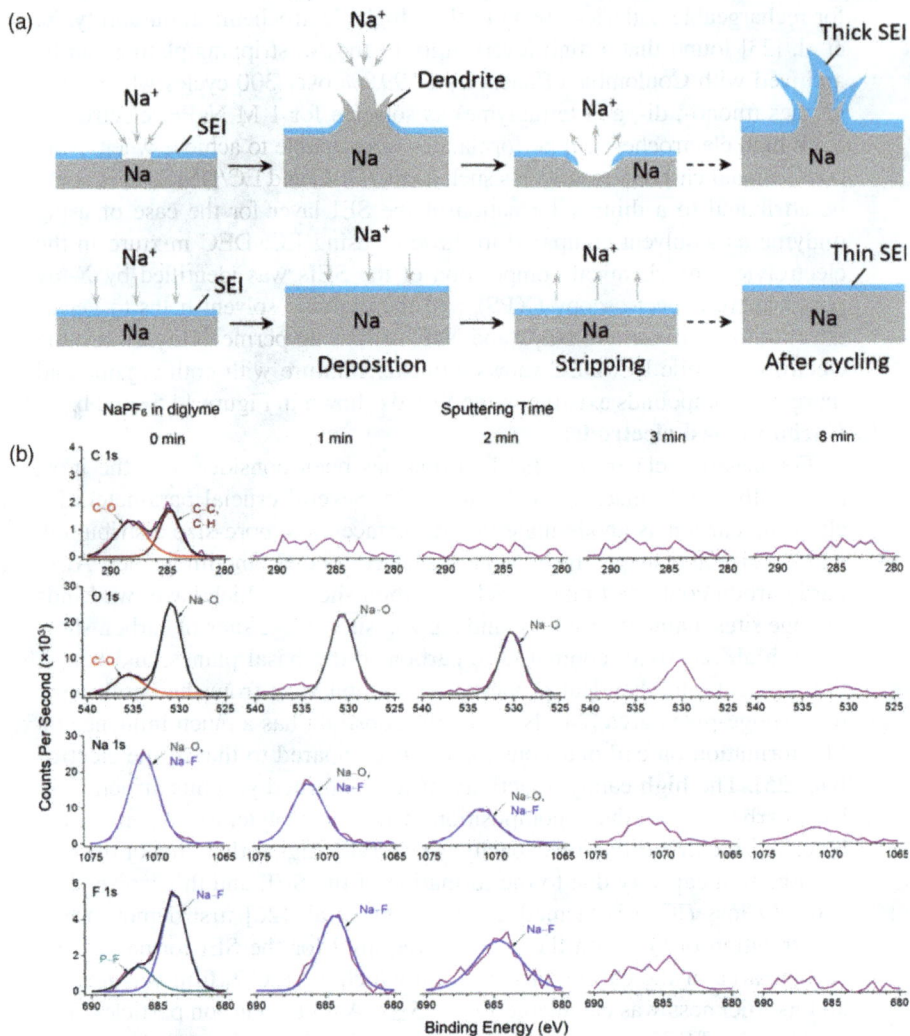

FIGURE 14.5 (a) Schematic illustration of stripping/plating process of Na-metal anode. (b) XPS analysis of SEI layers on Na-metal in diglyme and EC: DEC electrolytes. (Reproduced with permission [23].)

metal that renders low safety, while $NaPF_6$ exhibits poor thermostability with a few decompositions at 60°C–80°C [18,28–34]. When $NaPF_6$ was employed as the Na salt, NaF would take a great proportion in the components of SEI (Eqs. 14.1 and 14.2) [35]. As the inorganic NaX is thermodynamically stable, the induced SEI can be protected during the cycling.

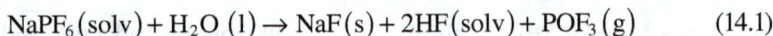

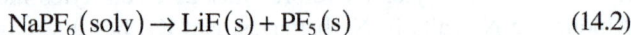

$$NaPF_6(solv) + H_2O\,(l) \rightarrow NaF(s) + 2HF(solv) + POF_3\,(g) \qquad (14.1)$$

$$NaPF_6(solv) \rightarrow LiF(s) + PF_5(s) \qquad (14.2)$$

TABLE 14.1

Literature XPS Data of the Main SEI Components Formed in Selected Na⁺-Based Electrolytes (Binding Energies in eV)

Core Level	Electrolyte Systems								Peak
	NaClO₄					NaPF₆			
	PC [18]	EC$_{0.45}$/PC$_{0.45}$/DMC$_{0.1}$ [28]	EC/DEC [29]	PC+FEC [30]	PC [31]	EC/DMC+FEC [32]	EC/PC [33]	PC+FEC [34]	
Na1s	-	-	-	1,071.8	-	-	1,071	1,073.5	NaF, NaCMC
C1s	290.0–290.5		288.5,	289.0,	291.0–	288.2,	-	288.5,	ROCO₂Na,
	286.8		289.8	290.2	291.5	290.0		290.0	Na₂CO₃,
	285.5		286.7	286.5	286.1–	286.0		286.2,	–(CH₂–O)–
			285.0	285.0	286.5	284.2		285.0	–CHₓ–
			290.9					290.8	PVDF
			284.3						Carbon additive
O1s	532.0	-	-	-	532.8,	-	-	531.5	Na₂CO₃,
	533.7				532.9				ROCO₂Na
					533.0				–(CO)–
					533.8				ONaClO₄
F1s	-	-	-	687.7	688.2	684.7	684.5	688.0	PVDF
				684.8	685.0		687.5–	684.8	NaF
							687.0		NaPF$_y$O$_z$
Cl2p	-		198.5	-	-	-	-	-	NaCl
			206.5						NaClO₃
			208.4						NaClO₄
P2p	-	-	-	-	-	-	136.0	-	NaPF$_y$O$_z$

Investigations on the formation of the passivation layer for Na⁺-based storage materials are still very limited, and the most frequently used electrolyte salts (i.e., NaClO₄ and NaPF₆) are summarized in Table 14.1. Major SEI components such as NaCO₃R, Na₂CO₃, PEO (CH₂–CH₂–O)n, NaF, and NaCl have been evidenced by several methods such as XPS, HAXPES, and TOF-SIMS, etc.

14.3.3 OTHER FACTORS

The SEI composition highly depended on the operation temperature [36]. Ishiikawa and co-workers [37] investigated the role of temperature on SEI layer and revealed that SEI layer formed at –20°C had compact and stable morphology and low impedance, leading to the best cycling performance. This is due to the dissolution of SEI and the insertion of solvents at high temperatures and consequently poor stability of SEI. Such a phenomenon was consistent with the results reported on graphite anodes. With the deepening of the research, opposite opinions were still held. The SEI layer can reform at high temperatures, and the dissolution and re-depositing lead to a more compact and stable structure. Additionally, the operation current density is also an important parameter of SEI formation. During the charging process, the formation of SEI competes with Na deposition onto the current collector. The surface electrochemical reactions require additional electrons. Therefore, both the SEI morphology and structure depend on the current density [38].

14.4   METHODS OF CHARACTERIZING SEI/CEI

As indispensable as in LIBs, interphases such as SEI/CEI are believed to be the key for the successful implementation of high-energy SIBs, in which electrodes operate outside the stability window of the electrolytes. Indeed, the SEI composition, morphology, and structure with which the battery performance is presumably associated are essentially explored and characterized to deeply understand the mechanism and set up guidelines for achieving the long-cycling performance batteries. Recently, the increasing studies in SEI in SIBs become more apparent that help researchers in the field to reach the optimized electrolyte recipe or design the stable electrode-electrolyte interphase by different approaches (artificial SEI, coating, additives, etc.).

Generally, SEI/CEI layers easily react with ambient CO_2 and H_2O to form inorganic sodium-containing compounds. Hence, washing the electrode in electrolyte solvents for analysis can easily introduce artifacts in the morphology and chemical composition of the SEI/CEI layer. Thus, for *ex situ* analysis, a transfer or encapsulating mechanism is required, which allows the sample to be transported from the glovebox (inert atmosphere) to the analysis chamber of any analytical machine without exposure to air and humidity. Some machines that operate under ultra-high-vacuum (UHV) or high vacuum-like XPS, scanning electron microscopy (SEM), and transmission electron microscopy (TEM) are even more difficult to design.

Therefore, it is adequate to select a characterization method to predict the interphase properly and precisely and to reveal its properties. Over the past decade, much progress has been made in developing and using particle and photon-based spectroscopy, scattering, and imaging techniques, and a new modeling framework for understanding and predicting the mechanism of interphase. Nowadays, a large variety of techniques have been used for analyzing SEI ranging from spectroscopy to microscopy to diffraction and thermal analysis. Since SEI is a surface phenomenon, the surface analysis techniques like XPS, AES, atomic force microscopy (AFM), SIMS, ToF-MS, and STM are most frequently used. SEM and TEM are used for imaging the morphology and surface film. Vibration spectroscopies like FTIR, IRAS, Raman, and XANES also provide a valuable surface information regarding the functionality. The traditional electrochemical methods like electrochemical impedance spectroscopy (EIS) and CV are used to study the evolution of the interphase resistance. Diffraction techniques like XRD are used for determining the ordering and structure of SEI. Thermoanalytical techniques like DSC, ARC, and TPD are used to study various reactions like SEI formation and degradation. And finally, the techniques like NMR and AAS give bulk information on SEI components [39].

These techniques along with the corresponding references are listed in Table 14.2.

Meanwhile, as an emerging method, theoretical studies and modeling already afford important insights into the SEI revolution of the working cell. Thus, a

TABLE 14.2
Advanced Techniques Used for SEI/CEI Analysis in Different Aspects

Analysis	Tools/Techniques	Advantages
Structural and microstructural techniques	• FTIR spectroscopy [1,40] • Raman spectroscopy [41]	• Simple, fast, and easy sample preparation. • Detection of a functional group on the electrode surface. • *In situ* and *ex situ* techniques were developed
Core and high depth profile of the structure	• XPS (X-ray photoelectron) [42] • AES (Auger electron spectroscopy) [43,44] • AFM (atomic force microscopy) [45,46] • TOF-SIMS (time-of-flight-secondary ion mass spectroscopy) [47] • EELS (electron energy loss spectroscopy) [48] • XANES (X-ray absorption near-edge structure) [49]	• Detailed information on energy binding linking to the molecular binding environment and the composition existent on few interfacial layers
Morphology, structure, thickness	• XRD (X-ray diffraction) [50] • STM (scanning tunneling microscopy) [51] • SPM (scanning probe microscopy) [52] • SEM (scanning electron microscopy) [53] • TEM (transmission electron microscopy) [54]	• High-resolution image of SEI formation on the metallic electrode or other electrodes. • Mapping technique for evaluating the homogeneity of elemental distribution • Atomic-level scale • Accurate evaluation of SEI thickness • Cryoprotection has been developed to prevent sample damage under a high dose of the beamline.
Formation of SEI and thermal degradation	• DSC (differential scanning calorimetry) [55,56] • ARC (accelerated rate calorimetry) [57,58] • TPD (temperature-programmed desorption) [59,60]	• Supplementary information on SEI formation during the cycling process • Thermal properties
Bulk information of SEI layer	• NMR (nuclear magnetic resonance) [61] • AAS (atomic absorption spectroscopy) [40] • EQCM (electrochemical quartz crystal microbalance) [62]	• Accurate quantitative or semi-quantitative of elemental content in SEI layer
Interphase resistance	• Electrochemical impedance spectroscopy (EIS) [39,41]	• Complex impedance in all components of cell • *In situ* and *ex situ* methods • Impedance at low frequency represents the Li-ion transfer resistance through SEI

combination of the Monte Carlo (MC) and molecular dynamic (MD) methods enables simultaneous simulations of many complex chemical reactions associated with the formation of the SEI film over a wide scan.

14.4.1 Surface Chemistry

14.4.1.1 Fourier Transformation Infrared Spectroscopy (FTIR)

FTIR spectroscopy is a fast and versatile tool for the analysis of the chemical nature of a substance once the sample could absorb infrared radiation (IR). Beams of wavelength in the range 100 and 1 μm interact with the material during the experiment. When IR is passed through a sample, the molecules within the sample will absorb some of the energy and start to vibrate/or rotate. By detecting the transmitted IR of the sample, it differentiates between different functional groups based on their dipole moment. Thus, the vibrational energy of various bonds is characteristic of the corresponding functional group, and it is possible to determine the molecular composition of the sample in question. However, not all vibrational modes are IR active, as a change in dipole moment is essential. Typically, NaF or LiF is therefore one of the components not visible in FTIR.

In IR experiments, UHV conditions are not mandatory for these instruments, thus making FTIR spectroscopy comparatively cheap and adapt quickly to our manipulation, while the other techniques need UHV due to the use of high-energy sources like XPS. IR spectroscopy could be used in various modes like attenuated total reflectance (ATR), subtractively normalized interfacial Fourier transform infrared (SNIFTIR), transmission infrared, near-normal incidence reflectance, grazing incidence reflectance, double modulation infrared, and reflection absorption infrared.

There has been an uncountable publication on analyzing FTIR spectra of the SEI on Na (Li), carbonaceous materials, silicon anode, etc. In particular, the various components of the SEIs have a critical impact on the electrochemical performance of graphite materials [7], which is revealed by FTIR spectroscopy. In the carbonate- or ether-based electrolyte, several bands appear after a full discharge. The assignments of all the bands of the graphite interphase in different electrolytes are listed in Table 14.1. Typically, the peaks in the range of 2,950–2,800 cm^{-1} are divided into the stretching signals of CH_3 and CH_2 (consult Figure 5a of Ref. [63]) [64]. The symmetrical stretching vibration of CH_3 signals at 2,881 cm^{-1} in the ether-based electrolyte is stronger than that in the carbonate-based electrolyte. The bands located at 2,943 and 2,824 cm^{-1} are the asymmetrical and symmetrical stretching vibrations of CH_2, respectively. The molecular structures of the products are shown in Figure 5d, e of Ref. [63]. In addition, the contents of CH_3 and CH_2 in the ether-derived SEI are higher and lower, respectively, than those in the carbonate-derived SEI. The stretching vibration of CO_3^{2-} peak is located at 1,765 cm^{-1} in the carbonate-derived SEI (Figure 5b of Ref. [63]) [65–67] that comes from the decomposition of EC and corresponds to the formation of $(CH_2OCO_2Na)_2$ and $CH_3CH_2OCO_2Na$ (Figure 5d of Ref. [63]). The bending modes of CO_3^{2-} in the sodium carbonate salts are located at 671–791 cm^{-1} (Figure 5c of Ref. [63]). However, there have no significant peaks characteristic of CO_3^{2-} in the wavenumber range of 671–791 cm^{-1} for the ether-derived

TABLE 14.3

FTIR Vibration Modes and the Correspondent Peak Positions for the Carbonate-/Ether-Based Electrolytes and Carbonate-/Ether-Derived SEIs

Vibration Modes	Peak Positions (cm⁻¹)				
	$ROCO_2Na$	RCH_2ONa	Diglyme	EC/DEC	$NaPF_6$
CH_3 stretch	2,885	2,881	2,877, 2,922	2,877, 2,934	
CH_3 stretch	2,860, 2,926	2,943, 2,824	2,820, 2,980	2,991	
R–C=O stretch	1,765			1,745, 1,712, 1,806, 1,858, 1,961	
CH_3, CH_2 bend	1,451, 1,583	1,462, 1,586	1,460	1,483, 1,552	
C–O stretch	1,021, 1,084, 1,195	1,024, 1,105, 1,203	1,116, 1,202	1,077, 1,163	
C–C stretch	853	850	854	854	
C=O bend	671, 736, 791			717,7 74	
					561

SEI. The bending modes of CH_3 and CH_2 are located at 1,451–1,586 cm⁻¹. The stretching modes of C–O and C–C are found at 1,021–1,203 and 850–853 cm⁻¹. Besides, the peaks at 561 cm⁻¹ are ascribed to the salt of $LiPF_6$ (Table 14.3), which are observed in both carbonate- and ether-based electrolytes. The co-intercalation of the solvated Na ions is also detected in the ether-based electrolytes, about 1,143 cm⁻¹ (corresponding to the feature mode of diglyme). Based on these characteristic absorption bands of the graphite interphase, the major components of the SEIs formed in carbonate- and ether-based electrolytes are carbonate salts and sodium alkoxides, respectively.

However, through these works, FTIR spectroscopy demonstrates some limitations in detecting SEI formation and composition. As SEI is a thin surface layer with an inhomogeneous composition that may not give a very strong vibration signal, a mapping analysis would be needed for distinguishing between the components and their distribution. Moreover, the various components of the SEI consisting of similar functional groups, mainly carbonyls, alkoxides, C–H vibrations, etc. have overlapping vibration signals. As a consequence, the spectra interpretation is tricky as it is difficult to distinguish between the various components.

14.4.1.2 X-Ray Photoelectron Spectroscopy (XPS)

XPS is the most widely accepted technology to probe the sensitive surface chemistry on different kinds of materials. In XPS, a sample is irradiated with X-rays of high energy (≈1,200–1,500 eV), while the kinetic energy and the number of electrons escaping the surface of the material are measured simultaneously. Based on the quantum state of escaped electrons, their characteristic energies possibly allow the analysis of all elements (except H and He). The elements can be accurately detected when presented in concentrations of >0.1 atomic percentage and the outermost 10 nm of the surface [68–70]. Therefore, XPS is one of the best operations for exploring the composition and evolution of SEI layers [31]. It is a semi-quantitative technique with an error in the range ±10% and the resolution of the XPS being ~0.1 eV. Combined with argon-ion sputtering technology, the XPS spectra of the SEI layer can be collected at different depths (several hundreds of nm) to reveal the uneven distribution of the SEI components [71].

Both FTIR and XPS techniques can provide a facile identification of the functional groups of organic components and types of bonds on the electrode surfaces from the locations and strengths of the peaks in spectra. However, the FTIR technique is nondestructive, while XPS may be performed under destructive mode on the electrode surface (depth-analysis profile). There are significant research results that have been reported by employing XPS techniques. Tables 14.4 and 14.5 summarize typical XPS data of the SEI components characterized based on the hard carbon anode cycling in the ether-based electrolyte [63] and carbonate-based electrolyte [72].

Regardless of the binding energies, Cls spectrum can be divided into three parts with different energy levels: 284.4 eV (C–C); 286.7 eV (C–O), and 289.2 eV (O–C(=O)–O). The energy binding of sp^2C [73] consists of sodium alkoxides (RCH_2ONa) and carbonate (Na_2CO_3) [74]. The Ols spectrum reveals mainly the products of sodium alkoxides (RCH_2ONa) or $CH_3CH_2OCO_2Na$ or Na_2CO_3 [39,74,75]. The F 1s and Na 1s spectra show peaks at ≈684.3 eV (Na – F) and ≈1072.1 eV (Na –O)/Na-F) or other sodium compounds (Na_2CO_3, RCH_2ONa) [73,74,76,77]. The peak of the F 1s spectrum at 687.4 eV (P – F) corresponds to the surplus of Na_xPF_y [23] (Table 14.5).

It is worth noting that depth-profile analysis using Ar-ion beam sputtering helps to estimate approximatively the atomic concentration at certain atomic levels, which

TABLE 14.4

XPS Binding Energy Values and Peak Assignments of the SEI Components Formed on the Interphase Using 1 M $NaPF_6$ in Diglyme

$NaPF_6$- Diglyme	Binding Energy (eV)	FWHM (eV)	Lorentzian -Gaussian (%)	Binding Energy (eV)	FWHM (eV)	Lorentzian -Gaussian (%)	Peak Assignment	Species
	0s			**50s**				
Cls	284.8	1.9	20	284.8	1.8	20	C–C	SP^2-C
	286.7	1.9	28	286.8	2.3	20	C–O	RCH_2ONa
	289.2	1.7	20	289.2	2	25	O–C(=O)O	Na_2CO_3
Ols	531	1.8	15	531	1.94	30	C=O (inorganic)	Na_2CO_3
	533	1.6	20	533	1.8	20	C–O	RCH_2ONa
Fls	684.3	2	28	684.1	2	30	Na-F	NaF
	687.4	2.1	25	687.5	2.2	27	P-F	Na_xPF_y
Nals	1,072.2	1.73	22	1,072.4	2.51	23	Na-O/Na-F	NaF RCH_2ONa Na_2CO_3

TABLE 14.5

XPS Binding Energy Values and Peak Assignments of the SEI Components Formed on the Interphase Using 1 M NaPF$_6$/EC-DEC (1:1)

NaPF$_6$. EC/ DEC	Binding Energy (eV) [0s]	FWHM (eV)	Lorentzian -Gaussian (%)	Binding Energy (eV) [50s]	FWHM (eV)	Lorentzian -Gaussian (%)	Peak Assignment	Species
	284.8	1.75	20	284.8	1.6	10	C–C	SP^2-C
C1s	286.7	2.2	30	286.5	2	30	C–O	ROCO$_2$Na
	290.1	2	25	289.2	1.8	20	O–C(=O)	ROCO$_2$Na
							O	Na$_2$CO$_3$
O1s	530.9	1.5	20	530.9	1.56	17	C=O (inorganic)	Na$_2$CO$_3$
	531.7	1.45	15	531.8	1.5	20	C=O	RCH$_2$ONa
	533	1.4	15	533	1.8	20	C-O	ROCO$_2$Na
	536.0	2	20	537.1	2	20		Na Auger
F1s	684	2	20	684.5	1.72	18	Na-F	NaF
	687.2	1.9	21	687.7	2.02	22	P-F	Na$_x$PF$_y$
Na1s	1071.8	2.03	20	1,071.9	1.75	24	Na-O/	NaF
							Na-F	RCH$_2$ONa
								Na$_2$CO$_3$

is also meaningful for element analysis or determining the surface layer thickness. Consulting Figure 6 of Ref. [63], after 50 s of sputtering, the upper surface is eliminated, so that the peaks of C and O representing the organic component become weak and vague. However, the peak that represents Na– O and NaF remains strong and sharp after 50 s and 100 s of sputtering, respectively. These results imply that the organic components such as sodium alkoxide are mainly distributed close to the surface of the SEI. The interior of the SEI is dominated by the inorganics such as Na$_2$CO$_3$ and NaF with sparsely distributed organics. The carbonate-derived SEI is also analyzed under the same conditions to compare them with the ether-derived SEI (consulting Figure S11 of Ref. [63]). After 100 s of sputtering, the C 1s spectrum of the graphite (after cycling) in the NaPF$_6$-EC/DEC electrolyte shows a strong peak at 290 eV. The EC decomposition paths induce different organic products revealing the functional group O–C(=O)–O [78,79] and the organic C=O in the O 1s spectrum mainly CH$_3$CH$_2$OCO$_2$Na and (CH$_2$OCO$_2$Na)$_2$[23,79]. The O 1s, Na 1s, and F 1s spectra obtained from XPS analysis with depth sputtering demonstrate that the inorganic products (Na$_2$CO$_3$, NaF) in the carbonate-derived SEI are similar to those in the ether-derived SEI. Moreover, the O 1s spectrum displays a Na auger electron peak, thereby illustrating that a majority of the organic species are sodium-containing compounds. Therefore, the coexistence of both the organic (CH$_3$CH$_2$OCO$_2$Na, (CH$_2$OCO$_2$Na)$_2$) and inorganic (Na$_2$CO$_3$, NaF) products is also observed for NaPF$_6$ in EC/DEC solvents [63].

Through XPS results, we could assume that the multilayer structure mode of the ether-derived SEI comprises flexible organic compounds such as sodium alkoxides and stiff inorganic compounds such as NaF and Na_2CO_3. In contrast, the relative surface carbonate-derived SEI comprises a mixture of different organic and inorganic compounds (Consulting Figure 7 of Ref. [63]).

Due to the use of high energetic photons for excitation of the sample, sample damage by radiation is possible in the XPS technique. In particular, the highly energetic beam spot may lead to the degradation of the SEI components and may change their chemical nature. When the substrate and the surface have the same elements, quantification becomes complicated. Regardless of the nonconductive sample, the spectrum shifts to lower kinetic energy; this is called "charging." The calibration of the specter in this case becomes tricky. The shifting depends on the conductivity and the microstructure of the component and thus rectifies it accurately. In the depth analysis of inhomogeneous films, the etching rate is different for hard and soft components. The hard material needs a longer sputtering time to go deeply as compared to the soft one. Etching may lead to unexpected modification of the surface and reactions of the active species to form undesirable products [39].

14.4.2 Surface Morphology and Structure

The electron microscope (EM) is the most convenient and useful choice to directly reveal the morphology and fine structure of the SEI layer in high resolution. The cross section of the electrode could also give some detailed information on the thickness and the presence of the SEI layer during the cycling process. However, during SEM observation, both the pristine materials and reaction mediates (like Na/Li metal, lithium or sodium peroxides, etc.) may be damaged by high electron dose due to its high reactivity. This fragile nature requires the modification and improvement of the well-established characterization tools that have been dedicated to a variety of electrodes for obtaining intrinsic information without any artifacts [80].

Recently, cryogenic EM (cryo-EM) has achieved great success in the materials research fields, such as nanoparticles, polymer, metal-organic frameworks, battery materials even though this tool was initially developed for the biology community [81,82]. Cryo-EM uses the cryo-protection, which not only helps to minimize damage from the air and the beamline but also keeps in preserving the intrinsic structure of the samples, which enables us to image the structure of the light and sensitive compounds at the atomic level, such as Li, Na, and its solid-state electrolyte interphase (CEI) [83]. In the early stage of cryo-EM development, high-resolution TEM, as one of the main structural techniques complementary to X-ray crystallography, has become a versatile tool in integrating the cryo-plunging of thin solution into a potent cryogen such as liquid ethane [81,84]. Subsequently, TEM images were to produce 2D projection of the object, and then 3D was then developed till now. Recently, the introduction of cryo-EM in battery material research has been expected to give new insights into this field. Although cryo-TEM is still a new technique for battery materials, profound progress has been made, which leads to discoveries and breakthroughs in the future (Table 14.6).

In the cryo-EM experiment, the sample preparation is a prerequisite step to mitigate the potential damage and avoid the artifacts, which are specific for high-energy

TABLE 14.6
Summary of the Recent Work on Cryo-EM in Battery Research [85]

Objective	Transfer Protection Holder Type	Characterization Techniques	Important Findings	References
Plated Li metal	Ar and vacuum	Cryo-TEM Cryo-XRD	Amorphous Li Crystalline LiF	[83]
Plated Li metal	Ar cooling holder	Cryo-TEM	LMC coated on the Li metal	[86]
C/S composite	Cryo Cryo transfer holder	Cryo-STEM	Not all the S in the porous carbon	[87]
TiS$_2$/S composite	Cryo Cryo transfer holder	Cryo-STEM	S embedded in the Fe$_3$O$_4$ nanosphere	[87]
Na metal–NaBr interface	Cryo	Cryo-FIB	Uniform NaBr coating on Na anode	[88]
Na metal interface	Cryo	Cryo-TEM	SEI layer on Na metal	[76]
Cathode interphase	Cryo	Cryo-TEM	CEI layer on the sodium cathode material	[89]

battery materials. To prepare a pristine and clean battery material sample without any contamination, scientists developed the methods to mount the sample on a grid, load it onto the holder, and transfer the holder into a TEM chamber without any air exposure [85]. Ion beam can mount the target particle of interest on the grid and control the thickness by milling. There are two commercial holders recently developed for cryo-TEM: the cryo transfer holder and the cooling holder. Depending on the sample characteristic, the sample holder should be developed to ensure the self-healing function and faster cooling to avoid any contamination. The ion flux used for milling contains plenty of high-energy Ga ions, which easily cause sample damage (local heating, surface construction, etc.). To reduce the above damage, cryo-FIB is used instead of using a focused ion beam (FIB) to enable the sample preparation for EM [85].

Na metal is the promising anode material beyond Li metal for rechargeable batteries due to its relatively low electrode potential (−2.91 V versus a standard hydrogen electrode) and high theoretical specific capacity (mAh g^{-1}). Because Na metal is so reactive and brittle that a slightly harsh environment (e.g., exposure to air and a strong beam) can contaminate and damage the samples leading to artifacts, misunderstanding, or controversial conclusions. In that case, conventional TEM at room temperature fails to image Na metal at the nanoscale because it quickly drifts, melts, and evaporates under the focusing beamline. In contrast, at a cryo-temperature, the Na metal is stable under the same exposure dose and can allow the higher electron dose needed for high-resolution imaging. Through cryo-TEM, Phung Le et al. [76] first revealed the nanoscopic origin of SEI growing on the Na metal electrodes, which lead to the uniform sodium deposition and high reversible stripping in TEGDME-based electrolyte. Different sodium salts (NaTFSI, NaFSI, NaBF$_4$, NaOT$_f$, NaClO$_4$, NaPF$_6$) dissolved in TEGDME are investigated as electrolyte for Na plating/stripping in Cu‖Na cells. In different sodium anions-TEGDME electrolytes, the SEI characteristics (structure, thickness, composition) could be distinguished by using cryogenic electron microscopy (cryo-TEM) combined with energy dispersive spectroscopy (EDS) on TEM images. In Figure 14.6, cryo-TEM elucidates the morphology and structure as well as the chemical composition and spatial distribution of the SEI on electrodeposited sodium metal (EDNa) at the nanoscale. The nanostructure of SEI on the plated sodium chunky is shown in Figure 14.6a and b. The identical contrast in the bulk demonstrates the uniformity of electrochemically deposited sodium (EDNa), while the characteristic bright rings and spots ensure its crystalline state (Figure 14.6a and c). Along the edges of sodium crystalline particles, an uneven SEI layer is a continuous dash line in Figure 14.6b showing a consistent SEI thickness of ~214 ($\pm$15 nm). In parallel, the EDS on TEM image inserted in Figure 14.6d proves that SEI is rich in inorganic species (that is, there is a much higher level of O^-, F^--containing compound derived from $NaBF_4$ than C-containing organic species; and negligible content of boron). Interestingly, these inorganic compounds in this SEI were not crystallized due to the absence of a bright spot or ring. Indeed, they were distributed uniformly in an amorphous state of this SEI, as shown by the selected area electron diffraction in Figure 14.6c. To our knowledge, this SEI layer is first reported for sodium-metal batteries utilizing cryo-TEM.

On the cathode side, a stable CEI is important to inhibit the continuous reaction between the cathode material and the electrolyte, especially at a high working voltage. Yan Jin et al. [89] reported non-flammable LHCE (sodium bis(fluorosulfonyl)imide (NaFSI)-triethyl phosphate (TEP)/1,1,2,2-tetrafluoroethyl-2,2,3,3-tetrafluoropropyl ether (TTE) (1:1.5:2 in mole) for SIBs, in which non-flammable TEP and thermally stable TTE diluent solvent ensures the safety and long cycling performance of NIBs.

Through cryo-TEM, the evolution of the CEI layers and structure transformation of Na-CNFM cathodes could be observed (Figure 14.7). High-resolution cryo-TEM images of cathode after first cycle in the carbonate-based electrolyte and NaFSI-TEP/TTE electrolyte are shown in Figure 14.7a and b. After the initial cycle, a CEI layer of 3 nm was formed on the surface of Na-CNFM cathode particles in carbonate electrolyte, and the layered structure with the sharp interplanar spacing for (002) crystal planes was maintained. However, after only 10 cycles, the CEI-layer thickness quickly increased to 9 nm, and a phase transition region (from layer to rock salt) with a thickness of ~7 nm in Na-CNFM cathode was observed (Figure 14.7c). It seems that the thin CEI layer formed during the first cycle was not stable and cannot prevent subsequent electrolyte decomposition and transition metal dissolution from cathodes. In contrast, in NaFSI-TEP/TTE electrolyte, although the thickness of the initial CEI layer formed on Na-CNFM cathode was only ~2–3 nm (Figure 14.7b), this thickness was almost not changed after 10 cycles as shown in Figure 14.7 d. Thus, this stable CEI layer could prevent further electrolyte decomposition and continuous growth of the interfacial layer. Moreover, it can also prevent transition metal dissolutions and minimize surface reconstructions (Figure 14.7d). As a result, the excellent cycling stability of HC‖Na-NCFM full cells with NaFSI-TEP/TTE electrolyte was achieved, which reaches a CE of 99.93% compared with 99.6% in conventional carbonate electrolyte [89].

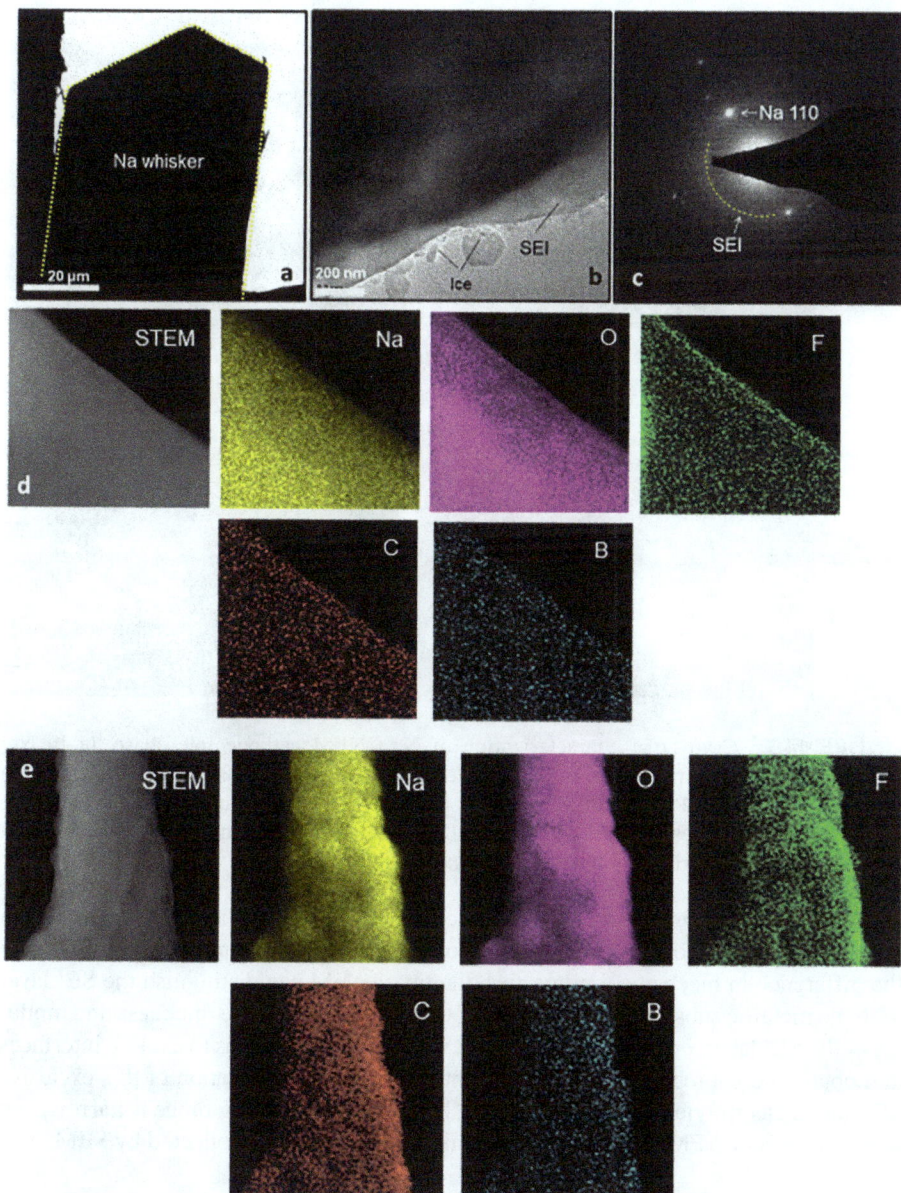

FIGURE 14.6 SEI on Na anode (a) Cryo-TEM images of Na deposited on a TEM grid; (b) Cryo-TEM image of SEI layer; (c) Selected-area electron diffraction pattern; (d) STEM image and energy-dispersive spectroscopy (EDS) spectra of the SEI layer; (e) STEM image of inactive sodium (Na^0) "leftover" and EDS spectra of residual SEI layer. (Reproduced with permission [76].)

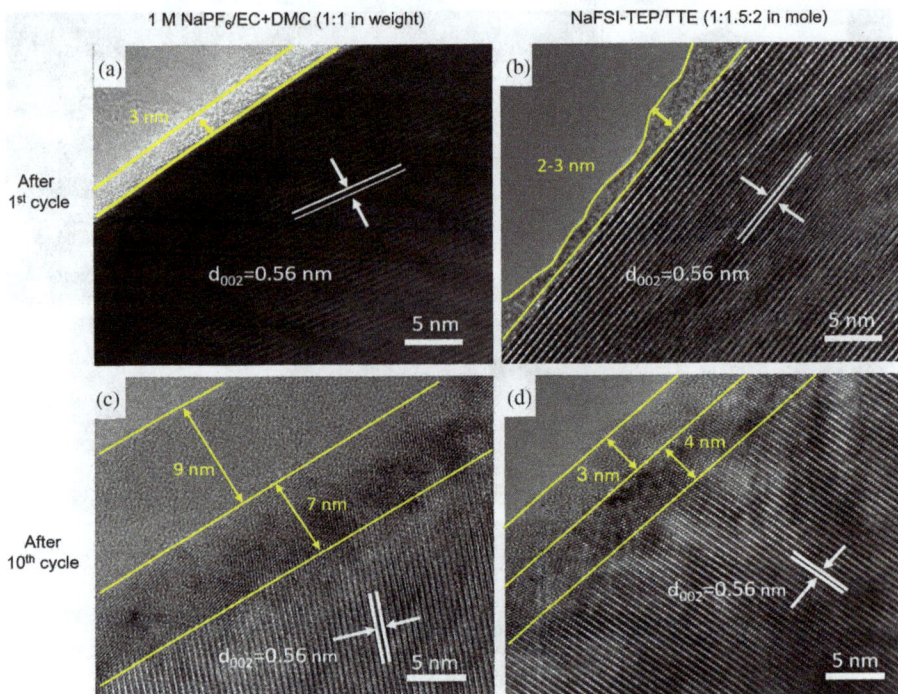

FIGURE 14.7 Cryo-TEM characterization of Na-CNFM cathode interphase. (a–b) Na-CNFM cathode after first cycle in different electrolytes: (a) 1 M $NaPF_6$/EC+DMC (1:1 in weight), (b) NaFSI-TEP/TTE (1:1.5:2 in mole); (c and d) Na-CNFM cathode after tenth cycles in different electrolytes: (c) 1 M $NaPF_6$/EC+DMC (1:1 in weight) and (d) NaFSI-TEP/TTE (1:1.5:2 in mole). (Reproduced with permission [89].)

Besides, to investigate the structure, thickness, and mechanical property of SEI, an AFM has been employed as a powerful tool in rechargeable batteries [90–93]. Due to the difference in mechanical properties, the tip of AFM can distinguish the SEI layer with the metallic substrate (or other substrates) and determine the thickness and modulus of the SEI layer. *In situ* AFM combined with *ex situ* XPS helps to explore interfaces and obtain a comprehensive exploration into the real-time formation of this exclusive SEI due to electrolyte decomposition [46]. However, the imaging range is narrow, and the resolution of AFM images is lower compared with images collected by SEM.

14.4.3 ELECTROCHEMICAL PROPERTIES

The electrochemical-related tests have been widely used as a nondestructive, convenient, and powerful technique for diagnosing and evaluating the SEI layer. For instance, frequency-domain EIS provides the complex impedance in all of the cell components (electrode, electrolyte, current collectors, contact interfaces) [94–100]. In fitting the EIS spectra with a physical equivalent circuit, the impedance from different processes occurring simultaneously at the electrode surface could be distinguished. In particular, Archer and co-worker [101] demonstrated the relationship between the resistance and cell working temperature. Both the interfacial and bulk impedances (related to the lower intercept of the spectra) rapidly increase with the rise of temperature. Qianye et al. [102] applied potential electrochemical impedance spectroscopy (PEIS) as a comprehensive and straightforward approach to determine the variation in impedance as a function of cycle number to demonstrate the influence of the few-layer graphene on the resistance magnitude of Si. Staircase PEIS (SPEIS) is a useful technique to evaluate the impedance at different voltage steps within a lithiation/de-lithiation cycle in a cell. For each measurement, it is easy to distinguish the diffusion-related Warburg impedance by splitting the frequency range (selected range from 1 Hz to 500 kHz). In general, high-frequency impedance responses (at 10–500 kHz) are attributable to the SEI resistance and the series. Impedance responses between 10 Hz and 10 kHz represent the interphase electronic contact and the charge transfer resistance. Impedance responses with frequencies less than 10 Hz are termed as "Warburg impedance," which is mainly determined by diffusion coefficient, is also termed as diffusion impedance in this study [103]. With PEIS and SPEIS, it was shown that FLG helps Si to maintain an overall lower series resistance during cycling. Also, the performance benefits could be through some prevention of the Si particles from agglomerating through electrochemical fusion.

EIS experiments in SEI-related publications have become a routine and easy characterization reported in recent papers. Morales-Ugarte et al. [104] reported the use of vinyl carbonate doped in imidazolium-based ionic liquid electrolyte enhanced the electrochemical properties of graphite-based lithium cell. In this study, the step-by-step EIS and XPS were performed during the first reduction cycle to follow the formation of the SEI. EIS spectra evolution, recorded at different cutoff voltages, indicated a two-step SEI thin film formation. These results were supported by XPS measurement at the same cutoff voltage. Hence, it was pointed out that VC was decomposed at 0.8 V to form an organic layer constituted of lithium alkyl carbonates and this film was developed up to 0.2 V. From 0.8 to 0.2 V, a slight decomposition of IL solvent was detected. This process was driven by a progressive sulfone decomposition reaction through the formation of polyoxysulfone, Li_2S, Li_3N, and LiF. As a result, the formation of the second film of inorganic nature was originated beyond 0.2 V.

The most troublesome step of EIS experiments is the interpretation of the original spectroscopy because of limited models that can clearly describe and obtain the theoretical fitting between calculated and experimental impedances. The interpretation of EIS is mostly empirical. The fitting model should be selected through professional software to convert the measured spectra into electrochemical signals. Sometimes, the selected model gives a very good fitting between the calculated and experimental impedances, but without clear physicochemical significance. Therefore, EIS data analysis using a basic fitting model becomes the toughest step of EIS characterization, which limits the wide application in SEI study and related battery research [104].

14.4.4 SEI UNDERSTANDING AND MODELING BY THEORETICAL CALCULATION

Numerous studies have been carried out to investigate and explain the SEI layer during the last decades. However, the formation and growth mechanism of the nanometer-thick SEI films are yet to experimentally identify and completely understand the working cells owing to their complex structure and lack of reliable *in situ* experimental techniques. In contrast, significant advances in computational methods have made it possible to predictively model the fundamentals of SEI and afford some insights into the SEI revolution of a working cell.

The state-of-the-art modeling has been making much progress in the investigation of SEI/CEI films on the anodes/cathodes, ranging from electronic structure calculations to mesoscale modeling, covering the thermodynamics and kinetics of electrolyte reduction reactions, SEI formation, modification through electrolyte design, correlation of SEI properties with battery performance, and the artificial SEI design. In addition, multiscale simulations have been applied to reach approximatively with the experiments. Computational details of the fundamental properties of SEI, such as electron tunneling, Li-ion transport, chemical/mechanical stability of the bulk SEI, and electrode/(SEI/) electrolyte interfaces, have been investigated and reported [105].

In 2000, Li and Balbuena [106] initially applied quantum chemistry to investigate the experimental ethylene carbonate reduction mechanism proposed by Aurbach et al. [107]. This step is a prerequisite step since the majority of SEI reaction mechanisms in the literature were deduced from experimentally observed products, and quantum chemistry helps to confirm these mechanisms by calculating the energies of the intermediate structures along the proposed reaction pathways. Following these initial calculations, the ethylene carbonate reduction pathways have been modeled extensively. Wang et al. [108] investigated the solvent effect in lithium-ion solvation and explicitly calculated the possible reduction processes of super-molecules of $Li^+(EC)_n$ ($n = 1$–5) using the high-level density functional theory (DFT) method in Gaussian 98. In this study, they treated the bulk solvent as a macroscopic and continuum medium using the polarized continuum models. In a later study by Yu et al. [109], they revealed that for the first time, all the possible products could be generated from EC decomposition, such as dilithium butylene dicarbonate (Li_2BDC), Li_2EDC, $LiO(CH_2)_2CO_2(CH_2)_2OCO_2Li$, $Li(CH_2)_2OCO_2Li$, and Li_2CO_3, some of which are just now being detected experimentally. Indeed, Li_2CO_3 may only be formed at low EC concentration.

In our days, ab initio molecular dynamic (AIMD) becomes feasible for larger systems with the development of supercomputers. Using AIMD, Leung and Budzie [110] tracked the initial decomposition process of liquid EC on graphite surfaces with different edge terminations and confirmed two EC decomposition pathways. In the meantime, due to limited size in DFT-based calculations, some support tools have been developed to enable classical MD simulations with a longer time scale and larger system size adapted for Li-ion battery systems. In particular, atomic force fields, COMPASS (Condensed-phase Optimized Molecular Potentials for Atomistic Simulation Studies), APPLE&P (Atomistic Polarizable Potential for Liquids, Electrolytes, and Polymers), ReaxFF reactive force field, etc. are integrating with MD to describe accurately the complex system with different components, in reality, the complicated interaction between molecules, the decomposition in many steps of solvents, the formation and growth of SEI layer and so on [111–115].

Though the theoretical calculation provides much atomic understanding of the SEI layer, the calculation is still mostly based on the ideal environment with some hypotheses, which is sometimes a broad gap from the working condition of the SEI/CEI layer on the electrodes in a real cell. In consequence, much more optimized methods should be continuously developed to simulate the working SEI layer.

14.5 ARTIFICIAL SOLID ELECTROLYTE INTERPHASE (SEI) LAYER: A FUTURE PERSPECTIVE

14.5.1 FEATURES OF IDEAL SEI LAYERS

In the conventional liquid electrolyte, the electrode/electrolyte interphase has become a vital aspect, particularly the SEI on the anode surface. Therefore, SEI serves as a medium film with enhanced stability, life cycle, and high CE which facilitates better battery performance.

The current obstacles related to SEI layer on Na-based anode:
- Lengthy formation process (i.e., up to 7 days for liquid electrolyte) leading to costly mass production
- High risk of capacity loss in irreversible Na^+ due to cathode side reaction with solvents as well as structure defect of active material, which leads to the reduction of CE
- Increasing ion-transport impedance from the bulk electrolyte into electrode due to unstable SEI composition structures during long-term cycling
- Competitive reactions from various components onto the electrode surface leading to the formation of an excessively complex structure SEI layer

Recently, research efforts have been extensively conducted to design stabilizing anodes by robust SEI layers. Such an attempt, hopefully, will serve to achieve better performance, lifetime as well as to overcome persistent safety problems commonly observed in the sodium-based anode.

Design features of a proper protective SEI layer on Na anode are summarized as follows:
- High selectivity, conductivity, and electrical resistance of the SEI layer for Na ion to promote the migration pathway and prevent the continuing electrolyte consumption or the continuous growth of SEI thickness, which is attributed to the homogeneous plating and stripping process of Na^+ ions
- Electrochemical, thermal, and chemical stabilities on the Na anode surface over the wide range of window voltage and operating temperature
- Restraining the side reactions of the dissolved ions from the cathode side allows diffusing to the anode host
- Certain flexibility (i.e., proper elastic module) to hinder the fracture/distortion of formed layer and the suppression of dendritic growth penetration from the Na host structure
- The rapid reaction rate with homogeneous layer structure leading to the formation of the compact thin film composition
- The higher electrochemical potential of the formation process in comparison with the Na deposition process

Generally, the electrochemical operation of SIBs can be controlled by creating an efficient passivation layer on the electrode and electrolyte interfaces. A proper coating layer blocks the further consumption of electrolyte species at the electrode surface by isolating electrons, which limits the oxidative and reductive reactions of solvated salt anions on the cathode and solvent molecules on anode sides, respectively.

14.5.2 Methods to Form Artificial SEI Layer

The natural SEI layer is instantaneously generated by a highly chemical-reactive Na anode with organic solvents molecules, salt anions, and additives when immersed in most organic electrolytes. As a result, the resulting layer is usually unstable and vulnerable to the infinite volume expansion/contraction of Na host, leading to the further decomposition of electrolytes on the negative electrode surface.

Therefore, designing a suitable artificial SEI layer with suitable approaches and unique techniques could control the formation of a desirable protective film in terms of the structural stability and composition to ensure high performance, lifetime, low cost, and safety of battery to shorten the path to commercially industrial production and application. There are some strategies to fabricate with acceptable interfacial stability of SEI films to upgrade the cyclability of Na as an anode. The efficient way to approach artificial SEI formation is the modification of the surface of the active material or separator with the designed passivation coatings [104]. It is commonly assumed that the high modulus strength of the formed interfacial architectures can effectively resist the dendritic growth at the anode side. Based on this idea, the solid electrolyte (SE) is one of the promising candidates, which possesses many advantages such as a wide electrochemical window, a broad working temperature, no leakage, and the ability to hinder unwanted chemical reaction with Na metal anode, which provides longer life and higher internal safety than other electrolytes. However, the presence of defects of SEI can lead to short-circuiting by dendritic growth penetrated from anode to cathode. For that reason, it has been widely studied in high energy density batteries, with extremely high strength of artificial SEI films compared with the conventional weak in polymer separators. Besides, the interface contact and low ionic conductivity of SE are major challenges for SE to be mass-produced on an industrial scale [116]. However, the issue when using the SE that is a high impedance of electrode/electrolyte interface and grain boundary defects of organic particles in the SEI. Hence, a variety of methods have been focused to construct an artificial interlayer to minimize the interface resistance such as *in situ* synthesis of the artificial film by the reaction of pure melting Na metal with NASICON at elevated temperature (380°C) [104].

14.5.2.1 Coated Protective Artificial Film SEI/Separator

Since the direct formation of a passivation layer (SEI) forms onto the active material that can be complicated for manufacturing, another effective way for engineering SEI layers is to design an SEI/separator, which is only initially synthesized on the separators. The common Celgard or PP membranes are used in commercial cells nowadays; however, the low thermal stability and weak dendritic resistance lead to their limited application. Ceramic coating separators have been widely commercialized

FIGURE 14.8 The SEM images of differential coating layers of ceramic coating, bohemite coating, and aramid fiber coating from left to right, respectively.

in batteries possessing high thermal resistance and excellent prohibition of dendritic penetration. Generally, to minimize the membrane thickness for optimizing the migration path of carrier ions, the ceramic layer is coated on one side of a separator where it directly contacts with an anode; the SEM images of different coating components of the separator are shown in Figure 14.8.

The ceramic and bohemite function as a dendrite-proof and chemically passivation layer, while the high thermal resistance and porosity properties of aramid fiber help the separator to obtain sufficient mechanical strength. This combination of these multilayer composites on membranes with the battery systems/batteries' structure can achieve the critical requirements for high-energy batteries such as high migration ions for high rate performance, minimal chances of short-circuit from dendritic growth, and high-temperature tolerance.

14.5.2.2 Chemical Pretreatment

Another method that has been widely applied for constructing the passivation layers is surface pretreatment by hybrid liquids to stabilize the Na, owing to its simplicity and low cost for mass production. The choice of reagents is the key to construct the desirable characteristic of generated SEI film in terms of structure, composition, and thickness under different pretreatment conditions such as temperature, reaction time, and reactants concentration.

Wang *et al.* proposed a simple method to modify the surface of Na via the reaction between SbF_3 and Na in DMC solvent, in which the generated film of two different layers of hybrid mixture polymer and NaF acts as the upper layer, while the alloy Na_3Sb layer is contacted on Na side. This hierarchically structured SEI exhibits the improving cycle life in carbonate liquid electrolyte with $NaClO_4$ 1 M, in which no increasing polarization is observed over 600 cycles [117]. ALD is one of the promising physical technologies for building advanced ultrathin artificial organic SEI layers on Na as the anode, which can precisely control the thickness of angstrom-level and smooth surface, enhancing the ion transport and preventing the electron transfer from the electrolyte to the electrode. However, the unstable chemical nature of highly reactive Na metal, the critical condition of inert gas (e.g., argon) environment, and low temperature (below the melting point of Na metal) need to be considered for deposition processes.

Wei Luo's group proposed a low-temperature plasma-enhanced ALD (PEALD) method to form an Al_2O_3 extreme thin layer as a protective artificial SEI onto Na metal as an electrode substrate. The chosen TMA (trimethylaluminum) and O_2 plasma as precursors produce a super-thin ALD-Al_2O_3 coating layer, in which the PEALD process was performed at 75°C under argon gas (Consulting 2a of Ref. [118]). In this work, the thin film of Al_2O_3 (≈2.8 nm in thickness) on the Na metal surface deposited by low-temperature PEALD method exhibits a stable deposition/ stripping cycling and cell impedance at a high current density of 0.25 mA cm⁻² (0.5 mAh cm⁻²) over 900 cycles in the same carbonate electrolyte system when compared to the formed traditional SEI films with bare Na. They found that a generated arti- ficial ultrathin Na-Al-O layer through the initial sodiated process possesses a high Na⁺ ions conductivity and an isolated electrode/electrolyte interface. Moreover, by analyzing the Na anode after the deposition/stripping process, they confirmed that the Al_2O_3-coated layer can deliver a protective and smooth surface, thus avoiding further electrolyte decomposition and dendritic growth [118].

14.6 CONCLUSION

SIBs offer cheaper alternatives to LIBs to address the need for potential large-scale energy storage projects that cannot be served by LIBs due to the limited natural abundance of lithium and expensive prices. In this study, we outlined exhaus- tively the main observations, recent development, and current understanding of the sodium-ion SEI/CEI based on different forms of electrodes. The formation of a stable SEI is very important to achieve high-cycling CE and the long-cycling performance of SIBs. Based on our deep literature assessment, the research com- munity seems to realize gradually that the key issue to the practical advancement of SIBs is the electrolyte along with the interfacial phenomena and safety-related issues. Therefore, the study of the SEI for the understanding of its dynamic nature should be given priority, which is the basis for guiding the strategies of SEI-layer design and modification. Herein, we looked at the existing state and potential possi- bilities of electrolytes for Na-based rechargeable batteries. An in-depth understand- ing of the electrolyte components that influence the electrochemical behavior of SIBs would allow the design of electrolytes that lead to a superior electrochemical performance in SIBs. Hence, a systematic way of designing electrolytes is needed for future Na-based rechargeable batteries. Furthermore, the controllable modifica- tion of the SEI is one of the effective strategies that have been proposed to modify the thickness, density, ion conductivity, etc., which are related to the Na-ion inser- tion/extraction processes as well as the performance of SIBs. Overall, the strong electrolytes interphase is the key problem in a rechargeable battery for practical application.

REFERENCES

1. S.J. An, J. Li, C. Daniel, D. Mohanty, S. Nagpure, D.L. Wood (2016), The state of understanding of the lithium-ion-battery graphite solid electrolyte interphase (SEI) and its relationship to formation cycling, *Carbon N. Y.* **105** 52–76.
2. G. Nazri, R.H. Muller, Composition of Surface Layers on Li Electrodes in PC (1985), LiClO₄ of Very Low Water Content, *J. Electrochem. Soc.* **132** 2050–2054.

3. M. Gauthier, T.J. Carney, A. Grimaud, L. Giordano, N. Pour, H.-H. Chang (2015), D.P. Fenning, S.F. Lux, O. Paschos, C. Bauer, F. Maglia, S. Lupart, P. Lamp, Y. Shao-Horn, Electrode–electrolyte interface in li-ion batteries: Current understanding and new insights, *J. Phys. Chem. Lett.* **6** 4653–4672.
4. F. Du, P. Sun, Q. Zhou, D. Zeng, D. Hu, Z. Fan, Q. Hao, C. Mei, T. Xu, J. Zheng (2020), Interlinking Primary Grains with Lithium Boron Oxide to Enhance the Stability of $LiNi_{0.8}Co_{0.15}Al_{0.05}O_2$, *ACS Appl. Mater. Interfaces.* **12** 56963–56973.
5. F.A. Soto, A. Marzouk, F. El-Mellouhi, P.B. Balbuena (2018), Understanding ionic diffusion through SEI components for lithium-ion and sodium-ion batteries: Insights from first-principles calculations, *Chem. Mater.* **30** 3315–3322.
6. M. Winter (2009), The solid electrolyte interphase – The most important and the least understood solid electrolyte in rechargeable Li batteries, *Zeitschrift Für Phys. Chemie.* **223** 1395–1406.
7. K. Xu (2014), Electrolytes and interphases in li-ion batteries and beyond, *Chem. Rev.* **114** 11503–11618.
8. P. Peljo, H.H. Girault (2018), Electrochemical potential window of battery electrolytes: The HOMO–LUMO misconception, *Energy Environ. Sci.* **11** 2306–2309.
9. S.J. An, J. Li, C. Daniel, D. Mohanty, S. Nagpure, D.L. Wood (2016), The state of understanding of the lithium-ion-battery graphite solid electrolyte interphase (SEI) and its relationship to formation cycling, *Carbon N. Y.* **105** 52–76.
10. Y. You, A. Manthiram (2018), Progress in high-voltage cathode materials for rechargeable sodium-ion batteries, *Adv. Energy Mater.* **8** 1701785.
11. S. Das, P.M. Attia, W.C. Chueh, M.Z. Bazant (2019), Electrochemical kinetics of SEI growth on carbon black II: Modeling, *ArXiv.* **166** E97.
12. M.B. Pinson, M.Z. Bazant (2013), Theory of SEI formation in rechargeable batteries: Capacity fade, accelerated aging and lifetime prediction, *J. Electrochem. Soc.* **160** A243–A250.
13. P.M. Attia, S. Das, S.J. Harris, M.Z. Bazant, W.C. Chueh (2019), Electrochemical kinetics of SEI growth on carbon black: Part I. Experiments, *J. Electrochem. Soc.* **166** E97–E106.
14. G. Gachot, P. Ribière, D. Mathiron, S. Grugeon, M. Armand, J.-B. Leriche, S. Pilard, S. Laruelle (2011), Gas chromatography/mass spectrometry as a suitable tool for the li-ion battery electrolyte degradation mechanisms study, *Anal. Chem.* **83** 478–485.
15. T. Hosaka, K. Kubota, A.S. Hameed, S. Komaba (2020), Research development on k-ion batteries, *Chem. Rev.* **120** 6358–6466.
16. G.G. Eshetu, T. Diemant, M. Hekmatfar, S. Grugeon, R.J. Behm, S. Laruelle, M. Armand, S. Passerini (2019), Impact of the electrolyte salt anion on the solid electrolyte interphase formation in sodium ion batteries, *Nano Energy.* **55** 327–340.
17. Y. Huang, L. Zhao, L. Li, M. Xie, F. Wu, R. Chen (2019), Electrolytes and electrolyte/electrode interfaces in sodium-ion batteries: From scientific research to practical application, *Adv. Mater.* **31** 1–41.
18. S. Komaba, W. Murata, T. Ishikawa, N. Yabuuchi, T. Ozeki, T. Nakayama, A. Ogata, K. Gotoh, K. Fujiwara (2011), Electrochemical Na insertion and solid electrolyte interphase for hard-carbon electrodes and application to Na-ion batteries, Adv. *Funct. Mater.* **21** 3859–3867.
19. B. Yuan, L. Zeng, X. Sun, Y. Yu, Q. Wang (2018), Enhanced sodium storage performance in flexible free-standing multichannel carbon nanofibers with enlarged interlayer spacing, *Nano Res.* **11** 2256–2264.
20. P. Poizot, S. Laruelle, S. Grugeon, L. Dupont, J.-M. Tarascon (2000), Nano-sized transition-metal oxides as negative-electrode materials for lithium-ion batteries, *Nature.* **407** 496–499.

21. M. Zarrabeitia, F. Nobili, M.Á. Muñoz-Márquez, T. Rojo (2016), M. Casas-Cabanas, Direct observation of electronic conductivity transitions and solid electrolyte interphase stability of $Na_2Ti_3O_7$ electrodes for Na-ion batteries, *J. Power Sources.* **330** 78–83.
22. D.I. Iermakova, R. Dugas, M.R. Palacín, A. Ponrouch (2015), On the comparative stability of Li and Na metal anode interfaces in conventional alkyl carbonate electrolytes, *J. Electrochem. Soc.* **162** A7060–A7066.
23. Z.W. Seh, J. Sun, Y. Sun, Y. Cui (2015), A highly reversible room-temperature sodium metal anode, *ACS Cent. Sci.* **1** 449–455.
24. H. P. Pierson (1993), *Handbook of Carbon, Graphite, Diamond and Fullerenes*, Noyes Publications, Park Ridge, NJ.
25. J. T. Cookson (1978), *Carbon Adsorption Handbook, Ann Arbor Science.*
26. G.G. Eshetu, S. Grugeon, H. Kim, S. Jeong, L. Wu, G. Gachot, S. Laruelle, M. Armand, S. Passerini (2016), Comprehensive insights into the reactivity of electrolytes based on sodium ions, *ChemSusChem.* **9** 462–471.
27. J. Fondard, E. Irisarri, C. Courrèges, M.R. Palacin, A. Ponrouch, R. Dedryvère (2020), SEI composition on hard carbon in Na-ion batteries after long cycling: Influence of salts (NaPF 6, NaTFSI) and additives (FEC, DMCF), *J. Electrochem. Soc.* **167** 070526.
28. A. Ponrouch, R. Dedryvère, D. Monti, A.E. Demet, J.M. Ateba Mba, L. Croguennec, C. Masquelier, P. Johansson, M.R. Palacín (2013), Towards high energy density sodium ion batteries through electrolyte optimization, *Energy Environ. Sci.* **6** 2361.
29. B. Philippe, M. Valvo, F. Lindgren, H. Rensmo, K. Edström (2014), Investigation of the electrode/electrolyte interface of Fe_2O_3 composite electrodes: Li vs Na batteries, *Chem. Mater.* **26** 5028–5041.
30. L. Bodenes, A. Darwiche, L. Monconduit, H. Martinez (2015), The solid electrolyte interphase a key parameter of the high performance of Sb in sodium-ion batteries: Comparative X-ray photoelectron spectroscopy study of Sb/Na-ion and Sb/Li-ion batteries, *J. Power Sources.* **273** 14–24.
31. M.A. Muñoz-Márquez, M. Zarrabeitia, E. Castillo-Martínez, A. Eguía-Barrio, T. Rojo, M. Casas-abanas (2015), Composition and evolution of the solid-electrolyte interphase in $Na_2Ti_3O_7$ electrodes for Na-ion batteries: XPS and auger parameter analysis, *ACS Appl. Mater. Interfaces.* **7** 7801–7808.
32. A. Darwiche, L. Bodenes, L. Madec, L. Monconduit, H. Martinez (2016), Impact of the salts and solvents on the SEI formation in Sb/Na batteries: An XPS analysis, *Electrochim. Acta.* **207** 284–292.
33. J. Song, G. Jeong, A.-J. Lee, J.H. Park, H. Kim, Y.-J. Kim (2015), Dendrite-free polygonal sodium deposition with excellent interfacial stability in a $NaAlCl_{4-2}SO_2$ inorganic electrolyte, *ACS Appl. Mater. Interfaces.* **7** 27206–27214.
34. S. Doubaji, B. Philippe, I. Saadoune, M. Gorgoi, T. Gustafsson, A. Solhy, M. Valvo, H. Rensmo, K. Edström (2016), Passivation layer and cathodic redox reactions in sodiumion batteries probed by HAXPES, *ChemSusChem.* **9** 97–108.
35. K. Edström, A.M. Andersson, A. Bishop, L. Fransson, J. Lindgren, A. Hussénius (2001), Carbon electrode morphology and thermal stability of the passivation layer, *J. Power Sources.* **97–98** 87–91.
36. S.K. Heiskanen, J. Kim, B.L. Lucht (2019), Generation and evolution of the solid electrolyte interphase of lithium-ion batteries, *Joule.* **3** 2322–2333.
37. M. Ishikawa, Y. Tasaka, N. Yoshimoto, M. Morita (2001), Optimization of physicochemical characteristics of a lithium anode interface for high-efficiency cycling: An effect of electrolyte temperature, *J. Power Sources.* **97–98** 262–264.
38. M. Dollé, S. Grugeon, B. Beaudoin, L. Dupont, J.-M. Tarascon (2001), In situ TEM study of the interface carbon/electrolyte, *J. Power Sources.* **97–98** 104–106.
39. P. Verma, P. Maire, P. Novák (2010), A review of the features and analyses of the solid electrolyte interphase in Li-ion batteries, *Electrochim. Acta.* **55** 6332–6341.

40. V. Zorba, J. Syzdek, X. Mao, R.E. Russo, R. Kostecki, (2012) Ultrafast laser induced breakdown spectroscopy of electrode/electrolyte interfaces, *Appl. Phys. Lett.* **100** 234101.

41. S.S. Zhang, K. Xu, T.R. Jow (2004), Electrochemical impedance study on the low temperature of Li-ion batteries, *Electrochim. Acta.* **49** 1057–1061.

42. C. Xu, B. Sun, T. Gustafsson, K. Edström, D. Brandell, M. Hahlin (2014), Interface layer formation in solid polymer electrolyte lithium batteries: An XPS study, *J. Mater. Chem. A.* **2** 7256–7264.

43. A. Kominato, E. Yasukawa, N. Sato, T. Ijuuin, H. Asahina, S. Mori (1997), Analysis of surface films on lithium in various organic electrolytes, *J. Power Sources.* **68** 471–475.

44. K. Morigaki, A. Ohta (1998), Analysis of the surface of lithium in organic electrolyte by atomic force microscopy, Fourier transform infrared spectroscopy and scanning auger electron microscopy, *J. Power Sources.* **76** 159–166.

45. D. Aurbach, Y. Cohen (1996), The application of atomic force microscopy for the study of Li deposition processes, *J. Electrochem. Soc.* **143** 3525–3532.

46. A.V. Cresce, S.M. Russell, D.R. Baker, K.J. Gaskell, K. Xu (2014), In situ and quantitative characterization of solid electrolyte interphases, *Nano Lett.* **14** 1405–1412.

47. I.V. Veryovkin, C.E. Tripa, A.V. Zinovev, S.V. Baryshev, Y. Li, D.P. Abraham (2014), TOF SIMS characterization of SEI layer on battery electrodes, *Nucl. Instruments Methods Phys. Res. Sect. B Beam Interact. with Mater. Atoms.* **332** 368–372.

48. R. Santos-Ortiz, T. Rojhirunsakool, J.K. Jha, S. Al Khateeb, R. Banerjee, K.S. Jones, N.D. Shepherd (2017), Analysis of the structural evolution of the SEI layer in FeF2 thin-film lithium-ion batteries upon cycling using HRTEM and EELS, *Solid State Ionics.* **303** 103–112.

49. T. Akai, H. Ota, H. Namita, S. Yamaguchi, M. Nomura (2005), XANES study on solid electrolyte interface of li ion battery, *Phys. Scr.* **2005** 408.

50. M.C. Smart, B. V. Ratnakumar, S. Surampudi, Y. Wang, X. Zhang, S.G. Greenbaum, A. Hightower, C.C. Ahn, B. Fultz (1999), Irreversible capacities of graphite in low-temperature electrolytes for lithium-ion batteries, *J. Electrochem. Soc.* **146** 3963–3969.

51. L. Seidl, S. Martens, J. Ma, U. Stimming, O. Schneider (2016), In situ scanning tunneling microscopy studies of the SEI formation on graphite electrodes for Li$^+$ -ion batteries, *Nanoscale.* **8** 14004–14014.

52. S.Y. Luchkin, S.A. Lipovskikh, N.S. Katorova, A.A. Savina, A.M. Abakumov, K.J. Stevenson (2020), Solid-electrolyte interphase nucleation and growth on carbonaceous negative electrodes for Li-ion batteries visualized with in situ atomic force microscopy, *Sci. Rep.* **10** 8550.

53. S.-H. Lee, H.-G. You, K.-S. Han, J. Kim, I.-H. Jung, J.-H. Song (2014), A new approach to surface properties of solid electrolyte interphase on a graphite negative electrode, *J. Power Sources.* **247** 307–313.

54. P. Guan, L. Liu, X. Lin (2015), Simulation and experiment on solid electrolyte interphase (SEI) morphology evolution and lithium-ion diffusion, *J. Electrochem. Soc.* **162** A1798–A1808.

55. A. Du Pasquier, F. Disma, T. Bowmer, A.S. Gozdz, G. Amatucci, J.-M. Tarascon (1998), Differential scanning calorimetry study of the reactivity of carbon anodes in plastic li-ion batteries, *J. Electrochem. Soc.* **145** 472–477.

56. Z. Zhang, D. Fouchard, J.R. Rea (1998), Differential scanning calorimetry material studies: implications for the safety of lithium-ion cells, *J. Power Sources.* **70** 16–20.

57. I.A. Profatilova, S.-S. Kim, N.-S. Choi (2009), Enhanced thermal properties of the solid electrolyte interphase formed on graphite in an electrolyte with fluoroethylene carbonate, *Electrochim. Acta.* **54** 4445–4450.

58. M.N. Richard, J.R. Dahn (1999), Accelerating rate calorimetry study on the thermal stability of lithium intercalated graphite in electrolyte. II. Modeling the results and predicting differential scanning calorimeter curves, *J. Electrochem. Soc.* **146** 2078–2084.

59. S.H. Ng, C. Vix-Guterl, P. Bernardo, N. Tran, J. Ufheil, H. Buqa, J. Dentzer, R. Gadiou, M.E. Spahr, D. Goers, P. Novák (2009), Correlations between surface properties of graphite and the first cycle specific charge loss in lithium-ion batteries, *Carbon N. Y.* **47** 705–712.

60. H. Ota, T. Sato, H. Suzuki, T. Usami (2001), TPD–GC/MS analysis of the solid electrolyte interface (SEI) on a graphite anode in the propylene carbonate/ethylene sulfite electrolyte system for lithium batteries, *J. Power Sources.* **97–98** 107–113.

61. L. Gireaud, S. Grugeon, S. Laruelle, S. Pilard, J.-M. Tarascon (2005), Identification of li battery electrolyte degradation products through direct synthesis and characterization of alkyl carbonate salts, *J. Electrochem. Soc.* **152** A850.

62. N. Ehteshami, L. Ibing, L. Stolz, M. Winter, E. Paillard (2020), Ethylene carbonate-free electrolytes for Li-ion battery: Study of the solid electrolyte interphases formed on graphite anodes, *J. Power Sources.* **451** 227804.

63. Z. Wang, H. Yang, Y. Liu, Y. Bai, G. Chen, Y. Li, X. Wang, H. Xu, C. Wu, J. Lu (2020), Analysis of the stable interphase responsible for the excellent electrochemical performance of graphite electrodes in sodium-ion batteries, *Small.* **16** 2003268.

64. J. Madejová (2003), FTIR techniques in clay mineral studies, *Vib. Spectrosc.* **31** 1–10.

65. K. Hongyou, T. Hattori, Y. Nagai, T. Tanaka, H. Nii, K. Shoda (2013), Dynamic in situ fourier transform infrared measurements of chemical bonds of electrolyte solvents during the initial charging process in a Li ion battery, *J. Power Sources.* **243** 72–77.

66. S. Pérez-Villar, P. Lanz, H. Schneider, P. Novák (2013), Characterization of a model solid electrolyte interphase/carbon interface by combined in situ Raman/Fourier transform infrared microscopy, *Electrochim. Acta.* **106** 506–515.

67. C. Wu, Y. Bai, F. Wu (2009), Fourier-transform infrared spectroscopic studies on the solid electrolyte interphase formed on Li-doped spinel $Li_{1.05}Mn_{1.96}O_4$ cathode, *J. Power Sources.* **189** 89–94.

68. V. Eshkenazi, E. Peled, L. Burstein, D. Golodnitsky (2004), XPS analysis of the SEI formed on carbonaceous materials, *Solid State Ionics.* **170** 83–91.

69. R. Dedryvère, S. Leroy, H. Martinez, F. Blanchard, D. Lemordant, D. Gonbeau (2006), XPS valence characterization of lithium salts as a tool to study electrode/electrolyte interfaces of Li-ion batteries, *J. Phys. Chem. B.* **110** 12986–12992.

70. D. Rehnlund, F. Lindgren, S. Böhme, T. Nordh, Y. Zou, J. Pettersson (2017), U. Bexell, M. Boman, K. Edström, L. Nyholm, Lithium trapping in alloy forming electrodes and current collectors for lithium based batteries, *Energy Environ. Sci.* **10** 1350–1357.

71. R. Xu, X.-B. Cheng, C. Yan, X.-Q. Zhang, Y. Xiao, C.-Z. Zhao, J.-Q. Huang, Q. Zhang (2019), Artificial interphases for highly stable lithium metal anode, *Matter.* **1** 317–344.

72. T. Miyayama, N. Sanada, M. Suzuki, J.S. Hammond, S.-Q.D. Si, A. Takahara (2010), X-ray photoelectron spectroscopy study of polyimide thin films with Ar cluster ion depth profiling, *J. Vac. Sci. Technol. A Vacuum, Surfaces, Film.* **28** L1–L4.

73. X. Shen, H. Liu, X.-B. Cheng, C. Yan, J.-Q. Huang (2018), Beyond lithium ion batteries: Higher energy density battery systems based on lithium metal anodes, *Energy Storage Mater.* **12** 161–175.

74. K. Li, J. Zhang, D. Lin, D.-W. Wang, B. Li, W. Lv, S. Sun, Y.-B. He, F. Kang, Q.-H. Yang, L. Zhou, T.-Y. Zhang (2019), Evolution of the electrochemical interface in sodium ion batteries with ether electrolytes, *Nat. Commun.* **10** 725.

75. D. Aurbach (2000), Review of selected electrode–solution interactions which determine the performance of Li and Li ion batteries, *J. Power Sources.* **89** 206–218.

76. P.M.L. Le, T.D. Vo, H. Pan, Y. Jin, Y. He, X. Cao, H. V. Nguyen, M.H. Engelhard, C. Wang, J. Xiao, J. Zhang (2020), Excellent cycling stability of sodium anode enabled by a stable solid electrolyte interphase formed in ether-based electrolytes, *Adv. Funct. Mater.* **30** 2001151.

77. B. Xie, L. Wang, H. Li, H. Huo, C. Cui, B. Sun, Y. Ma, J. Wang, G. Yin, P. Zuo (2021), An interface-reinforced rhombohedral Prussian blue analogue in semi-solid state electrolyte for sodium-ion battery, *Energy Storage Mater.* **36** 99–107.
78. J. Zhou, X. Kong, M.C. Sekhar, J. Lin, F. Le Goualher, R. Xu, X. Wang, Y. Chen, Y. Zhou, C. Zhu, W. Lu, F. Liu, B. Tang, Z. Guo, C. Zhu, Z. Cheng, T. Yu, K. Suenaga, D. Sun, W. Ji, Z. Liu (2019), Epitaxial synthesis of monolayer PtSe 2 single crystal on MoSe 2 with strong interlayer coupling, *ACS Nano.* **13** 10929–10938.
79. H. Kim, K. Lim, G. Yoon, J.-H. Park, K. Ku, H.-D. Lim, Y.-E. Sung, K. Kang (2017), Exploiting lithium-ether co-intercalation in graphite for high-power lithium-ion batteries, *Adv. Energy Mater.* **7** 1700418.
80. J. Conder, C. Marino, P. Novák, C. Villevieille (2018), Do imaging techniques add real value to the development of better post-Li-ion batteries? *J. Mater. Chem. A.* **6** 3304–3327.
81. M. Adrian, J. Dubochet, J. Lepault, A.W. McDowall (1984), Cryo-electron microscopy of viruses, *Nature.* **308** 32–36.
82. R. Henderson (2018), From electron crystallography to single particle CryoEM (Nobel Lecture), *Angew. Chemie Int. Ed.* **57** 10804–10825.
83. X. Wang, M. Zhang, J. Alvarado, S. Wang, M. Sina, B. Lu, J. Bouwer, W. Xu, J. Xiao, J.-G. Zhang, J. Liu, Y.S. Meng (2017), New insights on the structure of electrochemically deposited lithium metal and its solid electrolyte interphases via cryogenic TEM, *Nano Lett.* **17** 7606–7612.
84. J. Dubochet, J.-J. Chang, R. Freeman, J. Lepault, A.W. McDowall (1982), Frozen aqueous suspensions, *Ultramicroscopy.* **10** 55–61.
85. X. Wang, Y. Li, Y.S. Meng (2018), Cryogenic electron microscopy for characterizing and diagnosing batteries, *Joule.* **2** 2225–2234.
86. H. Liu, X. Wang, H. Zhou, H.-D. Lim, X. Xing, Q. Yan, Y.S. Meng, P. Liu (2018), Structure and solution dynamics of lithium methyl carbonate as a protective layer for lithium metal, *ACS Appl. Energy Mater.* **1** 1864–1869.
87. X.-C. Liu, Y. Yang, J. Wu, M. Liu, S.P. Zhou, B.D.A. Levin, X.-D. Zhou, H. Cong, D.A. Muller, P.M. Ajayan, H.D. Abruña, F.-S. Ke (2018), Dynamic hosts for high-performance Li–S batteries studied by cryogenic transmission electron microscopy and in situ X-ray diffraction, *ACS Energy Lett.* **3** 1325–1330.
88. S. Choudhury, S. Wei, Y. Ozhabes, D. Gunceler, M.J. Zachman, Z. Tu, J.H. Shin, P. Nath, A. Agrawal, L.F. Kourkoutis, T.A. Arias, L.A. Archer (2017), Designing solid-liquid interphases for sodium batteries, *Nat. Commun.* **8** 898.
89. Y. Jin, Y. Xu, P.M.L. Le, T.D. Vo, Q. Zhou, X. Qi, M.H. Engelhard, B.E. Matthews, H. Jia, Z. Nie, C. Niu, C. Wang, Y. Hu, H. Pan, J.-G. Zhang (2020), Highly reversible sodium ion batteries enabled by stable electrolyte-electrode interphases, *ACS Energy Lett.* **5** 3212–3220.
90. K.A. Hirasawa, K. Nishioka, T. Sato, S. Yamaguchi, S. Mori (1997), Investigation of graphite composite anode surfaces by atomic force microscopy and related techniques, *J. Power Sources.* **69** 97–102.
91. D. Alliata (2000), Electrochemical SPM investigation of the solid electrolyte interphase film formed on HOPG electrodes, *Electrochem. Commun.* **2** 436–440.
92. Z. Ogumi, S.-K. Jeong (2003), SPM analysis of surface film formation on graphite negative electrodes in lithium-ion batteries, *Electrochemistry.* **71** 1011–1017.
93. J. Zhang, X. Yang, R. Wang, W. Dong, W. Lu, X. Wu, X. Wang, H. Li, L. Chen (2014), Influences of additives on the formation of a solid electrolyte interphase on MnO electrode studied by atomic force microscopy and force spectroscopy, *J. Phys. Chem. C.* **118** 20756–20762.
94. S. Xiong, Y. Diao, X. Hong, Y. Chen, K. Xie (2014), Characterization of solid electrolyte interphase on lithium electrodes cycled in ether-based electrolytes for lithium batteries, *J. Electroanal. Chem.* **719** 122–126.

95. C.-F. Chen, P.P. Mukherjee (2015), Probing the morphological influence on solid electrolyte interphase and impedance response in intercalation electrodes, *Phys. Chem. Chem. Phys.* **17** 9812–9827.

96. E. Radvanyi, K. Van Havenbergh, W. Porcher, S. Jouanneau, J.-S. Bridel, S. Put, S. Franger (2014), Study and modeling of the solid electrolyte interphase behavior on nano-silicon anodes by electrochemical impedance spectroscopy, *Electrochim. Acta.* **137** 751–757.

97. P. Lu, C. Li, E.W. Schneider, S.J. Harris (2014), Chemistry, impedance, and morphology evolution in solid electrolyte interphase films during formation in lithium ion batteries, *J. Phys. Chem. C.* **118** 896–903.

98. Y.-C. Chang, H.-J. Sohn (2000), Electrochemical impedance analysis for lithium ion intercalation into graphitized carbons, *J. Electrochem. Soc.* **147** 50.

99. H. Schranzhofer, J. Bugajski, H.J. Santner, C. Korepp, K.-C. Möller, J.O. Besenhard, M. Winter, W. Sitte (2006), Electrochemical impedance spectroscopy study of the SEI formation on graphite and metal electrodes, *J. Power Sources.* **153** 391–395.

100. A.S. Bandarenka (2013), Exploring the interfaces between metal electrodes and aqueous electrolytes with electrochemical impedance spectroscopy, *Analyst.* **138** 5540.

101. Y. Lu, Z. Tu, L.A. Archer (2014), Stable lithium electrodeposition in liquid and nanoporous solid electrolytes, *Nat. Mater.* **13** 961–969.

102. Q. Huang, M.J. Loveridge, R. Genieser, M.J. Lain, R. Bhagat (2018), Electrochemical evaluation and phase-related impedance studies on silicon–few layer graphene (FLG) composite electrode systems, *Sci. Rep.* **8** 1386.

103. D.K. Kang, H.C. Shin (2007), Investigation on cell impedance for high-power lithium-ion batteries, *J. Solid State Electrochem.* **11** 1405–1410.

104. X.B. Cheng, R. Zhang, C.Z. Zhao, F. Wei, J.G. Zhang, Q. Zhang (2015), A review of solid electrolyte interphases on lithium metal anode, *Adv. Sci.* **3** 1–20.

105. A. Wang, S. Kadam, H. Li, S. Shi, Y. Qi (2018), Review on modeling of the anode solid electrolyte interphase (SEI) for lithium-ion batteries, *Npj Comput. Mater.* **4** 15.

106. T. Li, P.B. Balbuena (2000), Theoretical studies of the reduction of ethylene carbonate, *Chem. Phys. Lett.* **317** 421–429.

107. D. Aurbach, M.D. Levi, E. Levi, A. Schechter (1997), Failure and stabilization mechanisms of graphite electrodes, *J. Phys. Chem. B.* **101** 2195–2206.

108. Y. Wang, S. Nakamura, M. Ue, P.B. Balbuena (2001), Theoretical studies to understand surface chemistry on carbon anodes for lithium-ion batteries: Reduction mechanisms of ethylene carbonate, *J. Am. Chem. Soc.* **123** 11708–11718.

109. J. Yu, P.B. Balbuena, J. Budzien, K. Leung (2011), Hybrid DFT functional-based static and molecular dynamics studies of excess electron in liquid ethylene carbonate, *J. Electrochem. Soc.* **158** A400.

110. K. Leung, J.L. Budzien (2010), Ab initio molecular dynamics simulations of the initial stages of solid–electrolyte interphase formation on lithium ion battery graphitic anodes, *Phys. Chem. Chem. Phys.* **12** 6583.

111. K. Tasaki (2005), Solvent decompositions and physical properties of decomposition compounds in li-ion battery electrolytes studied by DFT calculations and molecular dynamics simulations, *J. Phys. Chem. B.* **109** 2920–2933.

112. O. Borodin, G.D. Smith (2009), Quantum chemistry and molecular dynamics simulation study of dimethyl carbonate: Ethylene carbonate electrolytes doped with LiPF 6, *J. Phys. Chem. B.* **113** 1763–1776.

113. D. Bedrov, G.D. Smith, A.C.T. van Duin (2012), Reactions of singly-reduced ethylene carbonate in lithium battery electrolytes: A molecular dynamics simulation study using the ReaxFF, *J. Phys. Chem. A.* **116** 2978–2985.

114. M.M. Islam, A.C.T. van Duin (2016), Reductive decomposition reactions of ethylene carbonate by explicit electron transfer from lithium: An eReaxFF molecular dynamics study, *J. Phys. Chem. C.* **120** 27128–27134.

115. S.-P. Kim, A.C.T. van Duin, V.B. Shenoy (2011), Effect of electrolytes on the structure and evolution of the solid electrolyte interphase (SEI) in Li-ion batteries: A molecular dynamics study, *J. Power Sources.* **196** 8590–8597.
116. W. Liu, P. Liu, D. Mitlin (2020), Review of emerging concepts in SEI analysis and artificial SEI membranes for lithium, sodium, and potassium metal battery anodes, *Adv. Energy Mater.* **10** 2002297.
117. Z. Zhang, Y. Zhao, Z. Zhao, G. Huang, Y. Mei (2020), Atomic layer deposition-derived nanomaterials: Oxides, transition metal dichalcogenides, and metal–organic frameworks, *Chem. Mater.* **32** 9056–9077.
118. W. Luo, C.-F. Lin, O. Zhao, M. Noked, Y. Zhang, G.W. Rubloff, L. Hu (2017), Ultrathin surface coating enables the stable sodium metal anode, *Adv. Energy Mater.* **7** 1601526.

15 Concluding Remarks

Ngoc Thanh Thuy Tran
National Cheng Kung University

Van An Dinh
Osaka University

Ming-Fa Lin
National Cheng Kung University

Hikari Sakaebe
National Institute of Advanced Industrial
Science and Technology (AIST)

Le My Loan Phung
University of Science, Viet Nam National
University – Ho Chi Minh city (VNU HCM)

Chin-Lung Kuo
National Taiwan University

*Jeng-Shiung Jan, Wen-Dung Hsu,
and Jow-Lay Huang*
National Cheng Kung University

CONTENTS

This book is devoted to the further development of the quasiparticle framework and the successful syntheses of cathode/anode/electrolyte materials in ion-based batteries [1–6]. Two kinds of quasiparticles, electrons and polarons, are predicted to exhibit the rich and unique phenomena through the first-principles numerical simulations and molecule dynamics. The concise mechanisms are thoroughly identified from the essential physical/chemical/material properties [7,8]. Furthermore, the experimental researches can enhance the main functions of core materials and reduce many disadvantages. By the delicate analyses, only a few theoretical predictions on the fundamental are consistent with the high-resolution measurements. However, most

DOI: 10.1201/9781003263807-15

inconsistencies mainly arise from the technical limits and intrinsic environments. A series of experimental examinations, which are absent up to now, are very useful in providing the full information for the theoretical studies. These need to be finished in the near-future researches under the great modifications of the updated equipment. Most importantly, a grand unified theory, corresponding to the stationary ion transports, is urgently required for the basic and applied sciences. That is, how to identify the continuous intermediate states during charging and discharging processes and establish their relations with the ion currents are the next studying focuses.

The main-stream electrolyte/anode/cathode materials are outstanding candidates in verifying the quasiparticle viewpoints of the first-principles simulations [9–11]. The significant multi-orbital hybridizations of chemical bonds in solid-state crystals, which are responsible for the essential properties, are thoroughly examined from the relevant physical quantities. The diverse bonding behaviors cover the $2p_z$-$2p_z$ π-bondings and $(2s, 2p_x, 2p_y)$-$(2s, 2p_x, 2p_y)$ σ-bondings in graphitic layers (carbon-honey lattices in Chapter 6), the interlayer $2p_z$-$(3s, 3p_x, 3p_y, 3p_z)$ and intramolecular $(d_{xy}, d_{yz}, d_{zx}, d_{x^2-y^2}, d_{z^2})$-$(3s, 3p_x, 3p_y, 3p_z)$ bondings for $FeCl_3$ graphite intercalation compounds (C–Cl and Fe–Cl bonds), the prominent evidence of $2s$-$(2s, 2p_x, 2p_y, 2p_z)$ and $(d_{xy}, d_{yz}, d_{zx}, d_{x^2-y^2}, d_{z^2})$-$(2s, 2p_x, 2p_y, 2p_z)$ orbital mixings for $LiFeO_2$ (Li-/Fe-O bonds in Chapter 3) and $(2s, 2p_x, 2p_y, 2p_z)$- $(2s, 2p_x, 2p_y, 2p_z)$ orbital for P_2S_5 (P-S bonds in Chapter 12). The different orbital mixings correspond to the specific energy ranges, according to their strengths and ionization energies. Furthermore, the active orbital hybridizations could be utilized to specify the optical absorption structures and the presence/absence of excitonic effect [8]. However, the important mechanisms, which dominate the quantum transports of electron Fermions/heavy boson ions, have not been included in the current quasiparticle framework yet. The close relations between the various asymmetric environments (e.g., the concentration- and distribution-dependent intermediate states; Refs) and the stationary ionic currents need to be fully established through the near-future basic researches. The complicated time-dependent chemical reactions are successfully investigated by the method of molecular dynamics, e.g., the lithiation processes of silicon anode (details in Chapter 7). Their direct combinations with the first-principles simulations will be very powerful in exploring the charging and discharging processes of ion-based batteries. This is under current investigations, e.g., those for very interesting Li-, P-, S-, and Cl-related compounds in lithium-sulfur batteries [12–14].

The experimental works are mainly focused on the enhanced functions and reduced dissipations of anode/electrolyte/cathode materials. The successful syntheses by chemical methods, the geometric symmetries (the X-ray patterns), and the optimal ion currents are examined simultaneously. Both lithium- and sodium-related batteries can provide the diverse quasiparticle phenomena from certain significant differences. The second part could be verified by the first-principles simulations. It is also noticed that the X-ray diffraction spectra are available in revealing the spatial charge distributions [15,16]. This will be very useful in verifying the theoretical predictions. Furthermore, how to develop the time-resolved X-ray scatterings in observing the various intermediate crystal structures becomes an emergent issue. The close relations among the electrode voltage, cycle number, and specific charge capacity are fully established through the high-precision measurements. On the other hand, the systematic investigations on the essential properties, as done for band structures [7], optical absorption spectra, and Coulomb excitations [8], are almost absent. Such studies are expected to play critical roles in greatly enhancing the battery performances and testing the quasiparticle framework. For example, they will be done for the various multicomponent materials of lithium-sulfur batteries.

In Chapter 2, the effect of small polaron during the alkali intercalation/de-intercalation is not neglectable. In principle, the small polaron binds with alkali vacancies/ions by strong binding energy to form polaron–alkali ion/vacancy complexes. Therefore, the diffusion mechanism can be explained via elementary diffusion processes of polaron–alkali ion/vacancy complexes. The advanced model of small polaron–alkali vacancy/ion complex diffusion has been used successfully to simulate the diffusion mechanism inside the cathode materials applicable for both Li-ion and Na-ion rechargeable batteries. Based on the density functional theory (DFT) results, it is concluded that polaron generally hinders the diffusion of alkali ions and then lowers the ionic conductivity [17–25].

In Chapter 3, the 3D ternary $LiFeO_2$ compound, a candidate for cathode materials of lithium-ion batteries (LIBs), is predicted to exhibit the rich and unique lattice symmetries, energy band structures, charge and spin density distributions, and atom-/orbital-projected density of states. The accurate analyses, which are made from the exact first-principles calculations, are available in identifying the significant single-/multi-orbital hybridizations and spin configurations in Li-O and Fe-O bonds. Specifically, the 3D $LiFeO_2$ compound presents the Moire superlattice with high atomic ordering and anisotropy geometric, the presence of 18 Li-O/18 Fe-O chemical bonds with identical lengths, the complicated spin-polarization band structure with various atom dominations, the sizable indirect gap of 1.9 eV, the spatial spin and charge density distribution, and a lot of van Hove singularities due to the spin polarizations and extreme point dispersions. As the consequence, the important single-/multi-orbital hybridizations in Fe-O and Li-O bonds that assign the optical excitations could be well characteristics. Regarded optical properties, the weak but important impart of excitonic effect is exhibited by slightly reduced optical gap and extremely enhanced the single-particle excitations. The optical response also involves low reflectance/high transmission coefficients at an energy lower than the optical gap. There exits various prominent excitation peaks due to the contribution of certain valence electrons which are displayed by the electron energy loss functions, the absorption and reflectance spectrum. Our prediction provides certain meaningful information about the critical physical/chemical pictures in LIBs. Such state-of-the-art analysis is very useful for fully comprehending the diversified properties in anode/cathode/electrolyte and other emerging materials. Furthermore, the theoretical predictions on the optimal geometries, the occupied bands, many van Hove singularities, and rich optical excitations could be verified from the high-resolution measurements of the X-ray diffraction, angle-resolved photoemission spectroscopy, scanning tunneling spectroscopy, and optical spectroscopy, respectively.

In Chapter 4, the surface structure, high-voltage charge/discharge characteristics, and Li-ion transfer kinetics at the electrode/electrolyte interface of Al- and Zr-oxide-coated $LiNi_{1/3}Co_{1/3}Mn_{1/3}O_2$ and $LiCoO_2$ electrodes were investigated. The solid-solution phase of $LiAlO_2$-$LiMO_2$ was formed uniformly at the surface of the Al-oxide-coated $LiNi_{1/3}Co_{1/3}Mn_{1/3}O_2$ and $LiCoO_2$. For Zr-oxide-coated $LiNi_{1/3}Co_{1/3}Mn_{1/3}O_2$, the oxide phase containing Zr, Ni, Co, and Mn was formed continuously in the surface region of the sea-like zones, where ZrO_2 grains were not deposited. The cycle performance in the high-voltage region was significantly improved by either Zr-oxide or Al-oxide coating. The increase in the resistance of Li-ion transfer at the electrode/electrolyte interface with the cycle was significantly suppressed by the coating. It can be considered that the oxide phases containing Al or Zr suppress the formation of rock-salt-like or spinel-like phases at the surface region. In the analyses using the composite electrode and the single-crystal-plane electrode, it was confirmed that the Zr-based coating not only stabilized the electrode/electrolyte interface, but also reduced the E_a value for Li-ion transfer at the interface. Surface modification is an essential factor that stabilizes the electrode/ electrolyte interface and controls the Li-ion-transfer kinetics at the interface, even in the high-voltage region.

Chapter 5 provided an overview of the layered structural materials for sodium-ion batteries (SIBs). The application and electrochemical performance of both P2- and O3-types have been systematically investigated. The application of in situ and ex situ techniques has been supported to understand the structural evolution and phase transition. Besides that, the advantages and disadvantages of the different synthesis methods utilized for powder cathode materials have been discussed in detail to prove the efficiency in developing the layered-structure materials. Hence, single metal-layered materials (Na_xMO_2), two metals ($Na_xMM'O_2$), and three metals ($Na_xMM'M''O_2$) also illustrated the structure and electrochemistry of sodium insertion materials. From there, it shows the importance and applicability of the layered structure cathode material for SIBs.

In Chapter 6, the stage-n $FeCl_3$-graphite intercalation compounds exhibit the diversified quasiparticle phenomena owing to large intercalants and distinct concentration cases. Based on the VASP calculation and analyses, the results including geometric structure/charge distributions/charge variations/band structures/density of states display the situations of active chemical bonding, such as intralayer/interlayer C–C orbital hybridizations, the interlayer C-intercalant interactions, and intra-intercalants, and other important essential properties like the shift of Fermi level. All the above-mentioned cases have been fully examined from stage-1 to stage-4. The calculated results cover crystal symmetries, C-/Fe-/Cl-dominated energy spectra at specific energy ranges, charge distributions after intercalations/de-intercalations, and atom-/orbital-projected van Hove singularities. The observable multi-/single-orbital hybridizations of C-C/C-Cl/Fe-Cl/Cl-Cl chemical bonds cover [2s, 2px, 2py]-[2s, 2px, 2py] &2pz-2pz/2pz-[3px, 3py, 3pz]/[3dxy, 3dyz, 3dz², 3dxz, 3dx²]-[3s, 3px, 3py, 3pz]/[3s, 3px, 3py, 3pz]-[3s, 3px, 3py, 3pz]. However, the evidence almost disappears for the C-Fe and Fe-Fe bonds. In addition, the stage-n systems, with $n \geq 2$ under the highest intercalant concentrations, are under the current investigations. There are certainly significant differences between molecule-related and alkali-atom graphite intercalation compounds in terms of featured properties, e.g., the different n- or p-type doping [26], single-particle and collective excitations of π electrons [27], and optical initial absorption structures. By the delicate analyses on the VASP simulations, the intralayer/interlayer C–C orbital hybridizations, the interlayer C-intercalant interactions, and the intra- and intercalant ones could be reached thoroughly. This is capable of providing very useful information about the suitable hopping integrals and site energies of active orbitals within the phenomenological tight-binding model. As for the complicated intercalation and de-intercalation during the charging and discharging processes of the high-performance batteries, the critical mechanisms on the large-ion transports are very important research topics and worthy of thorough studies, especially for the theoretical framework developments.

In Chapter 7, the anisotropic lithiation in silicon is studied by our newly developed ReaxFF [28] parameter. We have successfully reproduced the anisotropic lithiation as in the experiment [29–31] with the lithiation rate order as $Si(112) > Si(110) > Si(100) > Si(111)$. The anisotropy shows temperature dependence and becomes more evident at a lower temperature. The phase boundary thickness is found to be in reverse order with the lithiation rate. Stress analysis is performed, and it was found that $Si(100)$ has higher compressive stress on the phase boundary than $Si(110)$ and $Si(112)$. Furthermore, $Si(111)$ has an even higher compressive stress on the surface Si atoms than other facets. These compressive stress developments on the silicon surface significantly increase the Li insertion energy barrier into silicon. Therefore, it slows down the lithiation rate on $Si(100)$ and $Si(111)$ to a different degree and causes the anisotropic lithiation. Finally, the stress-related energy barrier difference shows that the anisotropic lithiation is a kinetic-dominated rather than a thermodynamic-dominated process.

As discussed in Chapter 8, investigating monolayer HfX_2 (X = S, Se, or Te) indicates the dependences of geometric, electronic, and optical properties. The trends in band structure energy and optical properties are caused by the species atoms replacing the chalcogen atoms. This finding is attributed to the distinct features of three compounds when moving from sulfur to selenium to tellurium. The characteristic of multi-orbital hybridization could also be explained by this trend. By investigating the optical properties, the relation of geometry, electronic and optical features is demonstrated that shows the effect of structural parameters of these materials on their properties. Furthermore, the absorption, reflectivity, and refractive consequences exhibit the dielectric constant dependence. These materials might be mounted on a substrate, which will influence the absorption or reflection properties. Due to their electronic and optical properties that are similar to MoS_2 on LIBs and SIBs [32], three compounds promote the enhancement and application on optoelectronics, particularly in batteries as an anode. The results also suggest further investigations in optical excitation or other properties of these structures to verify the probability of this application.

In Chapter 9, we have used room/low-temperature chemical reduction procedure to prepare the MnO_2/rGO and Mn_3O_4//rGO nanocomposites and used them as an anode in LIB. MnO_2/rGO nanocomposite has delivered a good capacity of 660.9 mAh g^{-1} after 50 cycles (@)123 mA g^{-1}). In comparison, Mn_3O_4/ rGO nanocomposite produced better energy storage with the capacitance of 677 mAh g^{-1} after 150 cycles and very good rate capability (640 mAh g^{-1}) at 1.2 A g^{-1} charge/discharge rate and excellent Li-ion diffusion coefficient. This indicates that the synergistic effect of the rGO with the metal oxide is the key to the enhancement of the electrochemical property of the metal oxide. Several other important parameters such as the conductivity of the rGO, size of the metal oxide and its distribution over rGO, available active sites, and the surface area of the anode also affect the anode characteristics and need serious investigation.

In Chapter 10, a series of polymer electrolytes were prepared through the in situ polymerization of thiosiloxane, poly(ethylene glycol) diacrylate (PEGDA), imidazolium-based cross-linker (C-VIm), and lithium-bis(trifluoromethyl-sulfnyl) amide (LiTFSI) in a poly(ethylene glycol) dimethyl ether (PEGDME) additive to obtain mechanically robust membranes. The electrolytes have high thermal stabilities, qualified ionic conductivities, and advantageous electrochemical stabilities. It was found that the mechanical properties can be improved by adjusting the ratio of PEGDA/C-VIm, and A-50% exhibits the best compression resistance. In addition, the method of in situ polymerization, as well as the solventless process, is responsible for the remarkable discharge capacity and the improved capacity retention of Li/LiFePO$_4$ cells based on the prepared electrolytes. In summary, this study demonstrates an efficient preparation for polymer electrolytes, and the present findings reveal that they are promising candidates for the next-generation LIB technologies.

In Chapter 11, the theoretical framework of quasiparticle viewpoints is further developed for the significant electrolyte/cathode/anode materials of ion-based batteries [11,33,34]. The 3D binary lithium-sulfur compound [Li_8S_4; [35]], with 12 atoms in a primitive unit cell, is chosen for a model study, which is useful in understanding the diverse quasiparticle behaviors. This study on the essential properties can provide the necessary information in establishing the grand theory of the stationary ion transports, or it can help in examining the critical mechanisms of ion currents in various emergent materials. By the delicate first-principles simulations and analyses [36]), the featured quasiparticle properties are achieved from the optimal crystal structure, atom-dominated band structure and wave function, spatial charge density distribution, and van Hove singularities. In addition, the spin-spit configuration is absent, i.e., Li_8S_4 is a non-magnetic system. The main features cover the cubic lattice symmetry with a uniform bond length, but a highly anisotropic chemical environment, the multi-fold-degenerate electronic states near the high-symmetry points, a direct energy gap of $E_g = 3.365$ eV at the Γ point, a giant energy spacing of $-12.2\text{eV} \leq E^v \leq -4.7\text{eV}$ through the absence of valence states [because of the onsite Coulomb energy difference between 3s and 3p orbitals in the sulfur atom; [37]], the asymmetric energy spectrum of the hole and electron bands versus the Fermi level, the lithium and sulfur codominance regardless of the specific energy range (the stronger contribution from the latter), the oscillatory wave-vector dependences, a plenty of band-edge states, the frequent subband crossings in the absence of anti-crossing behavior, the non-spherical charge density distribution around two bonding atoms, the zero density of states inside the gap and vacant regions, and the merged special structures due to the 2s-[3s, $3p_x$, $3p_y$, $3p_z$]-orbital-decomposed van Hove singularities. These rich characteristics lead to a conclusion: the critical mechanism of $s - sp^3$ chemical bonding in determining the fundament quasiparticle properties. It should be noticed that only the theoretical predictions [37] on crystal structure are consistent with the X-ray scatterings [37,38]. The time-resolved experimental technique will be very important in observing the dynamic transformation of the intermediate configurations during the charge and discharging processes [39]. There are no high-resolution ARPES and STS measurements for electronic properties up to now, mainly owing to the intrinsic limits [40,41] and material characteristics [41]. How to explore the close relations between the intermediate configurations and the stationary ion currents is worthy of the systematic investigations under an enlarged quasiparticle framework [11,33,34].

The first-principles simulations in Chapter 12 have identified very complicated multi-orbital hybridizations [3s, $3p_x$, $3p_y$, $3p_z$]-[3s, $3p_x$, $3p_y$, $3p_z$] for P=S bonds in 3D binary phosphorus sulfide compound. This critical quasiparticle behavior is achieved from the main features such as geometric and electronic properties: a large Moire superlattice with the unusual first Brillouin zone and the highly nonuniform/anisotropic chemical environment [42], many asymmetric holes and electron energy spectra about the Fermi level [43], the complicated atom dominances due to phosphorus and sulfur atoms in the absence of specific energy ranges [44], a direct bandgap of ~2.544 eV at the $L\Gamma$ points [Refs], the oscillatory/parabolic/linear/partially flat energy dispersions [Refs], the observable distortions in charge density distributions close to two nearest-neighbors atoms [associated with 3s and ($3p_x$, $3p_y$, $3p_z$) orbitals; [42,43], the unique emergences of sp^3-orbital-decomposed van Hove singularities within the specific energy ranges. Only the predicted crystal structure is confirmed by the high-precision elastic X-ray scatterings [38] and inelastic Raman spectra [45]. But both ARPES [40,41,46] and STS [41] measurements are not finished until now, mainly owing to the intrinsic limits [the non-conserved transferred momenta of surface effect [47] and the almost forbidden quantum currents because of a 3D thickness; [47]. The delicate VASP calculations [36] and reliable quasiparticle viewpoints [11,33] could be generalized to the isomer compounds of P_4S_x [$x \leq 10$; [48]] and the various Li-, P- and S-related multicomponent materials [48]. Systematic studies are urgently requested for the basic and applied sciences. This strategy is expected to be very helpful in greatly enhancing the efficiencies of lithium-sulfur batteries [39,49–51].

In Chapter 13, as the demand for LIBs is increasing, it is critical to increasing its specific energy. Archiving high-voltage LIBs has recently gained much attraction as one way to deal with this issue. There are a lot of theoretical and experimental studies to find out the candidate for high-voltage cathode materials [1–3]. Searching for electrolytes that are compatible with a low-voltage anode such as graphite and high-voltage cathode about 4.5–5V plays an important role. By using both the static and dynamic part of the first-principles approaches, the electrolyte system influences the stability of the cathode surface. By calculating the electron energy of both cathode surface and electrolyte, the result shows that the oxidation of electrolyte may be attributed to the transient deficit of electrons during charging, as electrons have higher mobility than lithium ions. On the other hand, the abstraction of hydrogen from the electrolyte molecule due to the catalytic interaction with cathode surface oxygen may also facilitate the oxidation and the following decomposition of the electrolyte. The use of first-principles calculation shows some hints to the possible improvement of existing materials, which may be complemented with experiments to achieve a satisfying result.

In Chapter 14, the anode/electrolyte interphase the solid electrolyte interphase (SEI) plays a key role in SIBs. In the early stage, we have had a basic understanding of the first lithium intercalation phase into hard carbon, as well as the processes that occur on the lithium-metal anode. To deeply analyze the processes occurring at the anode/electrolyte interphase, a combination of analytical instruments such as AIMD, XPS, EDS, SEM, XRD, FTIR, Raman spectroscopy, and EIS impedance measurements are used. In-depth research focused on advanced characterization, and combined strategies are urgently needed to support the continued production of SIBs.

REFERENCES

1. Lin M-F, Hsu W-D. *Green Energy Materials Handbook*: CRC Press; 2019.
2. Lin M-F, Hsu W-D, Huang J-L. *Lithium-Ion Batteries and Solar Cells: Physical, Chemical, and Materials Properties*: CRC Press.
3. Tsai P-C, Nasara RN, Shen Y-C, Liang C-C, Chang Y-W, Hsu W-D, et al. Ab initio phase stability and electronic conductivity of the doped-$Li_4Ti_5O1_2$ anode for Li-ion batteries. *Acta Materialia*. 2019;175:196–205.
4. Song J, Wang K, Zheng J, Engelhard MH, Xiao B, Hu E, et al. controlling surface phase transition and chemical reactivity of O_3-layered metal oxide cathodes for high-performance Na-ion batteries. *ACS Energy Letters*. 2020;5(6):1718–25.
5. Wood M, Li J, Ruther RE, Du Z, Self EC, Meyer III HM, et al. Chemical stability and long-term cell performance of low-cobalt, Ni-Rich cathodes prepared by aqueous processing for high-energy Li-Ion batteries. *Energy Storage Materials*. 2020; 24:188–97.
6. Xiao Z, Sheng L, Jiang L, Zhao Y, Jiang M, Zhang X, et al. Nitrogen-doped graphene ribbons/MoS_2 with ultrafast electron and ion transport for high-rate Li-ion batteries. *Chemical Engineering Journal*. 2021;408:127269.
7. Han NT, Dien VK, Tran NTT, Nguyen DK, Su W-P, Lin M-F. First-principles studies of electronic properties in lithium metasilicate ($Li_2\ SiO_3$). *RSC Advances*. 2020; 10(41):24721–9.
8. Dien VK, Pham HD, Tran NTT, Han NT, Huynh TMD, Nguyen TDH, et al. Orbital-hybridization-created optical excitations in $Li_2\ GeO_3$. *Scientific Repor ts*. 2021;11(1):1–10.
9. Lin S-Y, Liu H-Y, Nguyen DK, Tran NTT, Pham HD, Chang S-L, et al. *Silicene-Based Layered Materials*: IOP Publishing Limited; 2020.
10. Lin S-Y, Tran NTT, Chang S-L, Su W-P, Lin M-F. *Structure-and Adatom-Enriched Essential Properties of Graphene Nanoribbons*: CRC press; 2018.
11. Tran NTT, Lin S-Y, Lin C-Y, Lin M-F. *Geometric and Electronic Properties of Graphene-Related Systems: Chemical Bonding Schemes*: CRC Press; 2017.
12. Bhargav A, He J, Gupta A, Manthiram A. Lithium-sulfur batteries: attaining the critical metrics. *Joule*. 2020;4(2):285–91.
13. Huang L, Li J, Liu B, Li Y, Shen S, Deng S, et al. Electrode design for lithium–sulfur batteries: Problems and solutions. *Advanced Functional Materials*. 2020;30(22):1910375.
14. Zhao M, Li B-Q, Zhang X-Q, Huang J-Q, Zhang Q. A perspective toward practical lithium–sulfur batteries. *ACS Central Science*. 2020;6(7):1095–104.
15. Wolf M, May BM, Cabana J. Visualization of electrochemical reactions in battery materials with X-ray microscopy and mapping. *Chemistry of Materials*. 2017;29(8): 3347–62.
16. Ganapathy S, Adams BD, Stenou G, Anastasaki MS, Goubitz K, Miao X-F, et al. Nature of Li_2O_2 oxidation in a Li–O_2 battery revealed by operando X-ray diffraction. *Journal of the American Chemical Society*. 2014;136(46):16335–44.
17. Tran TL, Luong HD, Duong DM, Dinh NT, Dinh VA. Hybrid Functional Study on Small Polaron Formation and Ion Diffusion in the Cathode Material $Na_2Mn_3\ (SO_4)$ 4. *ACS Omega*. 2020;5(10):5429–35.

18. Luong HD, Dinh VA, Momida H, Oguchi T. Insight into the diffusion mechanism of sodium ion–polaron complexes in orthorhombic P2 layered cathode oxide NaxMnO$_2$. *Physical Chemistry Chemical Physics*. 2020;22(32):18219–28.

19. Luong HD, Pham TD, Morikawa Y, Shibutani Y, Dinh VA. Diffusion mechanism of Na ion–polaron complex in potential cathode materials NaVOPO$_4$ and VOPO$_4$ for rechargeable sodium-ion batteries. *Physical Chemistry Chemical Physics*. 2018;20(36):23625–34.

20. Bui KM, Dinh VA, Okada S, Ohno T. Hybrid functional study of the NASICON-type Na$_3$V$_2$(PO$_4$)3: Crystal and electronic structures, and polaron–Na vacancy complex diffusion. *Physical Chemistry Chemical Physics*. 2015;17(45):30433–9.

21. Bui KM, Dinh VA, Ohno T. Diffusion mechanism of polaron–Li vacancy complex in cathode material Li2FeSiO4. *Applied Physics Express*. 2012;5(12):125802.

22. Duong DM, Dinh VA, Ohno T. Quasi-three-dimensional diffusion of Li ions in Li$_3$FePO$_4$CO$_3$: First-principles calculations for cathode materials of Li-ion batteries. *Applied Physics Express*. 2013;6(11):115801.

23. Dinh VA, Nara J, Ohno T. A new insight into the polaron–Li complex diffusion in cathode material LiFe1-yMnyPO$_4$ for Li ion batteries. *Applied Physics Express*. 2012;5(4):045801.

24. Debbichi M, Debbichi L, Dinh VA, Lebègue S. First principles study of the crystal, electronic structure, and diffusion mechanism of polaron-Na vacancy of Na$_3$MnPO$_4$CO$_3$ for Na-ion battery applications. *Journal of Physics D: Applied Physics*. 2016;50(4):045502.

25. Bui KM, Dinh VA, Okada S, Ohno T. Hybrid functional study of the NASICON-type Na$_3$V$_2$(PO$_4$)3: crystal and electronic structures, and polaron–Na vacancy complex diffusion. *Physical Chemistry Chemical Physics*. 2015;17(45):30433–9.

26. Meng X, Tongay S, Kang J, Chen Z, Wu F, Li S-S, et al. Stable p- and n-type doping of few-layer graphene/graphite. *Carbon*. 2013;57:507–14.

27. Wu F, Das Sarma S. collective excitations of quantum anomalous hall ferromagnets in twisted bilayer graphene. *PhRvL*. 2020;124(4):046403.

28. van Duin ACT, Dasgupta S, Lorant F, Goddard WA. ReaxFF: A reactive force field for hydrocarbons. *The Journal of Physical Chemistry A*. 2001;105(41):9396–409.

29. Liu XH, Zheng H, Zhong L, Huang S, Karki K, Zhang LQ, et al. Anisotropic swelling and fracture of silicon nanowires during lithiation. *Nano letters*. 2011;11(8):3312–8.

30. Lee SW, McDowell MT, Choi JW, Cui Y. Anomalous shape changes of silicon nanopillars by electrochemical lithiation. *Nano Letters*. 2011;11(7):3034–9.

31. Liu XH, Wang JW, Huang S, Fan F, Huang X, Liu Y, et al. In situ atomic-scale imaging of electrochemical lithiation in silicon. *Nature Nanotechnology*. 2012;7(11):749–56.

32. Soares DM, Mukherjee S, Singh G. TMDs beyond MoS$_2$ for Electrochemical Energy Storage. *Chemistry-a European Journal*. 2020;26(29):6320–41.

33. Shih-Yang Lin H-YL, Nguyen DK, Tran NTT, Pham HD, Chang S-L, Lin C-Y, Lin M-F. *Silicene-Based Layered Materials: Essential Properties*: IOP Publishing 2020.

34. Lin SYT, NTT, Chang SL, Su WP, Lin MF. *Structure- and Adatom-Enriched Essential Properties of Graphene Nanoribbons*: CRC press 2018.

35. Liu Z, Hubble D, Balbuena PB, Mukherjee PP. Adsorption of insoluble polysulfides Li 2 S x (x= 1, 2) on Li 2 S surfaces. *Physical Chemistry Chemical Physics*. 2015;17(14):9032–9.

36. Hafner J. Ab-initio simulations of materials using VASP: Density-functional theory and beyond. *Journal of Computational Chemistry*. 2008;29(13):2044–78.

37. Zhang L, Sun D, Feng J, Cairns EJ, Guo J. Revealing the electrochemical charging mechanism of nanosized Li2S by in situ and operando X-ray absorption spectroscopy. *Nano Letters*. 2017;17(8):5084–91.

38. Li J, Sun JL. Application of X-ray diffraction and electron crystallography for solving complex structure problems. *Accounts of Chemical Research*. 2017;50(11):2737–45.

39. Ding B, Wang J, Fan Z, Chen S, Lin Q, Lu X, et al. Solid-state lithium–sulfur batteries: Advances, challenges and perspectives. *Materials Today*. 2020.

40. Damascelli A. Probing the electronic structure of complex systems by ARPES. *Physica Scripta*. 2004;2004(T109):61.

41. Bussolotti F, Chi D, Goh KJ, Huang YL, Wee AT. STM/STS and ARPES characterization—structure and electronic properties. *2D Semiconductor Materials and Devices*: Elsevier; 2020. p. 199–220.

42. Ozturk T, Ertas E, Mert O. A Berzelius reagent, phosphorus decasulfide (P4S10), in organic syntheses. *Chemical Reviews*. 2010;110(6):3419–78.

43. Tahri Y, Chermette H, Hollinger G. Electronic structures of phosphorus oxide P_4O_{10} and phosphorus sulfides P_4S_{10} and P_4S_7. *Journal of Electron Spectroscopy and Related Phenomena*. 1991;56(1):51–69.

44. Ertaş E, Öztürk T, Ösken I. Purification of phosphorus decasulfide (P_4S_{10}). Google Patents; 2018.

45. Cantarero A. Raman scattering applied to materials science. *Procedia Materials Science*. 2015;9:113–22.

46. Kordyuk A. ARPES experiment in fermiology of quasi-2D metals. *Low Temperature Physics*. 2014;40(4):286–96.

47. Kittel C. *Introduction to Solid State Physics*, John Wiley & Sons. New York. 1996:402.

48. Von Schnering HG, Hönle W. Chemistry and structural chemistry of phosphides and polyphosphides. 48. Bridging chasms with polyphosphides. *Chemical Reviews*. 1988;88(1):243–73.

49. Li G, Chen Z, Lu J. Lithium-sulfur batteries for commercial applications. *Chem*. 2018;4(1):3–7.

50. Ould Ely T, Kamzabek D, Chakraborty D, Doherty MF. Lithium–sulfur batteries: State of the art and future directions. *ACS Applied Energy Materials*. 2018;1(5):1783–814.

51. Kang W, Deng N, Ju J, Li Q, Wu D, Ma X, et al. A review of recent developments in rechargeable lithium–sulfur batteries. *Nanoscale*. 2016;8(37):16541–88.

16 Open Issues and Challenges

Ngoc Thanh Thuy Tran, Thi Dieu Hien Nguyen,
Thi Han Nguyen, and Ming-Fa Lin
National Cheng Kung University

CONTENTS

The various high-performance batteries, as clearly shown in Tables 1.1–1.5 of Chapter 1, are outstanding candidates in providing the basic science researches, the potential applications, and the commercial products. According to the up-to-date performances, there exists much progress in theoretical predictions [1–4] and experimental measurements [3,5–7] after plenty of scientists' efforts. For example, the systematic investigations have been successfully developed within unified quasiparticle viewpoints, as done for 3D bulk graphite [8], 2D layered graphene systems [9], 2D silicene-related emergent materials [10], 1D graphene nanoribbons [11], and 3D multicomponent core materials of ion-based batteries [1,3,12,13]. Furthermore, the diversified quasiparticle phenomena are clearly illustrated from the various crystal symmetries [12,13], electronic properties [2,14], magnetic quantization [12,13], optical absorption spectra [15], and quantum Hall transports [16]. Even certain significant predictions are achieved in the previous [12,13] and current works, while more delicate quasiparticle frameworks need to be required in the near-future investigations. On the experimental side, six kinds of available batteries (Tables 1.1–1.5) are available in observing and verifying the distinct quasiparticle phenomena. How to utilize the high-precision measurements to achieve this most important goal will be discussed later.

The first-principles method, which is assisted by molecular dynamics, might be very powerful in exploring the intermediate crystal structures during the charging and discharging process of ion-based batteries. These quasi-stable states will create the various chemical environments and thus are responsible for the stationary ion currents. In general, the current core materials clearly show two cases in dramatically changing the lattice symmetries: (I) intercalations/de-intercalations of ions under the same host structure and (II) release-/insertion-diversified crystal variations. For example, lithium graphite intercalation compounds [8] and ternary/quaternary lithium oxides, respectively, belong to types (I) and (II). Interestingly, both experimental measurements on the initial- and final-state properties, the cycle time, and their X-ray patterns can support the theoretical simulations. The numerical

DOI: 10.1201/9781003263807-16

calculations, with the boundary conditions under the extreme cases, need to qualitatively evaluate the specific phase-transition time in each chemical reaction and then delicately display the spatial charge density distribution. Obviously, the nonstable configurations possess very large Moire superlattices, corresponding to most of the intermediate ones. These lead to the great enhancement of the numerical compute time in understanding the fundamental quasiparticle behaviors, such as the absence of the direct combination among band structures, spatial charge density distributions, van Hove singularities, and optical excitations in achieving the critical mechanisms [2,17]. Specifically, the highly nonuniform Coulomb potentials, which are generated by the various crystal structures, are expected to dominate the ion transport. That is, they play critical roles in the creation of stationary currents in ion-based batteries [12,13]. How to develop a unified framework of ion transports is under the current investigations.

Interestingly, various heterojunctions frequently appear in the sample growth on a specific substrate (e.g., monolayer/bilayer silicene system on Ag(111) [18,19]) and the generation of composite materials (electrolyte/cathode/anode [12,13]). A boundary, with two distinct subsystems (Tables 1.1–1.5), is capable of creating the nonnegligible chemical bonding. Generally speaking, it will exhibit a commensurate lattice structure under the controlled fluctuations of chemical bond lengths, and it becomes a noncommensurate system. Apparently, a Moire superlattice, which possesses a plenty of atoms in a primitive unit cell, comes to exist for all the real cases, leading to very high barriers of numerical simulations. According to the previous quasiparticle framework of the VASP simulations [12,13], the original orbital hybridizations, as well as the extra ones near the heterojunction, are expected to totally change the low-energy fundamental properties, such as the diversified quasiparticle phenomena of geometric symmetries and electronic energy spectra. How to thoroughly examine the calculated results and then identify the active orbital hybridizations in each chemical bond or specific spin magnetism must be a tough work. It should be noticed that the optimal widths in two subsystems would greatly reduce the computer simulation time. There are many valence and conduction energy subbands, in which the main features are very rich and complicated, compared with those in independent subsystems. The delicate analyses are required for the position-dependent chemical bonds; the atom-dominated band structure at the specific energy ranges; the spatial charge density distribution and its variation; the ferromagnetic or antiferromagnetic spin arrangement; and the atom-, orbital-, and spin-projected density of states. To comprehend the battery performance, the close relations, which are associated with the intrinsic charge interactions and the ion current density, need to be built at the heterojunction and two subsystems.

The up-to-date VASP simulations are reliable for fully exploring the diverse optical transitions in cathode/electrolyte/anode materials of ion-based batteries (six kinds in Tables 1.1–1.5). They are successfully done for $LiFeO_2$ in Chapter 3. The real and imaginary parts of the transverse dielectric function, with an almost-vanishing moment transfer, are responsible for the dynamic charge responses under an electromagnetic wave perturbation. Furthermore, they can determine the energy loss function, the reflectance spectrum, the absorption coefficient, and the refractive index [15]. Their main features, the frequency, number, intensity, and form of special absorption structures, might be sensitive to the existence/absence of bound-state/quasi-stable excitons. Such effects mainly originate from the attractive electron-hole Coulomb potentials. How to link the excitonic effects, the prominent absorption structures, and the active orbital hybridizations/the significant spin arrangements in various chemical bonds needs to be delicately finished through the direct combination of the VASP electronic and optical properties [15]. As a result, each optical absorption structure, corresponding to the specific orbital hybridizations/magnetic configuration, could be achieved for any battery materials. The important similarities and differences among various cathodes, electrolytes, and anodes are worthy of the systematic investigations. This will clearly illustrate the diversified quasiparticle properties, thus leading to the further development of the theoretical quasiparticle framework. In addition, the photoluminescence spectra of unstable excitons cannot be solved by the first-principles calculation up to now.

The direct linking between the VASP band-structure simulations and the tight-binding model can open a series of phenomenological studies, such as the charged impurity screenings, Coulomb excitations, magneto-optical selection rules, and quantum Hall conductivities [20]. When cathode/electrolyte/anode materials have Moire superlattices, a lot of valence and conduction energy subbands near the Femi level, as done by the first-principles method, are difficult to achieve a good fitting. For example, as to $LiFeO_2$ in Chapter 3, it is almost impossible to get the reliable hopping integrals of the Hamiltonian matrix elements in the absence of delicate analyses. The critical factors cover the position-dependent chemical bonds and the complicated orbital hybridizations. On the other hand, the high-symmetry core materials only possess simple crystal structure, e.g., graphite/well-behaved graphite intercalation compounds [8], silicene/silicon nanotubes [21], and lithium/iron metals [3]. Their unusual low-lying energy spectra are expected to be well simulated through the concise intralayer and interlayer atomic interactions. Interestingly, the diversified magnetic quantization could be thoroughly identified from the generalized tight-binding model, which is assisted by the vector-potential-induced Peierls phases. The magnetic Moire superlattices come to exist, while their magneto-essential properties could be solved by the featured Landau levels, such as the successful framework development of quasiparticle quantization phenomena on 3D bulk graphites [8], 2D few-layer graphenes [22], and other group-IV and group-V layered materials [10]. It is also noticed that the direct combination of this model with the modified random-phase approximation/self-energy method [23,24] (the static/dynamic Kubo formula [25]) is very useful in comprehending single-particle and collective excitations/quasiparticle energy spectra and lifetimes (the electrical conductivities/vertical optical transitions). Obviously, the quasiparticle particles of cathode/electrolyte/anode materials will display the diverse behaviors.

The excited quasiparticle behaviors, which are associated with the time-dependent Coulomb fields, directly reflect the intrinsic electron–electron interactions. The quasi-particle energy spectra and their lifetimes (inelastic Coulomb scatterings) play the critical roles in the various fundamental properties. Generally speaking, the single- and many-particle charge screenings are well characterized by the longitudinal dielectric functions with the significant momentum and frequency dependences. The bare response function could be explored by the random-phase approximation (RPA) [23,24], in which the band-structure effects are fully included in the delicate calculations. Furthermore, screened response functions and electron energy loss spectra are very useful in comprehending the (momentum, frequency)-excitation phase diagrams. The diverse Coulomb excitation phenomena without/with the free carrier doping cover the rich electron-hole boundaries, the dimensionality-enriched acoustic/optical plasmon modes [26], and the unusual Landau dampings [27]. Most importantly, the modified RPA is successfully developed for 2D layered graphene systems [28] and 1D coaxial multiwalled carbon nanotubes [29], clearly illustrating the diversified Coulomb interactions. The regular RPA should be sufficient for 3D graphite and graphite intercalation compounds, in which the superlattice model might be suitable in investigating the doping-induced and π plasmon modes with the coherent oscillation ω_p frequencies of 0.5 and 5.0 eV, respectively [30,31]. When this model is utilized to study the electronic excitations of ternary/quaternary lithium oxides in ion-based batteries [31], the local-field effects are too strong to hinder the numerical calculations because of a very small Brillouin zone. That is, it is very difficult to observe the momentum (dependent coherent charge oscillations) in a Moire superlattice except for the long wavelength limit.

On the experimental side, a lot of challenges come to exist, mainly owing to Moire superlattices, intermediate configurations, nonnegligible drawbacks of measurement techniques, and intrinsic limits of condensed-matter systems [32]. During the significant charging/discharging processes in ion-based batteries (Figure 16.1), the reduction or enhancement of the specific ion concentration and arrangement is revealed at any position and time. This clearly indicates very rapid variations in crystal structures; that is, cathode, electrolyte, and anode are expected to exhibit a series of dramatic transformations in lattice symmetries. The up-to-date X-ray diffraction patterns (Figure 16.2) could be measured only for the initial and final configurations (the extreme cases under the chemical reactions) [33,34]. How to get more information about the dynamic processes of geometric structures using the femtosecond

(a) Zigzag nanotude   (b) Armchair nanotude   (c) Zigzag GNR   (d) Armchair GNR

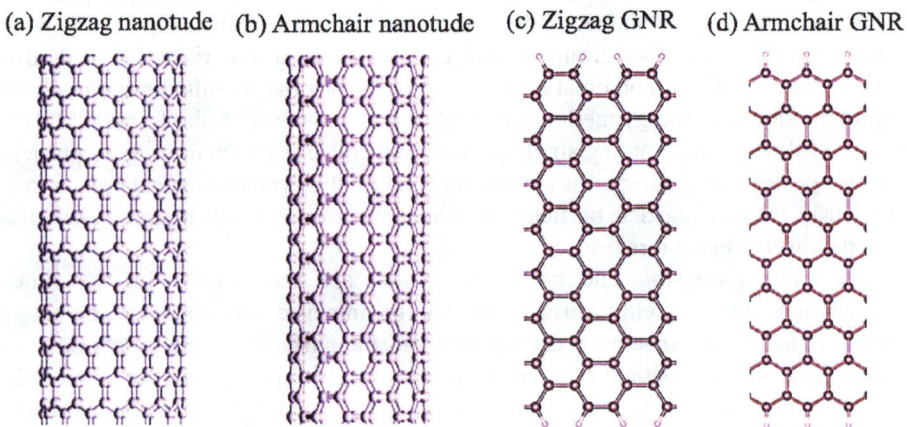

FIGURE 16.1   The charging and discharging processes in lithium-ion-related batteries.

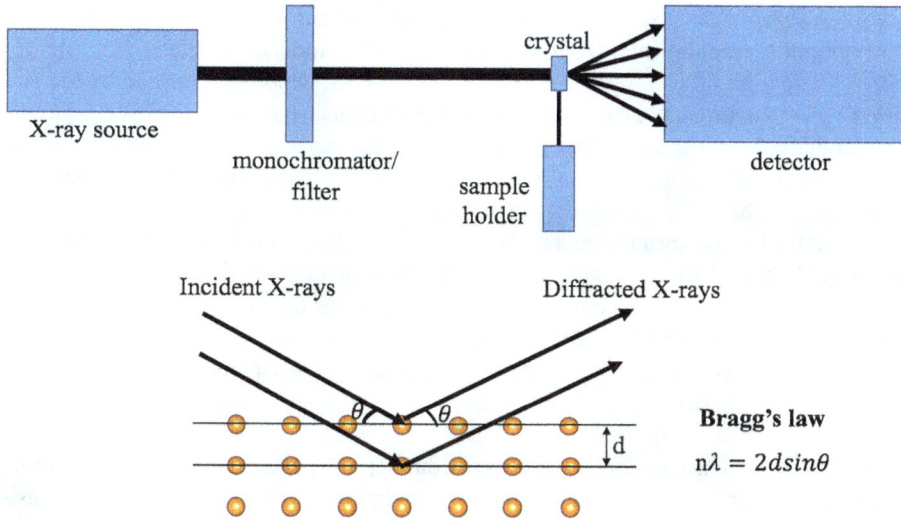

FIGURE 16.2 The X-ray diffraction apparatus.

techniques [35,36] will become important challenge. Such high-precision experiments are very useful in the further investigations of ion transports. In addition, scanning tunneling microscopy (STM) and tunneling electronic microscopy (TEM) cannot examine the time-dependent geometric structures of 3D bulk materials during the chemical reactions.

The predicted band structures, which are calculated along with the high-symmetry points within the first Brillouin zone, could be verified by the high-resolution angle-resolved photoemission spectroscopy (ARPES) as shown in Figure 16.3. The up-to-date measurements clearly show that ARPES is very powerful in examining the wave-vector dependences of quasiparticle energy spectra and lifetimes. They are very suitable for 2D-layered condensed-matted systems, e.g., AB- and ABC-stacked

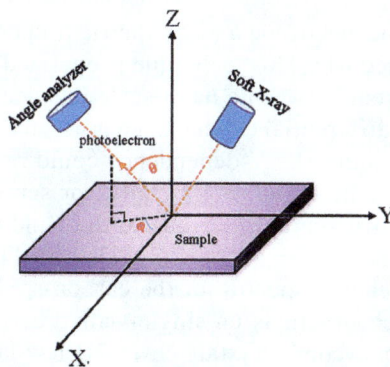

FIGURE 16.3 The angle-resolved photoemission spectroscopy.

few-layer graphene systems [37–39]. The outstanding examples even cover 3D bulk graphite/graphite intercalation compounds [38] and 1D graphene nanoribbons [40,41]. The high precisions of ARPES measurements, which are assisted by the featured band structures, are capable of solving the intrinsic barriers. However, a sample surface induces the nonconservation of a perpendicular momentum transfer, leading to the analysis difficulty, especially for Moire superlattices with a lot of valence subbands below the Fermi level. For example, the ARPES measurements, which are associated with the ternary/quaternary lithium oxides, with middle bandgaps, are never published in the previous works. Very complicated valence energy subbands, with the nonmonotonous behaviors, are responsible for the absence of delicate measurements, e.g., the VASP results for the occupied electronic energy spectra close to E_F in $LiFeO_2/LiGe_2O/LiTiO$ compounds (cathode/electrolyte/anode materials of the lithium-ion-based batteries). How to overcome such intrinsic measurements barriers is the next-step studying focus.

The optical measurements, which are conducted by the reflectance, absorption, transmission, and photoluminescence spectroscopies, can test the critical quasiparticle properties, the orbital-hybridization-induced absorption structures [17]. According to the single-particle and many-body VASP calculations, the excitonic effects are directly reflected in an obvious redshift of the threshold absorption frequency ω_1, and the observable changes of the spectral intensities, forms, peak numbers, and absorption frequencies [17]. The delicate examination and analysis, the linking of electronic energy spectra/active orbital hybridizations, and prominent absorption structures are absent up to now. For example, while the various optical spectra are measured for lithium oxides [42–44] within the wide-frequency ranges, the close relations between the important quasi-particle properties and vertically optical transitions cannot be reached from these previously published papers. Apparently, such very difficult works are closely related to a plenty of atoms in a Moire superlattice [45–47]. Also, the experimental photoluminescence spectra of [48,49] core materials, which come from the further excitonic decays, are unable to provide the concise physical pictures and the critical mechanisms. Furthermore, they do not get any supports from the theoretical predictions in the absence of VASP simulations and model calculations.

Scanning tunneling spectroscopy (STS), which is assisted by STM, is very powerful in verifying the electronic and geometric properties of the low-dimensional systems simultaneously. This technique is operated through a rather weak tunneling quantum current between a nano-scaled device and a sample surface. The voltage-dependent differential conductance is assumed to be roughly proportional to the density of states. Its V-dependence could reveal the band property near the Fermi level (the metallic, semimetallic, or semiconducting behaviors) and the featured van Hove singularities due to the band-edge states. The high-precision STS measurements have successfully identified the diverse lattice symmetries and electronic energy spectra for the carbon-sp^2 bonding systems since the first simultaneous observations on single-walled carbon nanotubes in 1998 [50,51]. Such carbon honeycomb crystals cover 2D few-layer graphenes [52–55] and 1D graphene nanoribbons [56–58]. The critical mechanisms are examined to

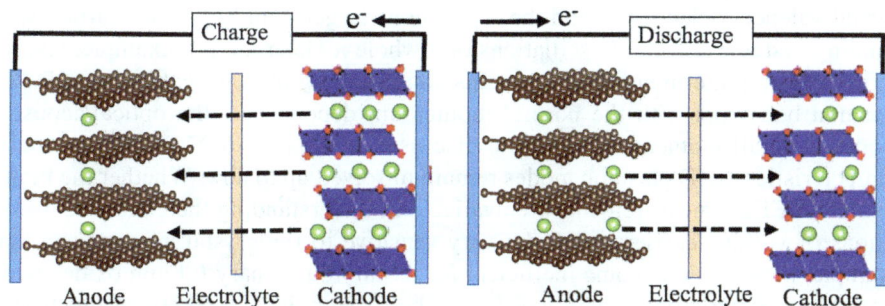

FIGURE 16.4 The low-dimensional carbon sp²-bonding systems: (a) Armchair, (b) zigzag carbon nanotube, (c) armchair, and (d) zigzag graphene nanoribbons.

be, respectively, stacking configuration and number of layers, and width and edge structure. According to geometric structures of (m, n) carbon nanotubes, there are three kinds of band properties: metals for (m, m) armchairs (Figure 16.4), narrow gaps for $2m + n = 3I$, and $m \neq, n$ (I an integer; curvature effects due to cylindrical surfaces), and middle gaps for others (the periodical boundary condition). The layered graphene systems show the unusual behaviors near the Fermi level, such as the semimetallic trilayer ABC stacking [59,60] and the narrow-gap AAB stacking [59]. Moreover, all graphene nanoribbons have energy gaps, mainly owing to the observable fluctuations of bond lengths near the passivated/open edges and spin arrangements at the edges. However, the main features of van Hove singularities, which mainly arise from most of the core materials in ion-based batteries, are very difficult to be examined by the tunneling current because of a sufficiently thick width, such as the absence of STS measurements on ternary and quaternary lithium oxides. The high-resolution STS experiments are worthy of the systematic investigations on low-dimensional electrode materials of rechargeable batteries.

Electronic energy loss spectroscopy (EELS) is very powerful for exploring the electron-hole pair excitations and the plasmon modes (momentum and frequency)-phase diagrams. Both reflection and transmission EELSs can measure the momentum-dependent dynamic carrier excitation behaviors due to the longitudinal Coulomb fields. However, the various optical spectra are measured only under the long-wavelength limit. These high-precision EELS measurements have been successfully done for graphite [61], stage-n graphite intercalation compound [62], layered graphene systems [63], and carbon nanotubes [64]. Interestingly, such results clearly clarify the dimensionality-, stacking-, layer-number-, and doping-enriched many-particle quasiparticle phenomena (the diversified Coulomb interactions). The up-to-date π-electronic excitations of EELS examinations could be classified into four kinds of plasmon modes according to their main features: the rather low-frequency plasmons of ~10–100 meVs in semimetallic graphite, free carrier plasmons of ~0.5–1.0 eV in donor-/acceptor-type graphite intercalation compounds and doped graphene ones, the inter-π-band plasmons of ~1.0–5.0 eVs in carbon nanotubes (the partial valence π electrons), and the π plasmons larger than 5 eV in all carbon-sp^2-bonding systems (coherent oscillations of a whole π electron). For example, alkali- and FeCl$_3$-graphite intercalation systems/alkali-dope graphenes exhibit the 3D/2D plasmon behaviors with the normal momentum dependence (the optical/acoustic modes at small momentum transfers) discussed in Chapter 6. However, the main characteristics of 1D plasmon modes remain unsolved up to now. Whether the high-resolution EELS measurements are available in understanding the collective oscillations of various carriers might be very sensitive to the crystal symmetries. For example, they would become inefficient for ternary/quaternary lithium oxides with large Moire superlattices, mainly owing to the strongly local-field effects due to the small first Brillouin zone. Such intrinsic limits also lead to the near-future serious challenges.

In addition to ion-based batteries, the other green energy materials have stirred a plenty of researches both theoretically and experimentally, such as hydrogen storages [65–67] and perovskite solar cells [68–70]. Interestingly, the direct combination of the half-occupied hydrogen atoms and the 2D morphologies are capable of generating the active chemical environments, greatly diversifying the quasiparticle phenomena, and thus largely enhancing the potential applications. For example, the hydrogenated group-IV materials are expected to exhibit the rich and unique properties, as done for graphene systems under the various hydrogenation cases [9]. The critical mechanisms, the significant couplings of the planar/buckling honeycomb crystals, chemical bonding, spin-orbital couplings, need to be clarified for any physical/chemical/material properties [9–11]. That is, according to the delicate VASP simulations and analyses, their close relations will be fully examined from the unusual geometric structures, H- & C-/Si-/Ge-/Sn-/Pb-dominated band structures within the distinct energy ranges, the spatial charge density distribution densities before and after hydrogenation reactions, and the atom- and orbital-projected density of states [71]. The important similarities and differences among them are under the current investigations. Most importantly, how to get a lot of merits and drawbacks on hydrogen storage/release, which are indicated from the above-mentioned VASP simulations, becomes a near-future challenge. The featured quasiparticle phenomena will be very helpful in green energy applications.

Interestingly, solar cells, which are friendly devices under the powerful conversion efficiency (transformation of photon energy into electricity), are one of the main-steam materials in basic sciences and engineering applications [72–75]. The well-characterized architecture of an unusual perovskite sample, as clearly shown in Figure 16.5a and b, covers the absorber layer sandwich between the electron- and the hole-transporting layers [72,73]. Most importantly, this core component efficiently absorbs the external electromagnetic waves and generates a lot of electron-hole pairs due to the vertically optical transitions. Apparently, the featured optical properties are strongly dependent on the main characteristics of electronic energy spectra and wave functions, such as the threshold absorption frequency and its relation with the bandgap, the prominent absorption structures associated with the joint van Hove singularities, and the excitonic effects. How to examine the close relations

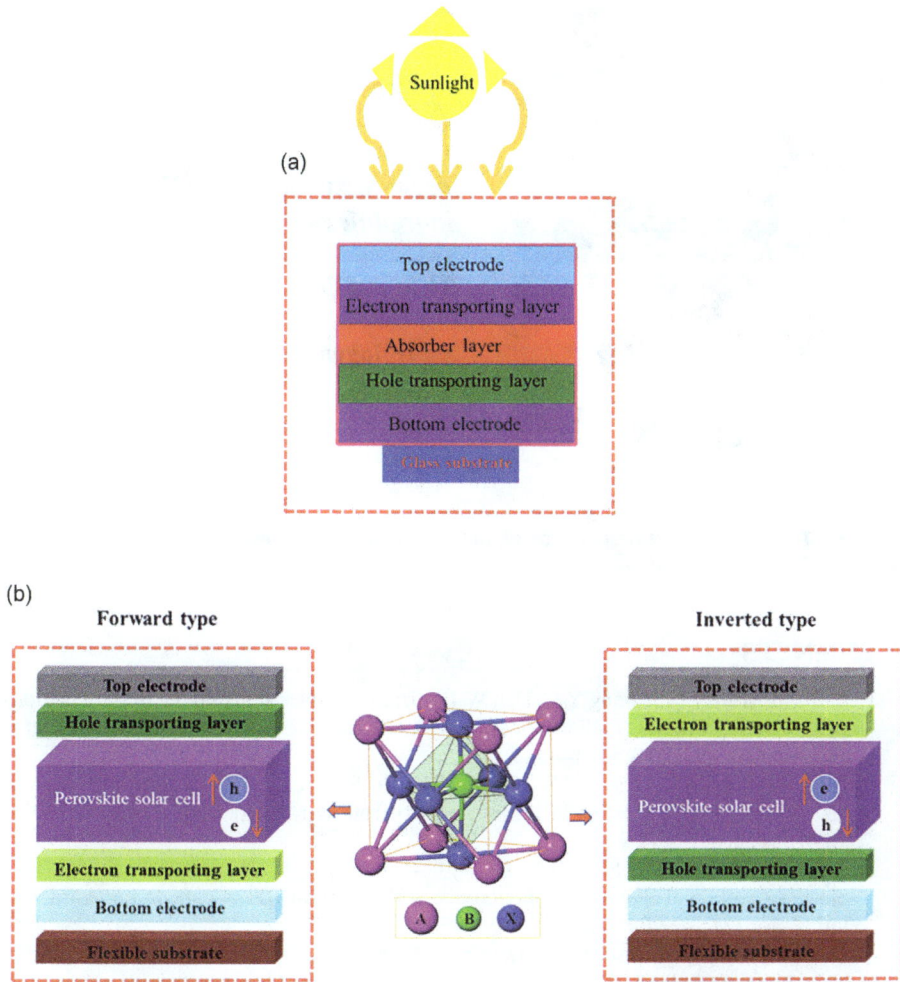

FIGURE 16.5   (a) The standard structure and (b) types of solar cells.

among the essential properties is worthy of the systematic investigations. Similarly, the high-precision experimental observations are urgently required for the consistent verifications of distinct physical quantities, e.g., the X-ray diffraction peaks due to specific crystal symmetry, the measurement of ARPES on the occupied energy spectrum, and the reflectance/absorption/transmittance/photoluminescence spectra by the corresponding optical spectroscopies. Recently, the metal halide perovskites (Figure 16.6), with a general formula of ABX_3 (A = cation, B = metal ion, and X = halogen anion), are expected to be an outstanding candidate in illustrating the diverse quasiparticle phenomena [76–78]. Such solar-cell materials should be included in the current development of the quasiparticle framework.

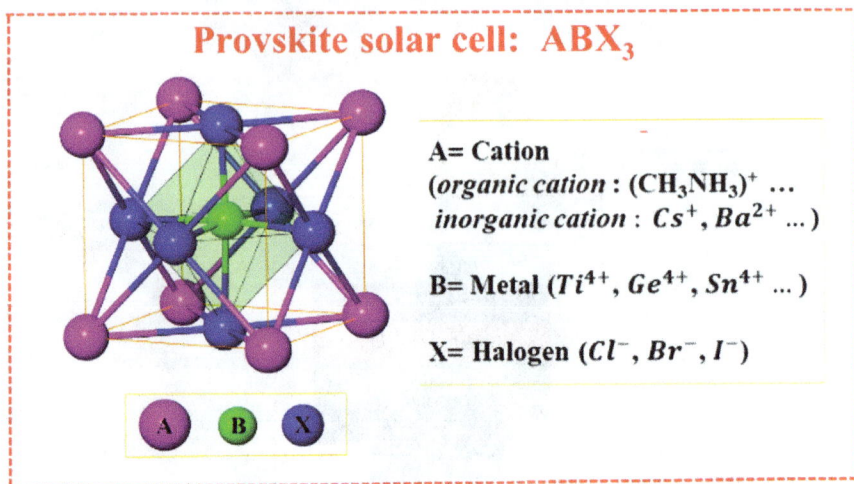

FIGURE 16.6 The crystal structure of the metal halide perovskites: AMX_3 (A = cation, M = metal ion, and X = halogen anion).

REFERENCES

1. Pan T-Y, Tran NTT, Chang Y-C, Hsu W-D. First-principles study on the initial reactions at $LiNi_{1/3}Co_{1/3}Mn_{1/3}O_2$ cathode/electrolyte interface in lithium-ion batteries. *Applied Surface Science*. 2020;507:144842.
2. Han NT, Dien VK, Tran NTT, Nguyen DK, Su W-P, Lin M-F. First-principles studies of electronic properties in lithium metasilicate ($Li_2 SiO_3$). *RSC Advances*. 2020; 10(41):24721–9.
3. Tsai P-C, Nasara RN, Shen Y-C, Liang C-C, Chang Y-W, Hsu W-D, et al. Ab initio phase stability and electronic conductivity of the doped-$Li_4Ti_5O_{12}$ anode for Li-ion batteries. *Acta Materialia*. 2019;175:196–205.
4. Lin F-W, Tran NTT, Hsu W-D. Effect of 1, 3-propane sultone on the formation of solid electrolyte interphase at Li-ion battery anode surface: A first-principles study. *ACS Omega*. 2020;5(23):13541–7.
5. Liu X, Wang Y, Yang Y, Lv W, Lian G, Golberg D, et al. A MoS_2/Carbon hybrid anode for high-performance Li-ion batteries at low temperature. *Nano Energy*. 2020;70:104550.
6. Wood M, Li J, Ruther RE, Du Z, Self EC, Meyer III HM, et al. Chemical stability and long-term cell performance of low-cobalt, Ni-Rich cathodes prepared by aqueous processing for high-energy Li-Ion batteries. *Energy Storage Materials*. 2020;24:188–97.
7. Song J, Wang K, Zheng J, Engelhard MH, Xiao B, Hu E, et al. Controlling surface phase transition and chemical reactivity of O3-Layered metal oxide cathodes for high-performance Na-ion batteries. *ACS Energy Letters*. 2020;5(6):1718–25.
8. Li W-B, Lin S-Y, Tran NTT, Lin M-F, Lin K-I. Essential geometric and electronic properties in stage-n graphite alkali-metal-intercalation compounds. *RSC Advances*. 2020;10(40):23573–81.
9. Tran NTT, Lin S-Y, Lin C-Y, Lin M-F. *Geometric and Electronic Properties of Graphene-Related Systems: Chemical Bonding Schemes*: CRC Press; 2017.
10. Lin S-Y, Liu H-Y, Nguyen DK, Tran NTT, Pham HD, Chang S-L, et al. *Silicene-Based Layered Materials*: IOP Publishing Limited; 2020.

11. Lin S-Y, Tran NTT, Chang S-L, Su W-P, Lin M-F. *Structure-and Adatom-Enriched Essential Properties of Graphene Nanoribbons*: CRC Press; 2018.
12. Lin M-F, Hsu W-D, Huang J-L. *Lithium-Ion Batteries and Solar Cells: Physical, Chemical, and Materials Properties*: CRC Press.
13. Lin M-F, Hsu W-D. *Green Energy Materials Handbook*: CRC Press; 2019.
14. Tran NTT, Nguyen DK, Lin S-Y, Gumbs G, Fa-Lin M. Fundamental properties of transition-metals-adsorbed graphene. *arXiv preprint arXiv:190501010*. 2019.
15. Dien VK, Pham HD, Tran NTT, Han NT, Huynh TMD, Nguyen TDH, et al. Orbital-hybridization-created optical excitations in $Li_2 GeO_3$. *Scientific Reports*. 2021;11(1):1–10.
16. Ramanayaka AN, Mani RG. Transport study of the Berry phase, resistivity rule, and quantum Hall effect in graphite. *Physical Review B*. 2010;82(16):165327.
17. Dien VK, Pham HD, Tran NTT, Han NT, Huynh TMD, Nguyen TDH, et al. Orbital-hybridization-created optical excitations in Li8Ge4O12. *arXiv preprint arXiv:2009021 60*. 2020.
18. Galashev A, Ivanichkina K, Vorob'ev A, Rakhmanova O. Structure and stability of defective silicene on Ag (001) and Ag (111) substrates: A computer experiment. *Physics of the Solid State*. 2017;59(6):1242–52.
19. Feng B, Ding Z, Meng S, Yao Y, He X, Cheng P, et al. Evidence of silicene in honeycomb structures of silicon on Ag (111). *Nano Letters*. 2012;12(7):3507–11.
20. Kaloni TP, Modarresi M, Tahir M, Roknabadi MR, Schreckenbach G, Freund MS. Electrically engineered band gap in two-dimensional Ge, Sn, and Pb: A first-principles and tight-binding approach. *The Journal of Physical Chemistry C*. 2015;119(21): 11896–902.
21. Tran NTT, Gumbs G, Nguyen DK, Lin M-F. Fundamental properties of metal-adsorbed silicene: A DFT study. *ACS Omega*. 2020;5(23):13760–9.
22. Tran NTT, Lin S-Y, Glukhova OE, Lin M-F. Configuration-induced rich electronic properties of bilayer graphene. *The Journal of Physical Chemistry C*. 2015;119(19): 10623–30.
23. Avogadro P, Nakatsukasa T. Finite amplitude method for the quasiparticle random-phase approximation. *Physical Review C*. 2011;84(1):014314.
24. Hesselmann A. Random-phase-approximation correlation method including exchange interactions. *Physical Review A*. 2012;85(1):012517.
25. Crépieux A, Bruno P. Theory of the anomalous Hall effect from the Kubo formula and the Dirac equation. *Physical Review B*. 2001;64(1):014416.
26. Van Men N, Khanh NQ, Phuong DTK. Plasmon modes in double bilayer graphene heterostructures. *Solid State Communications*. 2019;294:43–8.
27. Li X, Xiao D, Zhang Z. Landau damping of quantum plasmons in metal nanostructures. *New Journal of Physics*. 2013;15(2):023011.
28. Allison K, Borka D, Radović I, Hadžievski L, Mišković Z. Dynamic polarization of graphene by moving external charges: Random phase approximation. *Physical Review B*. 2009;80(19):195405.
29. Roche S, Triozon F, Rubio A, Mayou D. Electronic conduction in multi-walled carbon nanotubes: Role of intershell coupling and incommensurability. *Physics Letters A*. 2001;285(1–2):94–100.
30. Lebègue S, Harl J, Gould T, Ángyán J, Kresse G, Dobson J. Cohesive properties and asymptotics of the dispersion interaction in graphite by the random phase approximation. *Physical Review Letters*. 2010;105(19):196401.
31. Lazar P, Otyepková E, Karlický F, Čépe K, Otyepka M. The surface and structural properties of graphite fluoride. *Carbon*. 2015;94:804–9.
32. Ryu YK, Frisenda R, Castellanos-Gomez A. Superlattices based on van der Waals 2D materials. *Chemical Communications*. 2019;55(77):11498–510.

33. Ganapathy S, Adams BD, Stenou G, Anastasaki MS, Goubitz K, Miao X-F, et al. Nature of Li_2O_2 oxidation in a Li–O_2 battery revealed by operando X-ray diffraction. *Journal of the American Chemical Society*. 2014;136(46):16335–44.
34. Wolf M, May BM, Cabana J. Visualization of electrochemical reactions in battery materials with X-ray microscopy and mapping. *Chemistry of Materials*. 2017;29(8):3347–62.
35. Xiong C, Wang Z, Peng X, Guo Y, Xu S, Zhao T. Bifunctional effect of laser-induced nucleation-preferable microchannels and in situ formed LiF SEI in MXenes for stable lithium-metal batteries. *Journal of Materials Chemistry A*. 2020;8(28):14114–25.
36. Li Q, Sun X, Zhao W, Hou X, Zhang Y, Zhao F, et al. Processing of a large-scale microporous group on copper foil current collectors for lithium batteries using femtosecond laser. *Advanced Engineering Materials*. 2020;22(12):2000710.
37. Ohta T, Bostwick A, McChesney JL, Seyller T, Horn K, Rotenberg E. Interlayer interaction and electronic screening in multilayer graphene investigated with angle-resolved photoemission spectroscopy. *Physical Review Letters*. 2007;98(20):206802.
38. Grüneis A, Attaccalite C, Pichler T, Zabolotnyy V, Shiozawa H, Molodtsov S, et al. Electron-electron correlation in graphite: A combined angle-resolved photoemission and first-principles study. *Physical Review Letters*. 2008;100(3):037601.
39. Coletti C, Forti S, Principi A, Emtsev KV, Zakharov AA, Daniels KM, et al. Revealing the electronic band structure of trilayer graphene on SiC: An angle-resolved photoemission study. *Physical Review B*. 2013;88(15):155439.
40. Sugawara K, Sato T, Souma S, Takahashi T, Suematsu H. Fermi surface and edge-localized states in graphite studied by high-resolution angle-resolved photoemission spectroscopy. *Physical Review B*. 2006;73(4):045124.
41. Ruffieux P, Cai J, Plumb NC, Patthey L, Prezzi D, Ferretti A, et al. Electronic structure of atomically precise graphene nanoribbons. *ACS Nano*. 2012;6(8):6930–5.
42. Tumėnas S, Mackonis P, Nedzinskas R, Trinkler L, Berzina B, Korsaks V, et al. Optical properties of lithium gallium oxide. *Applied Surface Science*. 2017;421:837–42.
43. Joshi Y, Hadjixenophontos E, Nowak S, Lawitzki R, Ghosh PK, Schmitz G. Modulation of the optical properties of lithium manganese oxide via Li-ion de/intercalation. *Advanced Optical Materials*. 2018;6(12):1701362.
44. J A, Sahoo T. Effect of Li doping on conductivity and band gap of nickel oxide thin film deposited by spin coating technique. *Materials Research Express*. 2019;7(1):016405.
45. Dien VK, Pham HD, Tran NTT, Han NT, Huynh TMD, Nguyen TDH, et al. Orbital-hybridization-created optical excitations in Li_2GeO_3. *Scientific Reports*. 2021; 11(1):4939.
46. Han NT, Dien VK, Thuy Tran NT, Nguyen DK, Su W-P, Lin M-F. First-principles studies of electronic properties in lithium metasilicate (Li_2SiO_3). *RSC Advances*. 2020;10(41):24721–9.
47. Nguyen TDH, Pham HD, Lin S-Y, Lin M-F. Featured properties of Li+-based battery anode: $Li_4Ti_5O_{12}$. *RSC Advances*. 2020;10(24):14071–9.
48. Cao R, Wang W, Zhang J, Jiang S, Chen Z, Li W, et al. Synthesis and luminescence properties of Li_2SnO_3:Mn4+ red-emitting phosphor for solid-state lighting. *Journal of Alloys and Compounds*. 2017;704:124–30.
49. Trukhin AN, Rogulis U, Spingis M. Self-trapped exciton in Li_2GeO_3. *JLum*. 1997;72–74:890–2.
50. Wilder JW, Venema LC, Rinzler AG, Smalley RE, Dekker C. Electronic structure of atomically resolved carbon nanotubes. *Nature*. 1998;391(6662):59–62.
51. Odom TW, Huang J-L, Kim P, Lieber CM. Atomic structure and electronic properties of single-walled carbon nanotubes. *Nature*. 1998;391(6662):62–4.
52. Luican A, Li G, Reina A, Kong J, Nair R, Novoselov KS, et al. Single-layer behavior and its breakdown in twisted graphene layers. *Physical Review Letters*. 2011;106(12):126802.

53. Cherkez V, De Laissardière GT, Mallet P, Veuillen J-Y. Van Hove singularities in doped twisted graphene bilayers studied by scanning tunneling spectroscopy. *Physical Review B*. 2015;91(15):155428.
54. Lauffer P, Emtsev K, Graupner R, Seyller T, Ley L, Reshanov S, et al. Atomic and electronic structure of few-layer graphene on SiC (0001) studied with scanning tunneling microscopy and spectroscopy. *Physical Review B*. 2008;77(15):155426.
55. Yankowitz M, Wang F, Lau CN, LeRoy BJ. Local spectroscopy of the electrically tunable band gap in trilayer graphene. *Physical Review B*. 2013;87(16):165102.
56. Huang H, Wei D, Sun J, Wong SL, Feng YP, Neto AC, et al. Spatially resolved electronic structures of atomically precise armchair graphene nanoribbons. *Scientific Reports*. 2012;2(1):1–7.
57. Söde H, Talirz L, Gröning O, Pignedoli CA, Berger R, Feng X, et al. Electronic band dispersion of graphene nanoribbons via Fourier-transformed scanning tunneling spectroscopy. *Physical Review B*. 2015;91(4):045429.
58. Chen Y-C, De Oteyza DG, Pedramrazi Z, Chen C, Fischer FR, Crommie MF. Tuning the band gap of graphene nanoribbons synthesized from molecular precursors. *ACS Nano*. 2013;7(7):6123–8.
59. Que Y, Xiao W, Chen H, Wang D, Du S, Gao H-J. Stacking-dependent electronic property of trilayer graphene epitaxially grown on Ru (0001). *Applied Physics Letters*. 2015;107(26):263101.
60. Pierucci D, Sediri H, Hajlaoui M, Girard J-C, Brumme T, Calandra M, et al. Evidence for flat bands near the Fermi level in epitaxial rhombohedral multilayer graphene. *ACS Nano*. 2015;9(5):5432–9.
61. Wilkes J, Palmer R, Willis R. Phonons in graphite studied by EELS. *Journal of Electron Spectroscopy and Related Phenomena*. 1987;44(1):355–60.
62. Yoshida K, Sugawara Y, Saitoh M, Matsumoto K, Hagiwara R, Matsuo Y, et al. Microscopic characterization of the C–F bonds in fluorine–graphite intercalation compounds. *Journal of Power Sources*. 2020;445:227320.
63. Bangert U, Bleloch A, Gass M, Seepujak A, Van den Berg J. Doping of few-layered graphene and carbon nanotubes using ion implantation. *Physical Review B*. 2010;81(24):245423.
64. Yase K, Horiuchi S, Kyotani M, Yumura M, Uchida K, Ohshima S, et al. Angular-resolved EELS of a carbon nanotube. *Thin Solid Films*. 1996;273(1–2):222–4.
65. Graetz J. New approaches to hydrogen storage. *Chemical Society Reviews*. 2009;38(1):73–82.
66. Jena P. Materials for hydrogen storage: Past, present, and future. *The Journal of Physical Chemistry Letters*. 2011;2(3):206–11.
67. Ma Y, Zhang T, He W, Luo Q, Li Z, Zhang W, et al. Electron microscope investigation on hydrogen storage materials: A review. *International Journal of Hydrogen Energy*. 2020;45(21):12048–70.
68. Kim JY, Lee J-W, Jung HS, Shin H, Park N-G. High-efficiency perovskite solar cells. *Chemical Reviews*. 2020;120(15):7867–918.
69. Park N-G, Zhu K. Scalable fabrication and coating methods for perovskite solar cells and solar modules. *Nature Reviews Materials*. 2020;5(5):333–50.
70. Luo D, Su R, Zhang W, Gong Q, Zhu R. Minimizing non-radiative recombination losses in perovskite solar cells. *Nature Reviews Materials*. 2020;5(1):44–60.
71. Lin S-Y, Chang S-L, Tran NTT, Yang P-H, Lin M-F. H–Si bonding-induced unusual electronic properties of silicene: A method to identify hydrogen concentration. *Physical Chemistry Chemical Physics*. 2015;17(39):26443–50.
72. Shabbir SA, Azher Z, Latif H. Comparison of efficiency analysis of perovskite solar cell by altering electron and hole transporting layers. *Optik*. 2020;208:164061.

73. Khan F, Rezgui BD, Kim JH. Analysis of PV cell parameters of solution processed Cu-doped nickel oxide hole transporting layer-based organic-inorganic perovskite solar cells. *Solar Energy*. 2020;209:226–34.

74. Ren H, Yu S, Chao L, Xia Y, Sun Y, Zuo S, et al. Efficient and stable Ruddlesden–Popper perovskite solar cell with tailored interlayer molecular interaction. *Nature Photonics*. 2020;14(3):154–63.

75. Jeong M, Choi I W, Go EM, Cho Y, Kim M, Lee B, et al. Stable perovskite solar cells with efficiency exceeding 24.8% and 0.3-V voltage loss. *Science*. 2020;369(6511):1615–20.

76. Hu M, Chen M, Guo P, Zhou H, Deng J, Yao Y, et al. Sub-1.4 eV bandgap inorganic perovskite solar cells with long-term stability. *Nature Communications*. 2020;11(1):1–10.

77. Quarti C, Katan C, Even J. Physical properties of bulk, defective, 2D and 0D metal halide perovskite semiconductors from a symmetry perspective. *Journal of Physics: Materials*. 2020;3(4):042001.

78. Heo JH, Im K, Lee HJ, Kim J, Im SH. Ni, Ti-co-doped MoO_2 nanoparticles with high stability and improved conductivity for hole transporting material in planar metal halide perovskite solar cells. *Journal of Industrial and Engineering Chemistry*. 2021;94:376–83.

17 Problems

*Thi Dieu Hien Nguyen, Ngoc Thanh
Thuy Tran, and Ming-Fa Lin*
National Cheng Kung University

CONTENTS

Problem 17.1: Obviously, the heterojunctions between electrolyte and cathode/anode (Figure 17.1) play the critical roles on the current transport within the ion-based batteries. For example, the multicomponent lithium oxides are frequently used as core materials [1–5]. The boundary of composite two subsystems is capable of inducing a Moire superlattice under a specific environmental match [1–5]. The sufficiently wide samples need to be utilized in the relaxation process of geometric structures, e.g., four- or five-layer widths in the Vienna *ab initio* simulation package (VASP) simulations, as done for substrate-related emergent materials [monolayer and bilayer silicene on Ag(111); [6]]. How to greatly reduce the computer time in a very large unit cell has to be overcome by the physical pictures. (1) Calculate an optimal crystal structure and explore its highly nonuniform chemical environment, especially for the large fluctuations of chemical bond lengths near the specific heterojunction. (2) Evaluate the atom-dominated electronic energy spectrum and wave functions in terms of main features. Such unusual results could reveal the energy-specified atom dominances. Verifying (3) the spatial charge density distributions and their variations after the boundary formation is expected to display the drastic changes, the rich orbital-hybridization behaviors will be qualitatively examined from the accurate results. Finally, more complicated atom- and orbital-projected density of states can

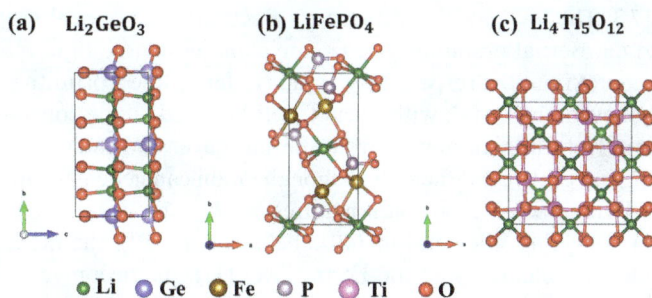

FIGURE 17.1 The heterojunction of electrolyte and cathode/anode for lithium-ion-based batteries, such as (a) Li_2GeO_3 (b) $LiFePO_4$, and (c) $Li_4Ti_5O_{12}$.

DOI: 10.1201/9781003263807-17

clearly identify the active orbital hybridizations of each chemical bond. The delicate analyses are strongly required in dealing with the merged special structures due to the van Hove singularities [1,2]. Moreover, the energy-dependent prominent peaks, corresponding to the significant chemical bondings, could be tested by the high-resolution scanning tunneling spectroscopy (STS) measurements [7,8].

Problem 17.2: Exploring the single- and many-particle optical excitations due to the external perturbation of an electromagnetic wave by the VASP simulations. The lithium-sulfide-related batteries are outstanding candidates for the VASP simulations. Both standard LiS_2/LiO_2 and P_2S_5 electrolytes [Figures 11.1 and 12.1] are expected to exhibit the rich optical properties in terms of the dielectric function [$s(\omega)$], reflectance spectrum [$R(\omega)$; [9]], absorption coefficient [$\alpha(\omega)$; [9]], refractive index [$n(\omega)$; [9]], and energy loss function [$Im[-1/s(\omega)]$; [9]]. The first-principles results could be further utilized to accurately analyze the significant vertical electronic excitations without any momentum transfers; therefore, the close relations between the electronic and optical properties could be achieved from the reliable data. As for the first-step VASP work, (1) evaluate the optimal crystal structures and discuss the chemical environments, (2) calculate electronic energy spectra and plot the important vertical excitation channels, and (3) qualitatively understand the joint van Hove singularities from the initial- and final-state density of states. For these traditional semiconductors, (4) energy gaps are in great contrast with optical threshold absorption frequencies in the presence of excitonic effects [the red-shift effects in [10]]. (5) Fully explore how many quasi-stable excitonic states are clearly revealed near the lowest absorption frequency [10]. Moreover, (6) the many-body electron-hole Coulomb scatterings might have strong modifications on the featured absorption peaks [the spectral forms, intensities, numbers, and frequencies; [10]]. Very interestingly, (7) the linking together of electronic and optical properties is capable of getting the most important quasiparticle properties. That is to say, each absorption structure is dominated by the specific single- or multi-orbital hybridizations of chemical bonds [10]. (8) It is worthy of making a detailed comparison between lithium oxides and sulfides in the geometric, electronic, and transport behaviors. This concise chemical picture will be generalized to other emergent battery materials. The above-mentioned theoretical predictions could be directly tested by the optical spectroscopies, reflectance [9,10], absorption [9,10], and transmission ones [9,10].

Problem 17.3: The low- and middle-energy electronic and optical properties, corresponding to the Bernal graphite systems, are explored by the tight-binding model and gradient approximation, respectively. First, (1) determine a primitive unit cell for this layered honeycomb crystal, with a set of well-behaved lattice constants. (2) Built the π-electronic Hamiltonian matrix from the intralayer and interlayer carbon-$2p_z$-orbital hopping integrals. And then, its diagonalization can get eigenvalues and wave functions. (3) Fully discuss their main features, such as the significant valence and conduction band overlap [the semi-metallic behavior in [11]], the great asymmetry of hole and electron bands about the Fermi level [11], the occupied π-band width, the strong energy dispersions [linear, parabolic, and weakly dispersive ones; [11]], the critical points in the energy-wave-vector space [van Hove singularities in [11]], and the non-crossing/crossing/anti-crossing behaviors [11]. The main features of electronic properties are responsible for the unusual optical excitations. Using the

Fermi golden rule [12], the vertical transitions, which mainly arise from the initial to final states with the same wave vectors, are determined by their joint density of states and dipole matrix elements [9]. The optical excitation probabilities are approximated by the first derivation of Hamiltonian versus k_x or k_y [details in [13,14]. (4) The π-electronic electron-hole pair excitations and plasmon modes are directly reflected in the frequency-dependent transverse dielectric function, energy loss function, and reflectance spectrum and absorption coefficient. That is to say, the single-particle and collective excitations can be clearly identified from the special structures in the calculated results [11]. (5) Similarly, the important differences between graphite and monolayer graphene [among three kinds of 3D bulk graphite systems [11]] could be investigated thoroughly. These results will clearly display the critical roles of dimensionalities [9,11] and stacking configuration [11].

Problem 17.4: Alkali graphite intercalation compounds (Figure 17.2) exhibit the rich and unique quasiparticle properties since their successful syntheses 4 years ago [15], especially for the lithium-atom/lithium-ion-related systems in batteries [16]. Both first-principles and phenomenological methods should be suitable in exploring the fundamental properties, mainly owing to the obvious atomic interactions due to the concise orbital hybridizations [17]. The former is utilized to investigate the optimal stage-n layered structures by calculating the lowest ground-state energy. (a) Evaluate all the available geometric parameters. (b) Propose the physical and chemical pictures to clearly illustrate the unusual chemical environments, such as the uniform or non-uniform chemical-bond lengths, the [x, y]-plane distribution arrangements, stacking configurations, and their dependences on n. The chemical intercalations/de-intercalations are expected to strongly modify the low-lying valence and conduction bands [17]. (c) Explore the main features of electronic properties, covering band property across the Fermi level [metal, semi-metal, or a zero-gap semiconductor; [17–19]], the asymmetry of electron and hole energy spectra, the alkali- and carbon-dominated band structure, the well-defined/slightly modified/highly distorted π-electronic structures, the spatial charge density distributions before and after chemical modifications, and more van Hove singularities with/without merged peaks due to the active orbital hybridizations. (d) Examine the active orbital hybridizations in determining the diverse quasiparticle behaviors. (e) Clarify the significant differences between the neutral and ionic guest cases. Also, (f) optical excitation spectra could provide another sharp contrast.

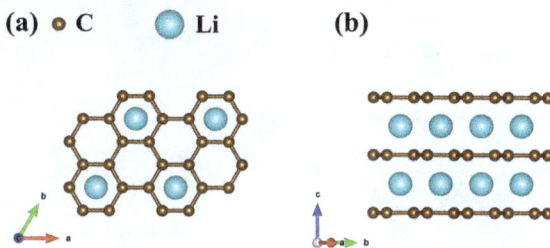

FIGURE 17.2 A stage-1 lithium-related graphite intercalation compound: (a) Top and (b) side views.

Problem 17.5: The large molecule/saturated ion intercalations/de-intercalations into graphitic layers (Figure 17.3) are able to provide another kind of chemical bonding phenomenon. The aluminum-related batteries are deduced to be driven by the large $AlCl_4^{-1}$ and $Al2Cl_7^{-1}$ transport [20]. Both $AlCl_4$ and $AlCl_4^{-1}$ are chosen for a model study in determining all the active orbital hybridizations of chemical bonds [20], as done for alkali graphite intercalation compounds in Prom.17.4. [20]. (a) Evaluate the periodical distance, the significant lengths for C–C, C–Cl, Al–Cl, and intra-/inter-molecule Cl–Cl bonds, as well as the intra-molecule bonding angle. (b) Discuss the top and side views. (b) Investigate the relative C-, Al-, and Cl-weights from the calculated energy spectra, indicating the distinct atomic contributions within the specific energy ranges. (c) Clearly identify the obvious orbital mixings from the spatial charge density distributions in the presence and absence of chemical modification [achieving the most prominent orbital hybridizations; [21]]. (d) Delicately analyze the C-, Al-, and Cl-decomposed density of states, as well the projections of their active orbitals. A lot of highly accurate figures cover thorough information about the merged structures of the atom- and orbital-dependent van Hove singularities. These results, which are assisted by (a)–c), are able to clearly examine all the intrinsic/significant atomic interactions: the intra-layer, inter-layer, intra-molecule, and inter-molecule ones. (e) Clarify the important similarities and differences between two distinct atomic configurations. (f) Discuss whether the stationary ion transport is strongly affected by the intercalation-induced or intercalation-destroyed free carrier density.

Problem 17.6: The rich Coulomb excitations in carbon-sp^2 bonding systems by the phenomenological methods, the diverse [momentum, frequency]-excitation phase diagram through the tight-binding model and the random-phase approximation [11,22,23]. Monolayer graphene, with a planar honeycomb lattice, exhibits the unusual Coulomb excitation phenomena. (a) The near-nearest Hamiltonian interactions, which only come from the carbon-2pz orbitals, are enough in fully exploring the low- and middle-frequency quasiparticle excitation behaviors [11,22,23]. This can reach the well-known Dirac-cone band structure near the K and Kj valleys [the corner of the hexagonal Brillouin zone; [22,24]], in which a pair of valence and conduction bands linearly intersect each other at the Dirac points. As for a pristine system, it belongs to a zero-gap semiconductor without free carriers, since its density of states is vanishing at the Fermi level [24]. (b) Evaluate the screening properties, the dielectric functions and energy loss functions, through the random phase

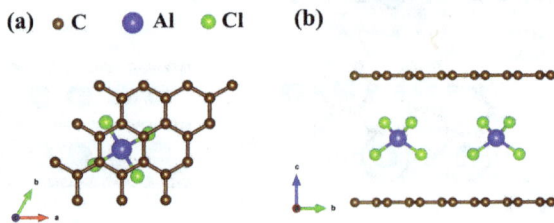

FIGURE 17.3 The stage-1 $AlCl_4$-molecule graphite intercalation compound, as shown in (a) top and (b) side views.

approximation (RPA). (c) Discuss the electron-hole excitation boundaries due to intra-band and inter-band channels. (d) Examine whether the low-frequency acoustic plasmon modes exist at very low temperature. (e) Investigate the critical temperatures and momenta in determining the collective excitations of temperature-induced free electrons and holes, and their close relations by using the plasmon frequency dispersions. By the efficient n- or p-type dopings, the single-particle and collective excitations are greatly diversified by the various carrier densities. (f) Identify the doping-enriched Coulomb excitations from the momentum and E_F dependences. However, (g) the π plasmons, which correspond to the coherent of all C-$2p_z$ orbitals in a honeycomb crystal, might hardly depend on the doping behaviors, such as their frequencies much high than 1 eV [about 5.0 eV in [11]]. In addition, the generalized tight-binding model is suitable in studying electronic properties and Coulomb excitations in the presence of a uniform perpendicular magnetic field [11,22,23]; that is, it is reliable for the diverse phenomena of magnetic quantization.

Problem 17.7: An armchair carbon nanotube, as clearly shown in Figure 17.5(a), exhibits the unique elementary Coulomb excitations. Their π-electronic states could be regarded as those from those of a planar honeycomb structure under a specific periodical condition [25]. (a) By using the band-structure calculations in Prob. 18.3 (a), identify the 1D metallic features from the gapless property at the Fermi level and a finite density of states there. (b) Calculate the bare and screened response functions, respectively, to examine the single-particle and collective excitation behaviors. (c) Based on the physical pictures of prominent peaks in energy loss, spectra, they could be classified into three kinds of plasmon modes: (I) 1D acoustic plasmons with frequencies lower than/comparable with 1 eV, (II) interband optical plasmons of ω_p ~ 1.0–5.0 eV, and (III) π plasmons with $\omega_p > 5.0$ eV in which they are, respectively, due to only free carriers, certain one valence subband, and all π-electronic states. (d) Explore the significant dependences on the discrete angular momentum transfer, radius, doping, and temperature by examining the electron-electron Coulomb interactions on a cylindrical surface and the unusual variations of the Fermion distributions under the thermal excitations. Moreover, electronic energy spectra and wave functions are easily modulated by the external electric and magnetic fields, such as the metal-semiconductor transition by the periodical Aharonov-Bohm effect [a magnetic field parallel to a nanotube axis in [26]]. It should be noticed that a transverse gate voltage and a longitudinal magnetic field, respectively, lead to the on-site Coulomb potential energies and the Peierls-phase-dependent hopping integrals [11,27]. (e) Thoroughly discuss their strong effects on magneto-electronic properties and magnetoplasmons with/without the Landau dampings. (f) The similar investigations could be generalized to narrow- and middle-gap semiconducting carbon nanotubes [types of II and III due to curvature effects and periodical boundary conditions, respectively; [26,28]]. (g) Clearly identify the important differences among three types of single-walled carbon nanotubes in the diverse quasiparticle phenomena.

Problem 17.8: In addition to vertical optical transitions, Bernal graphite is an outstanding candidate in illustrating the diversified Coulomb excitations. When RPA is assisted by the reliable tight-binding model [books], two kinds of longitudinal p-electronic excitation behaviors could be clarified from the delicate calculations and analyses. (a) Discuss the semi-metallic behavior along the KH line [$-\pi \leq k_z I_c \leq \pi$, I_c

the periodical distance in the perpendicular direction; [29]], in which free conduction electrons and valence holes come to exist simultaneously. The band-overlap opposite charges, which mainly arise from the interlayer carbon-$2p_z$ orbital interactions [13,14], make the same conductions to the single-particle and collective excitations, such as the temperature-dependent plasmon frequencies. (b) Thoroughly explore the dynamic screening behaviors of free carriers along the honeycomb plane and the perpendicular direction, respectively. This covers the suitable calculations on the magnitude and direction dependences of the 3D transferred moments, as well as the temperature ones, in terms of the bare and screened response functions. (c) Characterize plasmon behavior under a long wavelength limit for the various cases, and its critical momentum/temperature under a specific temperature/momentum. (d) Compare those of the electron energy loss spectroscopy (EELS) experiments in detail [9], especially for the low-frequency collective excitations. (e) Propose physical pictures to comprehend the diversified phenomena of the propagating plasma waves. (f) Illustrate the important differences among monolayer graphene, a single-walled carbon nanotube, and AB-stacked graphite in electron-hole excitations and plasmon modes.

Problem 17.9: Hydrogen atoms and 2D surface can provide very active chemical environments, especially for green energy resources [details in Chap. 17; [30,31]]. The 2D hydrogenated group-IV materials (Figure 17.4a–h) are worthy of the systematic investigations in basic sciences [32,33] and highly potential applications [30,31,34]. For example, the single- and double-side hydrogen absorptions, which come to exist under the highest concentrations [31,34,35], clearly illustrating for monolayer graphene [monolayer plumbene in (e)/(f) and (g)/(h)]. The former and the latter, respectively, possess the almost vanishing sp^3 bonding/the planar honeycomb crystal/the negligible spin-orbital coupling and the very obvious four-orbital hybridizations/the largest buckling structure/the highest spin-orbital interactions [31,36], as predicted from the VASP simulations [35,36] and the tight-binding model [37]. The first-principles method is very suitable for thoroughly exploring the fundamental quasiparticle properties [24]. First, consider hydrogenated graphene cases, (a) investigate the top-, bridge- and hollow-site positions and determine the optimal ones under the various concentration and distribution arrangements. (b) Discuss the existence of buckling structures and their dependences on chemical adsorptions. (c) Examine the clear evidences of sp^3 bondings in interlayer coupled H-C bonds, covering the atom-dominated band structures, the spatial charge density distribution before and after the chemical modifications, and the van Hoe singularities. The featured energy spectra need to be illustrated under the delicate analyses. They are directly reflected in other quasiparticle behaviors. (d) The main optical properties, the transverse dielectric function, reflectance, absorption coefficient, refractive index and energy loss function, are fully studied in the absence/presence of excitonic effects. These results can achieve the close relations between the prominent absorption structures and the active orbital hybridizations, as well as the discrete stable/quasi-stable exciton states. Very interestingly, plumbene/tinene/germanene/silicene materials are able to greatly diversify the hydrogenation reactions, owing to the significant sp^3 bonding, buckling structures, and spin-orbital interactions [6,38–41]. (e) Clearly illustrate their important differences in terms of essential properties [6,38–40] and hydrogen storage [6,38–43].

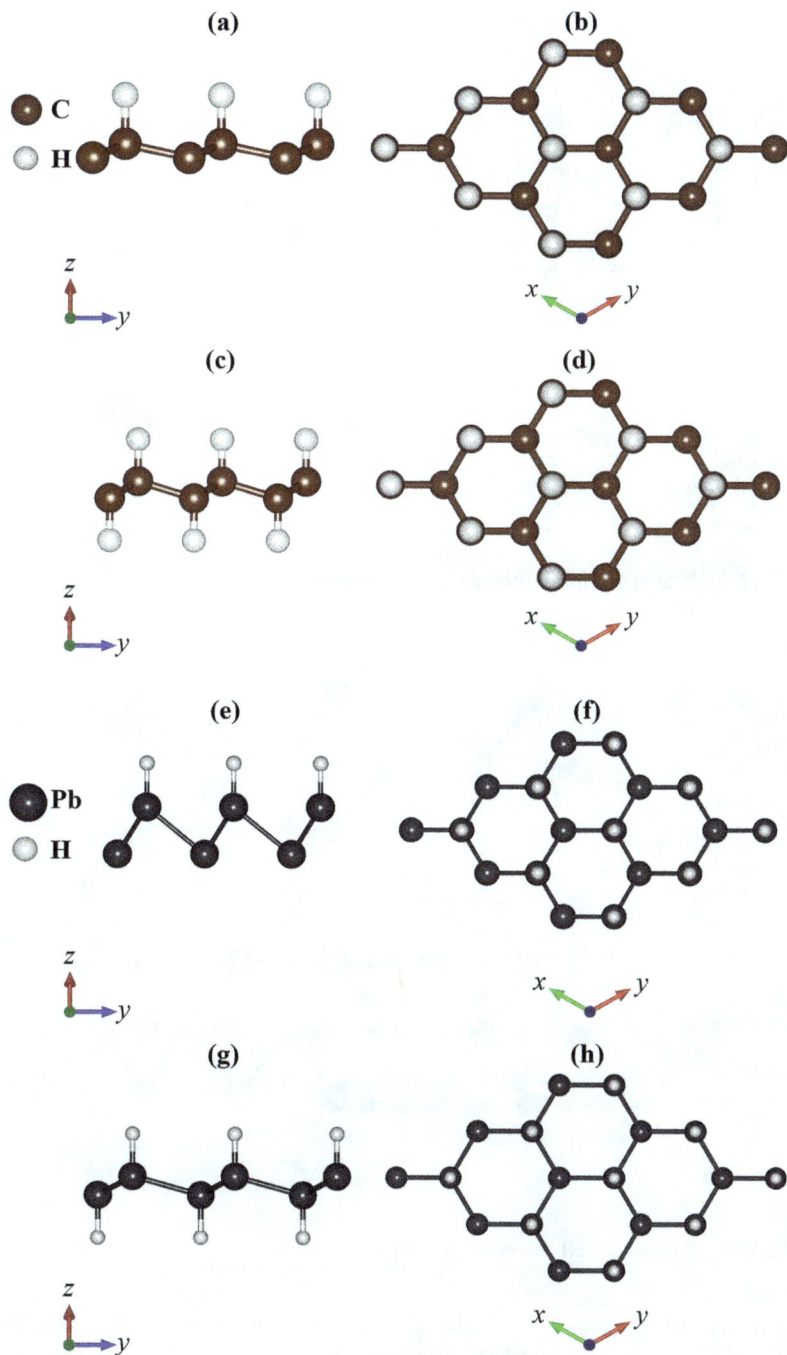

FIGURE 17.4 (a)/(b) The single- and (c)/(d) double-side hydrogenated graphene under the highest absorption under the top/side views; similarly for the monolayer Pb cases in (e)/(f) and (g)/(h).

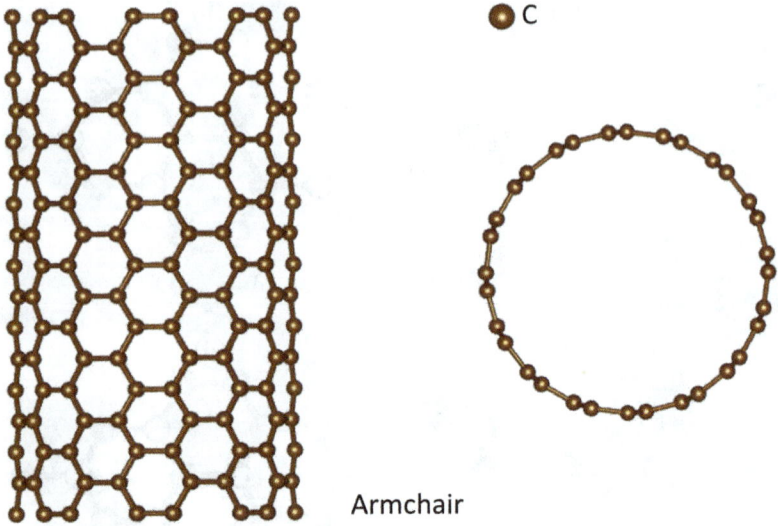

FIGURE 17.5   The armchair carbon nanotube under the top/side views.

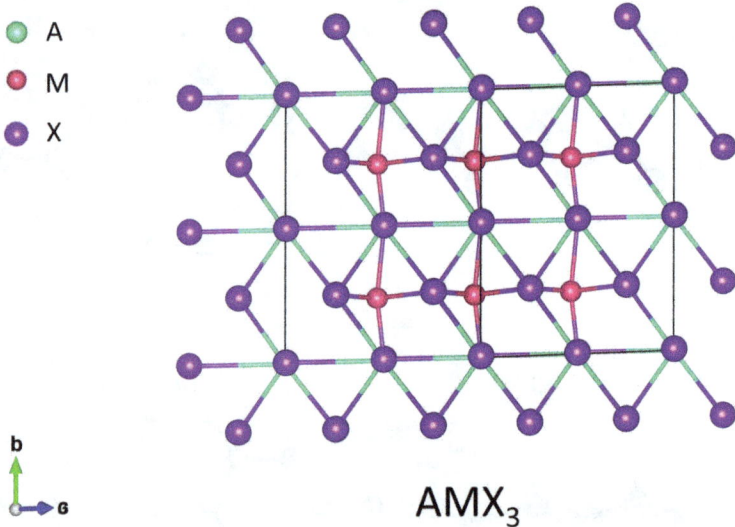

FIGURE 17.6   The solar cell material, AMX_3.

Problem 17.10: By using the VASP simulations, the metal halide perovskites of efficient solar cell materials [AMX_3 in Figure 17.6 [44,45]] will exhibit the unusual quasiparticle behaviors. (a) Investigate 3D crystal symmetries of $CsGeI_3$ from bond lengths and bonding angles. (b) Clearly illustrate the main features of valence and conduction energy spectra and wave functions, and distinct atom dominances within the specific energy ranges. (c) Thoroughly clarify the critical multi-orbital hybridization in Cs-I and Ge-I bonds through merged van Hove singularities and the spatial charge density distribution. After the perturbation of an external electromagnetic wave, the charge screening ability is represented by the transverse dielectric functions. (d) Discuss the optical threshold frequency in the presence/absence electron-hole coupling effects, as well as the other prominent absorption structures. (e) Specifically, establish the close relations between each significant transition channel and the active orbital hybridizations. Very interestingly, (f) examine the collective excitations from the strong peaks of energy loss spectra [9] and/or the plasmon edges of reflectance spectra [9].

REFERENCES

1. Nguyen T D H, Pham H D, Lin S Y and Lin M F 2020 Featured properties of Li+-based battery anode: $Li_4Ti_5O_{12}$ *RSC Advances* **10** 14071–9.
2. Khuong Dien V, Thi Han N, Nguyen T D H, Huynh T M D, Pham H D and Lin M-F 2020 Geometric and electronic properties of Li_2GeO_3 *Frontiers in Materials* **7** *arXiv:2009.02154.*
3. Wang L, He X, Sun W, Wang J, Li Y and Fan S 2012 Crystal orientation tuning of $LiFePO_4$ nanoplates for high rate lithium battery cathode materials *Nano Letters* **12** 5632–6.
4. Zhou F, Kang K, Maxisch T, Ceder G and Morgan D 2004 The electronic structure and band gap of $LiFePO_4$ and $LiMnPO_4$ *Solid State Communications* **132** 181–6.
5. Buyukyazi M and Mathur S 2015 3D nanoarchitectures of a-$LiFeO_2$ and alpha-$LiFeO_2$/C nanofibers for high power lithium-ion batteries *Nano Energy* **13** 28–35.
6. Shih-Yang Lin H-Y L, Nguyen D K, Tran N T T, Pham H D, Chang S-L, Lin C-Y, Lin M-F 2020 *Silicene-Based Layered Materials: Essential Properties*: IOP Publishing.
7. Hipps K 2006 *Handbook of Applied Solid State Spectroscopy*: Springer. pp 305–50.
8. Bussolotti F, Chi D, Goh K J, Huang Y L and Wee A T 2020 *2D Semiconductor Materials and Devices*: Elsevier. pp 199–220.
9. Fox M 2001 *Optical Properties of Solids*: Oxford University Press.
10. Dien V K, Pham H D, Tran N T T, Han N T, Huynh T M D, Nguyen T D H and Fa-Lin M 2020 Orbital-hybridization-created optical excitations in $Li_8Ge_4O_{12}$ *arXiv preprint arXiv:2009.02160.*
11. Chiun-Yan Lin J-Y W, Chiu C-W and Lin M-F 2019 *Coulomb Excitations and Decays in Graphene-Related Systems*: CRC Press.
12. Kittel C 1996 *Introduction to Solid State Physics*: John Wiley & Sons New York 402.
13. Lai Y, Ho J, Chang C and Lin M-F 2008 Magnetoelectronic properties of bilayer Bernal graphene *Physical Review B* **77** 085426.
14. Ho Y-H, Chiu Y-H, Lin D-H, Chang C-P and Lin M-F 2010 Magneto-optical selection rules in bilayer Bernal graphene *ACS Nano* **4** 1465–72.
15. Assouik J, Hajji L, Boukir A, Chaouqi M and Lagrange P 2017 Heavy alkali metal-arsenic alloy-based graphite intercalation compounds: Investigation of their synthesis and of their physical properties *Comptes Rendus Chimie* **20** 116–24.
16. Xu J, Dou Y, Wei Z, Ma J, Deng Y, Li Y, Liu H and Dou S 2017 Recent progress in graphite intercalation compounds for rechargeable metal (Li, Na, K, Al)-ion batteries *Advanced Science* **4** 1700146.
17. Li W-B, Lin S-Y, Tran N T T, Lin M-F and Lin K-I 2020 Essential geometric and electronic properties in stage-n graphite alkali-metal-intercalation compounds *RSC Advances* **10** 23573–81.

18. Lin W-B, Tran N T T and Lin S-Y 2019 Diverse fundamental properties in stage-n graphite alkali-intercalation compounds: anode materials of Li+-based batteries *arXiv preprint arXiv:2001.02042.*

19. Li Y, Lu Y, Adelhelm P, Titirici M-M and Hu Y-S 2019 Intercalation chemistry of graphite: Alkali metal ions and beyond *Chemical Society Reviews* **48** 4655–87.

20. Wu M, Xu B, Chen L and Ouyang C 2016 Geometry and fast diffusion of A_lCl_4 cluster intercalated in graphite *Electrochimica Acta* **195** 158–65.

21. Bhauriyal P, Mahata A and Pathak B 2017 The staging mechanism of A_lCl_4 intercalation in a graphite electrode for an aluminium-ion battery *Physical Chemistry Chemical Physics* **19** 7980–9.

22. Lin C-Y, Ho C-H, Wu J-Y, Do T-N, Shih P-H, Lin S-Y and Lin M-F 2019 *Diverse Quantization Phenomena in Layered Materials*: CRC Press.

23. Chen S, Wu J, Lin C and Lin M 2016 Theory of magnetoelectric properties of 2d systems *New Journal of Physics* **18** 103024.

24. Tran N T T, Lin S-Y, Lin C-Y and Lin M-F 2017 *Geometric and Electronic Properties of Graphene-Related Systems: Chemical Bonding Schemes*: CRC Press.

25. Ho Y, Chang C, Shyu F, Chen R, Chen S and Lin M-F 2004 Electronic and optical properties of double-walled armchair carbon nanotubes *Carbon* **42** 3159–67.

26. Shyu F-L, Chang C-P, Chen R-B, Chiu C-W and Lin M-F 2003 Magnetoelectronic and optical properties of carbon nanotubes *Physical Review B* **67** 045405.

27. Ho J-H, Lu C, Hwang C, Chang C and Lin M-F 2006 Coulomb excitations in AA-and AB-stacked bilayer graphites *Physical Review B* **74** 085406.

28. Shyu F L and Lin M 2002 Electronic and optical properties of narrow-gap carbon nanotubes *Journal of the Physical Society of Japan* **71** 1820–3.

29. García N, Esquinazi P, Barzola-Quiquia J and Dusari S 2012 Evidence for semiconducting behavior with a narrow band gap of Bernal graphite *New Journal of Physics* **14** 053015.

30. Lin S-Y, Chang S-L, Tran N T T, Yang P-H and Lin M-F 2015 H–Si bonding-induced unusual electronic properties of silicene: A method to identify hydrogen concentration *Physical Chemistry Chemical Physics* **17** 26443–50.

31. Tozzini V and Pellegrini V 2013 Prospects for hydrogen storage in graphene *Physical Chemistry Chemical Physics* **15** 80–9.

32. Pontes R B, Mançano R R, da Silva R, Cótica L F, Miwa R H and Padilha J E 2018 Electronic and optical properties of hydrogenated group-IV multilayer materials *Physical Chemistry Chemical Physics* **20** 8112–8.

33. Mohebpour M A, Mozvashi S M, Vishkayi S I and Tagani M B 2020 Prediction of hydrogenated group IV–V hexagonal binary monolayers *Scientific Reports* **10** 1–14.

34. Ohno S, Wilde M and Fukutani K 2014 Novel insight into the hydrogen absorption mechanism at the Pd (110) surface *The Journal of Chemical Physics* **140** 134705.

35. Crespo E A, Ruda M, de Debiaggi S R, Bringa E M, Braschi F U and Bertolino G 2012 Hydrogen absorption in Pd nanoparticles of different shapes *International Journal of Hydrogen Energy* **37** 14831–7.

36. Boukhvalov D, Katsnelson M and Lichtenstein A 2008 Hydrogen on graphene: Electronic structure, total energy, structural distortions and magnetism from first-principles calculations *Physical Review B* **77** 035427.

37. Konschuh S, Gmitra M and Fabian J 2010 Tight-binding theory of the spin-orbit coupling in graphene *Physical Review B* **82** 245412.

38. Das D K, Sarkar J and Singh S K 2018 Effect of sample size, temperature and strain velocity on mechanical properties of plumbene by tensile loading along longitudinal direction: A molecular dynamics study *Computational Materials Science* **151** 196–203.

39. Matthes L, Pulci O and Bechstedt F 2014 Optical properties of two-dimensional honeycomb crystals graphene, silicene, germanene, and tinene from first principles *New Journal of Physics* **16** 105007.
40. Matthes L, Pulci O and Bechstedt F 2013 Massive Dirac quasiparticles in the optical absorbance of graphene, silicene, germanene, and tinene *Journal of Physics-Condensed Matter* **25** 395305.
41. Vogt P 2018 Silicene, germanene and other group IV 2D materials. Beilstein-Institute.
42. Roome N J and Carey J D 2014 Beyond graphene: Stable elemental monolayers of silicene and germanene *ACS Applied Materials & Interfaces* **6** 7743–50.
43. Wei W, Dai Y, Huang B B and Jacob T 2013 Many-body effects in silicene, silicane, germanene and germanane *Physical Chemistry Chemical Physics* **15** 8789–94.
44. Zhou T, Zhang Y, Wang M, Zang Z and Tang X 2019 Tunable electronic structures and high efficiency obtained by introducing superalkali and superhalogen into AMX3-type perovskites *Journal of Power Sources* **429** 120–6.
45. Tang L-C, Chang C-S and Huang J Y 2000 Electronic structure and optical properties of rhombohedral $CsGeI_3$ crystal *Journal of Physics: Condensed Matter* **12** 9129.

Index

Note: **Bold** page numbers refer to tables; *italic* page numbers refer to figures.

For Product Safety Concerns and Information please contact our EU
representative GPSR@taylorandfrancis.com
Taylor & Francis Verlag GmbH, Kaufingerstraße 24, 80331 München, Germany

www.ingramcontent.com/pod-product-compliance
Lightning Source LLC
Chambersburg PA
CBHW060804220326
41598CB00022B/2534